D1348578

Telecommunications
Transmission
Engineering

Telecommunications Transmission Engineering is published in three volumes:

Volume 1 — *Principles*
Volume 2 — *Facilities*
Volume 3 — *Networks and Services*

Telecommunications Transmission Engineering

Third Edition

Technical Personnel
Bellcore and
Bell Operating Companies

Volume 3
Networks and Services

Operating and regional company employees may obtain copies of this volume, order the Bellcore Technical References cited in this volume, as well as subscribe to Bellcore's DIGEST of Technical Information by contacting their company documentation coordinators.

All others may obtain the same copies as well as subscribe to the DIGEST by calling Bellcore's documentation hotline 1–800–521–CORE or (201) 699–5802.

Order each *Telecommunications Transmission Engineering* volume by the ST number as follows.

Volume 1, *Principles*	ST–TEC–000051
Volume 2, *Facilities*	ST–TEC–000052
Volume 3, *Networks and Services*	ST–TEC–000053

Prepared for publication by
Bellcore Technical Publications.

This is the third edition of the *Telecommunications Transmission Engineering* three–volume series previously copyrighted by AT&T and AT&T Bell Laboratories.

ISBN 1–878108–03–4 (Volume 3)
ISBN 1–878108–04–2 (three–volume set)

Library of Congress Catalog Card Number: 90–62180

Telecommunications Transmission Engineering

Introduction

Telecommunications engineering is concerned with the planning, engineering, design, implementation, operation, and maintenance of the network of facilities, channels, switching equipment, and user devices required to provide voice and data communications between various locations. Transmission engineering is that part of telecommunications engineering that deals with the channels, the transmission facilities or systems that carry the channels or circuits, and the combinations of the many types of channels and facilities that help form a network. It is a discipline that combines skills and knowledge from science and technology with an understanding of economics, human factors, and system operations.

This three–volume reference book is written for the practicing transmission engineer and for the student of transmission engineering in an undergraduate curriculum. However, the material was planned and organized to make it useful to anyone concerned with the many facets of telecommunications engineering. Of necessity, it represents only the current status of communications technology being used and deployed today by the Bell operating companies in their intraLATA (local access and transport area) networks, which provide exchange telecommunications and exchange access services, as well as their official networks. The reader should be aware of the dynamic nature of the subject.

Volume 1, *Principles*, covers the transmission engineering principles that apply to communications systems. It defines the characteristics of various types of signal, describes signal impairments arising in practical channels, provides the basis for understanding the relationships between a communication network and its components, and provides an appreciation of how transmission objectives and achievable performance are interrelated.

Volume 2, *Facilities*, emphasizes the application of the principles of Volume 1 to the design, implementation, and operation of the transmission systems and facilities that are used to form the public networks.

Volume 3, *Networks and Services*, builds on the principles and facilities discussed in Volumes 1 and 2 and shows how the principles are applied to facilities to form networks that are used by the exchange carriers to provide services for various users.

The authors use a generic approach throughout all three volumes. However, they often use specific examples and illustrations that are most familiar to them to help clarify a concept. These examples are not intended to recommend any equipment or to imply that there is only one solution.

The material has been written, edited, prepared, and reviewed by a large number of technical personnel from the Bell operating companies and Bellcore. Thus the book represents the cooperative efforts and views of many people.

As a point of reference for the reader, a brief description of these organizations follows. At divestiture, on January 1, 1984, 22 Bell operating companies were transferred from AT&T to seven newly formed regional companies: Ameritech, Bell Atlantic, BellSouth, NYNEX, Pacific Telesis, Southwestern Bell Corporation, and U S WEST. These regional companies, through their operating telephone companies, were empowered to provide local exchange telecommunications and exchange access services. They were also called upon to provide a common central point to meet the requirements of national security and emergency preparedness. Finally, they were allowed to create and support a centralized organization for the provision of engineering, administrative, and other services. This *central point* and organization is Bellcore.

The regional companies are both the owners and the clients of Bellcore. Bellcore's mission—to provide research, technical support, generic requirements, technical analyses, and other services to the Bell operating companies—allows them to provide modern, high-quality yet low-priced services to their customers. Bellcore provides neither manufacturing nor supplier recommendations, as procurement is a function of the regional companies.

Irwin Dorros
Executive Vice President –
Technical Services
Bellcore

Volume 3 — Networks and Services

Preface

Overall exchange–carrier objectives are to provide high–quality, low–cost communications services with a fair return on investment. This volume presents transmission–related technical and administrative information needed to achieve these objectives.

Service quality depends on meeting transmission and reliability objectives. Networks and services must be engineered to meet design objectives; facilities and circuits must be constructed to meet them also. Facilities and circuits must be maintained so that deviations from the engineered objectives are controlled; the effects of failures are thus minimized. Transmission, maintenance, and reliability objectives are discussed throughout this volume as they relate to various networks and services.

The control of costs is an integral part of the process of deciding how to provide and maintain any network. The process is one of compromise, i.e., of striking the best balance among customer satisfaction, plant performance, and cost.

Volume 3 builds on the principles covered in Volume 1 and the facilities discussed in Volume 2. The definition and characterization of impairments, their effect on performance as measured by grade of service, the methods of setting objectives, and the physical plant used to provide service are basic to the objectives and maintenance methods covered in this volume. In essence, the provision of networks and services is a fundamental exchange–carrier objective.

Section 1 discusses the overall structure and features of the message network, which consists of loops, trunks, and switching machines configured for the efficient handling of telephone calls.

Local and toll portions of the network are discussed, with their transmission plan. New networking concepts (integrated services digital network, packet switching) are included.

Loops are the circuits that connect telephone stations to local central offices and thus to the rest of the message network. Their performance characteristics are important because each connection usually involves two loops. Section 2 discusses the characteristics of actual plant and design considerations for the provision of loops.

Trunks provide transmission paths to interconnect switching machines. Section 3 defines the various trunk types and then discusses traffic engineering concepts used to determine the required number of trunks. Design criteria are different for local, toll, and auxiliary–service trunks. Techniques used in the control of echo impairments are included, as are modern operator services systems.

The many types of special services are introduced and defined in Section 4. Design criteria for the principal switched and private–line types of special services are included, along with a discussion of centrex offerings and exchange access for interexchange carriers.

Transmission performance must be monitored to ensure that quality standards are met, to detect trends, and to develop plans for improvement. Section 5 covers the measurement plans, and the maintenance, planning, engineering, and management functions required in operating the complex facilities network for telecommunications services.

Contents

Contents

Contents

Contents

Contents

Contents

Contents

Contents

Contents

Figures

Figures

Tables

Tables

Tables

Telecommunications Transmission Engineering

Section 1

The Message Network

Section 1 is devoted to a review of the purposes and functions of the message network because of its fundamental importance and central role in meeting today's telecommunications needs. In addition, it provides background and understanding of the overall functions and transmission objectives that underly the design of the loop and trunk components of the network.

Chapter 1 discusses the overall message network, both long–haul and metropolitan, in its original hierarchical form and its later, evolved, structure. Significant changes in technology in a span of more than 30 years (abandonment of analog switching at upper levels of the hierarchy, wide use of digital facilities, adoption of common–channel signaling) led to a network only superficially like its original form.

Chapter 2 covers a simplified, digital–oriented transmission plan for intraLATA networks. It is intended to take advantage of digital transmission and switching to derive both transmission improvement and cost savings.

In Chapter 3, new structures for the telecommunications industry are introduced: the integrated services digital network, the packet switching networks that supply public data service, and the prospective broadband networks of a few years hence.

These chapters provide an overview of the various trunk networks that have evolved and the switching systems necessary for efficient interconnection and utilization of the complex array of transmission paths.

Chapter 1

The Network Plan for Message Service

The message telecommunications service (MTS) network provides connections among virtually all of the more than 140 million access lines in the United States, Canada, and some Caribbean islands. This network comprises the individual facilities of exchange carriers, facilities–based interexchange carriers (ICs), nonfacilities–based (resale) carriers, and other administrations.

An overview of the switching and transmission plans for the message network requires a description of the historic switching hierarchy, with its classes of switching offices, types of trunk, and features that permitted efficient call routing prior to the division of network responsibilities in 1984 [1]. It will include descriptions of network transmission requirements [2] and relate them to trunk loss and office balance. Completing conversion of the entire message network from analog to digital, while moving rapidly, will require some years. During this period, there will be a mix of network plans that will require a "transition plan" to make the conversion process transparent to the user. This chapter also covers the tandem switching plans that provide local and access service in small local access and transport areas (LATAs), plus toll service in the larger LATAs. A proposed ultimate plan for exchange carriers' use is included in Chapter 2.

1-1 THE HIERARCHICAL SWITCHING PLAN

Large amounts of traffic between any two central offices are generally routed most economically over direct trunks; however, when the volume of traffic between offices is small, use of direct trunks may not be economical. In these cases, traffic originating from several wire centers destined for one office may be concentrated at intermediate switching locations that connect two or more trunks to build up the required connections [3].

Conversely, where concentrating networks have been established, the amount of traffic between any two offices may become large enough to support direct trunks economically. Thus, an economic balance is maintained between the cost of trunks and the cost of switching equipment.

The original switching plan consisted of a hierarchy of switching offices interconnected by trunk groups in a pattern that provided rapid and efficient handling of long–distance traffic. The hierarchical routing discipline provided for the concentration of traffic and interconnected all offices in the network. The principle of automatic alternate routing provided a low incidence of call blockage with reasonable trunk efficiency. The hierarchical structure of the switching plan was as shown in Figure 1–1. Note that this was a model, subject to technical evolution and to sizable modification in specific cases.

Switching Offices

Under this switching plan, each office involved was designated according to its switching function, its interrelationship with other switching offices, and its transmission requirements. There were five ranks in the hierarchy, as shown in Figure 1–1; the rank of the office was given by its class number with class 1 being the highest. Offices that performed switching functions of more than one rank were assigned the classification for the highest function that they performed. Also, these offices needed to meet the transmission requirements of the highest classification.

End Offices. The central–office entity where loops are terminated is called an end office and was designated class 5. An end office was often located physically in the same building that housed an office of higher classification. In many cases, end–office and toll–office functions were performed by one switching system. A class 5 equipment entity may have been a subgroup of switching equipment, such as a control group in an electronic switching system. However, the offices were considered to be separate entities and loops were terminated at the class 5 office only. The class 5 function was also performed by remote switch units controlled by, and homed on, a host class 5 switch. Centrex switches were also considered to be class 5.

4

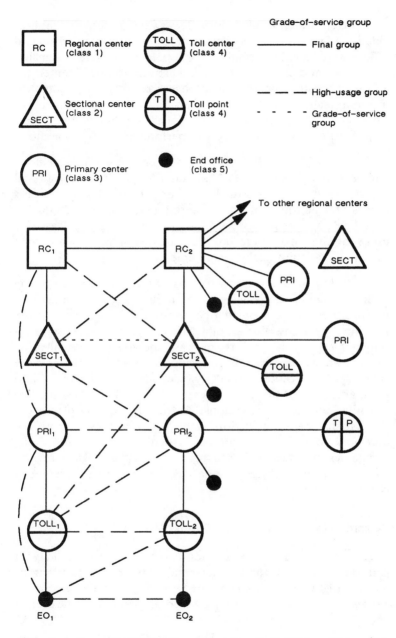

Figure 1–1.   Hierarchical network switching plan for distance dialing.

Toll Centers and Toll Points. The switching centers that provided the first stage of concentration for intertoll traffic from end offices were called toll centers or toll points and were designated as class 4C and class 4P offices, respectively. The toll center was an office at which operator assistance was provided to complete incoming calls in addition to other traffic operating functions. The toll point was an office where operators handled only outward calls or where switching was performed without operators.

Control Switching Points. Regional centers, sectional centers, and primary centers (classes 1, 2, and 3, respectively) constituted control switching points (CSPs). The CSPs were key switching offices at which intertoll trunks were interconnected. To qualify as a CSP, a switching office of a given rank had at least one office of the next lower rank homing on it and met certain switching and transmission requirements.

Switching Areas. The serving area of a switching office of any rank was composed of the areas of all the offices that homed on it. Thus, there were areas that corresponded to each rank in the switching hierarchy. For example, each regional center served a geographical region. Each region was subdivided into smaller areas known as sections, whose principal switching offices were called sectional centers. Similarly, sections were subdivided into small areas served by primary centers. Figure 1–2 shows the two Canadian and ten U.S. regions, and the numbering plan areas (NPAs) that were included in each. Note that these regions are unrelated to the U.S. territories assigned to regional holding companies at divestiture. The figure shows the original regions and the locations of the regional centers, but with present–day NPAs.

Classification of Trunks and Trunk Groups

Trunks were classified in several ways according to traffic types and uses or transmission characteristics. Traffic classifications indicated the way trunks were used in the switching hierarchy. Transmission classifications were based on positions in the hierarchy.

Basic Transmission Types. The network was made up of three types of trunk group distinguished by their respective

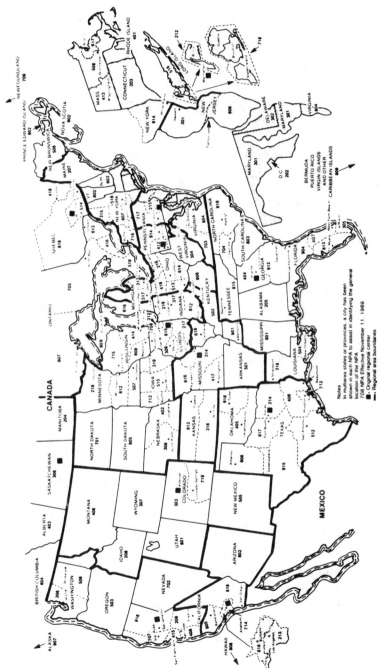

Figure 1-2. Numbering plan areas (including regional area boundaries).

transmission design requirements. A *toll–connecting trunk* connected a class 5 office to any office of higher rank, an *intertoll trunk* connected any class 1 through class 4 office with any other class 1 through class 4 office, and a *direct trunk* interconnected two class 5 offices. The direct trunks may have carried either local or toll traffic; if used for long–haul toll service (e.g., Los Angeles—San Francisco), they were termed *end–office intertoll trunks*.

Final Trunk Groups and Homing Arrangements. Final trunk groups are shown by the solid lines in Figure 1–1. One, and only one, final group was always provided from each office to an office of higher rank; the lower ranking office was said to home on the higher. Class 5, 4, and 3 offices always homed on an office of higher rank, but not necessarily the next higher rank, as shown at RC_2 in Figure 1–1.

Each final group was the route of last resort between its terminal offices; i.e., there was no alternate route and calls failing to find an idle trunk in the group were not completed. Consequently, each final trunk group in the network was engineered for a low probability of blocking, so that on the average no more than a small fraction of the calls offered to such a group in the busy hour would find all trunks busy. Objectives for final groups were that not more than one call in a hundred would be blocked by a no–circuit condition in the busy hour (B.01 grade of service). Final trunk groups interconnected all ten U.S. and the two Canadian regional centers.

A series of final trunk groups connected in tandem constituted a final–route chain. For example, the final–route chain between EO_1 and RC_1 had four final groups; the final–route chain between class 5 offices EO_1 and EO_2 in Figure 1–1 consisted of nine final groups, which represented the path of last resort of a call between these offices.

High–Usage Trunk Groups. In addition to the final trunk groups, direct high–usage trunks were provided between offices of any class where the volume of traffic and economics warranted and where the necessary alternate–routing equipment features were available. However, the choice of traffic carried by these trunks needed to be consistent with routing practices. High–usage trunk groups carried most, but not all, of the offered traffic in the busy hour. Overflow traffic was offered to an alternate route. The proportion of the offered traffic that was carried on a direct high-usage trunk group in each case was determined by the relative costs of the direct route and the alternate route, including the additional switching cost on the alternate route.

Grade-of-Service Group. A trunk group that would have normally been in the high-usage category, but for service or economic reasons was engineered for a low probability of blocking and not provided with an alternate route, was called a grade-of-service group. These groups effectively limited the hierarchical final-route chain for only certain items of traffic but did not change the homing arrangements of their terminal offices. The group shown in Figure 1-1 between $SECT_1$ and $SECT_2$ would have been in the final-route chain for only those end offices that homed on these sectional centers. Traffic destined for other locations would have been switched via the high-usage and final groups to RC_1 and RC_2.

Later Evolution of the Plan

The five-level hierarchy was always a model; in many cases, it was economical to handle traffic without ever overflowing to a sectional or regional center. The average number of trunks in tandem on a long-distance call, including toll-connecting trunks, was slightly over three. Even where the full five-level routing chain was available, only one call in several million ever used the full regional-center-to-regional-center route. In later years, as the cost of direct trunks fell relative to the cost of multiple stages of switching and as electromechanical switches were replaced by digital machines of much larger trunk capacity, the regional centers lost their importance. Some were actually removed. In the limit, one major IC is now using dynamic nonhierarchical routing which, as the name implies, does not follow a fixed hierarchy. It changes routing preferences in accordance with time of day, traffic load, and network management controls.

Network modernization also resulted in the deactivation of large numbers of class 4 offices. Consolidation of class 4 functions by rehoming of end offices made it possible to combine operator forces into larger but fewer teams. The former class 4 switches were "detolled" to class 5. Likewise, remote trunk arrangements for operator access to distant offices spurred the consolidation of operator work groups.

Also later in time, a subclass of office termed 4X came into limited use. Termed an "intermediate point," it was located between a class 4C/4P switch and an end office. It was used for concentration of trunking and access to remote operator positions, generally in rural areas. The switch had to be "transparent" in terms of transmission, and was usually digital.

Call Routing

Calls that were carried by the network were routed according to a plan or set of rules. Elements of the routing plan included the numbering plan, routing codes, and switching office capabilities as well as the network configuration.

Numbering Plan. An essential element of MTS network operation was and is the North American Numbering Plan (NANP) [4], whereby each main station telephone in the entire network is identified by a unique ten–digit number. The first three digits of this number are the NPA code. The remaining seven–digit number is made up of a three–digit central–office code and a four–digit station number.

Destination Code Routing. The NPA and office codes of the numbering plan constitute a unique designation or network address for each central office. A call can be routed from any location in the network to any office using the network address of the destination office. This process is known as destination code routing; the NPA and office codes are called routing codes.

There are other routing codes in addition to the NPA and central–office codes. System group codes are three–digit codes used for routing traffic where calls cannot be routed by NPA code. Nonsystem group codes are one–, two–, or three–digit codes that are used to meet special local needs such as police and fire calls. There are also standard three–digit service codes such as operator codes, test codes, and terminating toll center codes.

CSP Switching Requirements. From a routing code, a switching system must be able to interpret the address information, determine the route to or toward the destination, and manipulate the codes in order to advance the call properly. CSPs without common−channel signalling (CCS) must meet certain switching system requirements for efficient call routing, including storing of digits, variable spilling (deletion of certain digits when not required for outpulsing), prefixing of digits when required, code conversion (a combination of digit deletion and prefixing), translation of three or six received digits, and automatic alternate routing.

Call−Routing Pattern. In the following discussion, the term "final−route chain" is applied to the series of final groups in tandem between a class 5 office and its home class 1 office. The term "overall final−route chain" is applied to the final groups between two class 5 offices.

The routing pattern for a call between two points consisted of a combination of the overall final−route chain between the originating and terminating offices and high−usage groups between switching offices in the chain. A call was switched only at offices on the overall final−route chain. A call was routed only upward along the originating final−route chain shown in Figure 1−1 and only downward in the terminating final−route chain. It may have been offered to a high−usage trunk group that bypassed one or more switching centers along the chain, provided that the call progressed toward its destination. For transmission and administration reasons, calls originating in one final−route chain were not routed along a second final−route chain to destinations in a third chain. This normal routing pattern was sometimes suspended by network managers to accommodate network traffic conditions caused by natural or manmade emergencies, or such unusual loads as occur on Mother's Day.

Route Selection Guidelines. In addition to the fundamentals of call routing, other principles were applied in assigning routes for traffic on existing trunk groups or in establishing new high−usage groups. These guidelines were used to provide economical handling of traffic; they also had a favorable effect on network transmission performance. The guidelines were:

(1) As mentioned before, traffic was to be handled on a direct route whenever such a route was feasible and economical. The ability to overflow from the direct to an alternate route was to be provided.

(2) In general, a direct high–usage group was established between offices of any rank when there was a sufficient volume of traffic to support the group. High–usage trunking was to be developed to the maximum economical extent in order to reduce the requirements of intermediate switching by routing traffic as low in the hierarchy as possible. To achieve this objective, there was a restriction on establishing high–usage trunk groups and the traffic routed over them. By this rule, called the *one–level inhibit rule*, the switching functions performed for the first-routed traffic at either end of the high–usage trunk group were to differ from those at the other end by only one class number. For example, a trunk group may have been established between an end office (class 5 switching function) and a distant regional center, *but only* for the class 4 switching function performed by the regional center switching system. A regional center acted as a toll center for the end offices homing on it; as an extreme example, there were class 5 offices in many of the buildings that also housed regional centers, with class–5–to–class–1 trunk groups connecting them.

(3) In general, traffic between any pair of switching offices, class 1 through class 4, was to have the same first–choice route in both directions. This rule became less applicable as more metropolitan areas acquired switching networks that used directional alternate routing.

(4) The number of intermediate switches was to be kept at a minimum. When there was a choice of routes whose cost differences were not significant, the route with the fewest switches was to be selected.

(5) When there was a choice of routes with an equal number of switches and insignificant cost differences, that route

was to be selected in which switching was done at the lowest level in the hierarchy.

Figure 1–3 illustrates a routing pattern that might have been involved in completing a call from EO_1 to EO_2. In this example, $TOLL_1$ had trunks to PRI_1 only; hence, the call was routed to

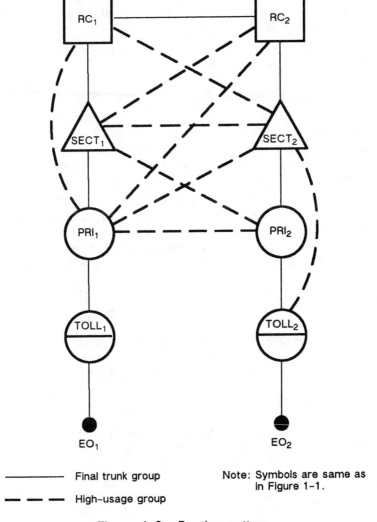

	Final trunk group
– – –	High-usage group

Note: Symbols are same as in Figure 1–1.

Figure 1–3. Routing pattern.

that primary center. At PRI_1 the call was offered first to the high-usage group to PRI_2. At PRI_2 the switching equipment selected an idle trunk in the final group to $TOLL_2$ and the call was routed to the called customer at EO_2.

If all the trunks in the high-usage group between PRI_1 and PRI_2 were busy, the call was next offered to the high-usage group between PRI_1 and $SECT_2$. At $SECT_2$ there was a choice of two routings: (1) via high-usage trunks to $TOLL_2$ or, if all trunks were busy, (2) over the two final trunk groups, $SECT_2$-to-PRI_2 and PRI_2-to-$TOLL_2$.

In the event all trunks in the group between PRI_1 and $SECT_2$ were busy, the call was next offered to the final group to $SECT_1$. There were available at PRI_1 other high-usage groups to RC_2 and RC_1; however, these were intended for certain other traffic items (e.g., terminal) that were to be so routed. Traffic routed via PRI_1 was not to be offered directly to regional centers if there were other lower ranking switching centers in the final-route path, to which the traffic had not yet been offered. It was desirable to restrict the switched load to centers of lower rank, even at the cost of foregoing other alternate-route possibilities. At $SECT_1$ there was a choice of four routings in the following sequence:

(1) via the $SECT_1$-to-PRI_2 high-usage group

(2) via the $SECT_1$-to-$SECT_2$ high-usage group

(3) via the $SECT_1$-to-RC_2 high-usage group

(4) via the final group from $SECT_1$ to RC_1.

Automatic Alternate Routing. The trunking network was so designed that direct high-usage groups were provided as a first choice for traffic between switching offices when the traffic load warranted such groups. These high-usage groups were engineered so that a predetermined portion of the busy-hour traffic was forced to another route where it could be carried at less cost with little or no blocking. A call that found an all-trunks-busy condition on the first route tested was automatically offered in sequence to one or more alternate routes for completion, with the last choice being a final group.

The number of trunks in a direct high—usage group depended on the offered load, the efficiency of added trunks in the alternate route, and the cost ratio of the alternate route to the direct route. The cost ratio is the average incremental annual costs for transmission and switching facilities for one added trunk path in the alternate route, divided by like costs for a trunk in the direct route.

Signalling in the Message Network

Signalling information, such as supervisory and address, was originally conveyed over the trunks themselves. This method is called circuit—associated signalling. As technical advances were incorporated into the message network, common—channel interoffice signalling (CCIS) came into use in the intertoll network in the late 1970s. All three major ICs now use one form of CCS or another. Within today's LATAs, CCS is being deployed to carry call—control signals over a separate signalling network [5]. The CCS network carries supervisory information between elements of the "intelligent network" for call—control and data base queries and responses. Signalling information is transmitted in packets called message signal units and conforms to the Signalling System No. 7 protocol developed by the International Telegraph and Telephone Consultative Committee (CCITT). Switching systems that generate and receive CCS information for call setup are the CCS network's interface to the message network through end offices, tandems, and operator services systems. Signalling transfer points are the packet switches that route supervisory information through the signalling network. Service control points [6] are sites that provide data base access for routing calls in accordance with customer needs and the overall network plan.

1-2   TRANSMISSION PLAN

The switching plan was provided for the handling of most traffic with a minimum of switching. Nevertheless, the most serious impact of the switching plan on transmission was that different combinations of trunks were often used on successive calls (between the same two telephones) and that as many as nine trunks could theoretically be connected. If satisfactory performance were to be provided, the transmission characteristics of every trunk had to be controlled and the plan had to accommodate the varying numbers of analog trunks used without introducing large transmission differences (contrast) on successive calls.

The transmission design of the network had to provide low trunk losses if the requirements of satisfactory speech volumes and low contrast were to be met. However, other factors, such as the provision of margin against singing and echo, tended to make trunk losses high. A compromise design required the design and operation of every trunk at the lowest loss consistent with echo and singing control. The design also required the implementation of a program of transmission maintenance designed to ensure that trunks met their requirements and were kept as uniform as possible.

Network Transmission Design

The transmission design of the network assumed the use of 500–type or equivalent station sets connected to class 5 offices by means of two–wire loops. This two–wire loop operation created conditions of low return loss that limited the minimum loss at which network trunks could operate without echo or singing. These problems were controlled by the via net loss (VNL) design and in some cases by the use of echo suppressors and echo cancellers.

Via Net Loss Design. The relationship between the minimum loss required to control echo and the round–trip delay between class 5 offices is shown in Figure 1–4. This relationship was the basis for VNL design (see Volume 1, Chapter 26, Part 1). Inspection of Figure 1–4 shows that as the number of analog trunks is increased, an increase in loss of 0.4 dB per added trunk is required. This increment compensated for the greater loss variability that occurs with an increased number of analog trunks in the connection. The VNL design rules were applied to all trunks in a connection when the round–trip delay in the overall

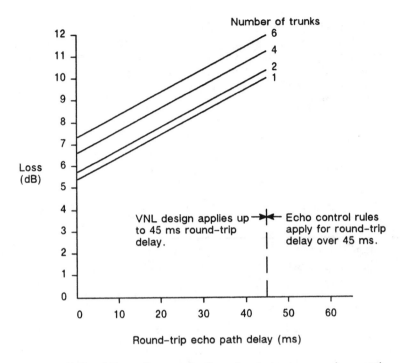

Figure 1-4.   Overall connection loss versus echo path delay between class 5 offices.

connection was less than 45 milliseconds (ms). The 45-ms restriction was imposed to limit the maximum trunk loss to a value that permits satisfactory received speech volume. When the round-trip delay was more than 45 ms, one of the trunks in a connection was equipped with echo control. Interregional trunks equipped with echo control operated at zero loss.

Via Net Loss and Via Net Loss Factors. The overall connection loss (OCL) between class 5 offices, shown in Figure 1-4, is given by the expression

$$OCL = 0.102D + 0.4N + 5.0 \text{ dB} \qquad (1-1)$$

where D is the round-trip echo path delay in ms and N is the number of analog trunks in the connection. In order that each analog trunk operated at the lowest practical loss, 2.5 dB of the 5.0-dB constant was assigned to each toll-connecting trunk in the connection. The 2.5–dB loss included an allowance of 0.5 dB for the loss, formerly allocated to loops, in battery supply equipment. The remaining loss was assigned to all trunks in the connection, including the toll–connecting trunks. This remainder is called via net loss and is expressed as follows:

$$VNL = 0.102D + 0.4N \quad dB. \qquad (1\text{-}2)$$

Then, for each trunk in a connection

$$VNL = 0.102D_t + 0.4 \quad dB \qquad (1\text{-}3)$$

where D_t is the round–trip echo path delay in ms for the trunk.

Since the echo path delay of a trunk is proportional to its length, Equation 1–3 is usually given in terms of trunk length and a via net loss factor (VNLF) for the trunk facility type as

$$VNL = VNLF \times \text{trunk length in miles} + 0.4 \quad dB \qquad (1\text{-}4)$$

where VNLF = (2 × 0.102 ÷ velocity of propagation in miles per ms) dB per mile. Equation 1–4 is used in VNL calculations. For example, assume that the VNL of a 600–mile intertoll trunk on carrier facilities is to be determined. The VNLF for carrier facilities is 0.0015 dB per mile. Therefore, for this trunk,

$$VNL = (0.0015 \times 600) + 0.4 \quad dB = 1.3 \ dB. \qquad (1\text{-}5)$$

The VNL concept is largely obsolete for the message network, but remains embedded in industry loss plans for special–services networks.

Echo Control. Echo suppressors are four–wire signal–activated devices that insert a high loss in the return echo path when speech signals are transmitted in the direct path. They give an effective echo improvement of about 30 dB. Since tandem echo suppressors produce degradation in received speech, application rules permitted only one suppressor in a connection. This restriction was readily met because of the hierarchical structure of the network and the finite size of the regional center areas. The maximum echo path delay within a region was usually low enough that echo suppression was not required for intraregional trunks.

However, it was possible to exceed 45 ms delay for connections between points in different regional center areas. Therefore, echo suppressors were required on roughly half of the regional–center–to–regional–center trunks. In addition, echo control was used on interregional high–usage intertoll trunks, on interregional toll–connecting and end–office toll trunks more than 1850 miles long, and on any trunk routed via a communication satellite.

Originally, echo was controlled by echo suppressors applied circuit by circuit. Later on, digital echo suppressors became available as part of the switch frames in digital toll switches, and echo cancellers came into use on both a per–circuit and a DS1 (24–channel) basis.

A national test survey made in the peak years of the VNL plan [7] indicated that echo suppressors were present on 0.4 percent of connections between 360 and 725 miles in length, 25.4 percent between 725 and 1450 miles, and 90.1 percent between 1450 and 2900 miles. The mean echo path delays for the same three mileage bands were 16.4, 24.8, and 37.3 ms, respectively. These results allowed increasing the distance for application of echo control to 1850 miles from a previous value of 1565.

Analog Trunk Loss Objectives with VNL Design. The VNL objectives for analog trunk losses were stated in terms of inserted connection loss (ICL), defined as the 1000–Hz loss inserted by switching the trunk into an actual operating connection. Figure 1–5 shows the ICL objectives for intertoll, toll–connecting, and end–office intertoll trunks. If an interregional toll–connecting trunk were more than 1850 miles long, it was equipped with echo control and operated at an ICL of 3.0 dB. The ICL design objective for end–office toll trunks between class 5 offices was VNL + 6.0 dB with a maximum of 8.9 dB.

Fixed Loss Plan. With the introduction of digital switching machines and their integration with digital transmission facilities, the loss plan for the switched message network was modified so that a fixed 6–dB loss was specified for all toll connections. In this fixed loss plan, each toll–connecting trunk is allocated 3 dB of loss and each intertoll trunk is to operate at 0–dB loss. These loss objectives were to be applied to the ultimate all–digital network; during the transition from the predominantly analog

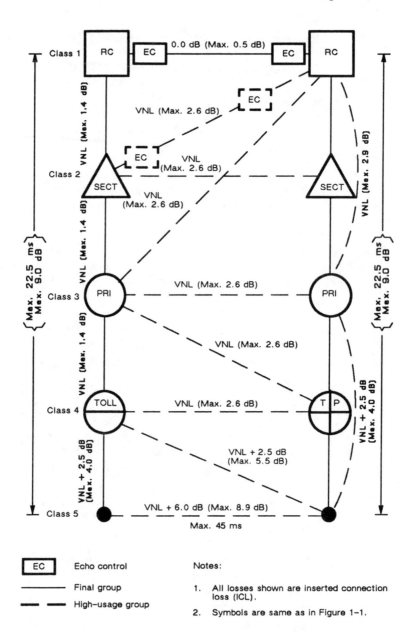

Figure 1-5.   Trunk losses and maximum round-trip delay with VNL design.

network to the ultimate digital network, compromise loss objectives were to be applied, e.g., 1 dB for trunks from an analog toll office to a digital one. This plan was not fully in place at the time of divestiture; however, it led to a modern successor, the region digital switched network (RDSN) plan described in Chapter 2.

Through and Terminal Balance. In the development of VNL design, the only reflections considered to be significant from an echo and singing standpoint were those at the class 5 offices. There was no intermediate echo if the entire connection between class 5 offices including the switching paths were four-wire. However, many class 4 offices and some CSPs employed two-wire switching systems. At two-wire offices, special procedures were implemented to reduce reflections to a point where they approximated four-wire operation. Also, at four-wire switching offices, reflections caused by two-wire toll-connecting and switchboard trunks were controlled.

Through Balance at Two-Wire CSPs. All intertoll trunks were provided on four-wire facilities. At two-wire CSPs, four-wire terminating sets were used to convert these trunks to two-wire for switching. On a connection of two intertoll trunks through a two-wire CSP, echoes arose due to the imbalance between the impedances of the balancing network and the two-wire side of each four-wire terminating set. This two-wire impedance is the two-wire input impedance of the other four-wire terminating set involved in the connection, as modified by the office equipment and cabling. Drop build-out capacitors on each trunk were used to give the same capacitance for all paths through the office; network building-out capacitors on the balancing network of each four-wire terminating set were applied to make that network match the office capacitance. Thus, a 27-dB through balance adequate for VNL operation could be achieved.

Terminal Balance. The balance at the point where an intertoll trunk was switched to a toll-connecting trunk was called terminal balance. Generally, it was the balance between the balancing network of a four-wire terminating set and a toll-connecting trunk appropriately terminated at the class 5 office. Terminal balance improvements were made by using impedance compensators in two-wire metallic toll-connecting trunks so that their impedance more closely resembled the relatively fixed impedance of the four–wire terminating set network. In addition, at two–wire offices the effects of office cabling had to be treated in a manner similar to that used for through balance.

Matching Office Impedance. The switching office impedances used were 900 ohms for all class 5 and crossbar tandem offices, and 600 ohms for all other tandem and toll offices. These impedance values were based on average impedances of trunk and loop facilities connected to the office.

Trunk impedances had to match both local and toll office impedances in order to meet terminal balance requirements. At class 1, 2, 3, and 4 offices, all intertoll and toll–connecting trunks were designed to the common office impedance. At class 5 offices, incoming and outgoing trunk circuits were designed to match the 900–ohm compromise value, representing a nominal value for loop facilities.

Maintenance Considerations

In the development of the VNL plan, only a small variation of trunk losses from assigned values was considered. To meet all the requirements of the VNL plan, a trunk was not placed in service unless it met all of the applicable circuit–order requirements; tests were performed often enough to detect transmission difficulties before they could have a significant effect on network performance. In addition, troubles found by tests had to be corrected promptly. Otherwise, the trunk was to be removed from service until remedial measures were taken.

1-3   METROPOLITAN NETWORK PLANS

Most exchange carriers serve metropolitan areas comprising complex trunk and   switching networks. These networks are complex because of the number of end offices to be interconnected, the volumes of point–to–point traffic to be carried, and the possible special routings required for call accounting, operator assistance, and signalling conversion. Also, a few remaining end offices are not equipped for alternate routing.

The substantial traffic load of large metropolitan areas and the variety of interlocal trunking arrangements in use have prompted recommendations for metropolitan networks for general use. Requirements for economy and better service under unusual traffic loads and the availability of switching systems capable of regulated alternate routing (dynamic overload control) were motivating factors in the development of the metropolitan network plan. The preferred arrangement, called the *multialternate routing* (MAR) arrangement, has the the general characteristics of multistage automatic alternate routing, multitandem switching in the final route, optional integration of local and toll traffic, and the use of a high–volume or directional tandem office where needed to augment the basic network.

Studies are made to anticipate the characteristics of a metropolitan network. Strategies must be developed for growing in this network by reflecting the variable cross–section requirements year by year, exhaustion of switching machines, etc. Capital and expense requirements are developed for each plan for comparison purposes.

There are a number of metropolitan tandem arrangements in addition to the MAR configuration. An understanding of call routing for each arrangement provides a background in metropolitan network trunk switching patterns. The three local and LATA access switching–office combinations that use the MAR arrangement have different but related transmission designs, each having slightly different performance characteristics.

Metropolitan Tandem Networks

With modern high–capacity digital switching systems, a metropolitan area is often served by only one tandem switching system. The tandem office also provides access to operator services, as covered more fully in Chapter 10. Where more tandem systems are required, the area may be subdivided into smaller areas called *sectors*. A sector is composed of the serving areas of a number of end offices. These offices are not necessarily contiguous but offer a blend of traffic such that advantage can be taken of the noncoincidence of busy–hour local– and toll–traffic loads. Each sector is served by a *sector tandem*, which is a local area switching center used as an intermediate switching point for traffic between other offices. The trunks used in these networks may be operated one–way or two–way; i.e., address signalling be-tween switches may be in one or both directions. Four types of trunk are used in the local networks: *direct* trunks, which inter-connect end offices; *tandem* trunks, which connect end offices to tandem offices; *intertandem* trunks, which interconnect tandem offices; and LATA access trunks.

There are special local conditions where network configura-tions other than the MAR arrangement have service or economic advantages. In addition, it may be appropriate for a specific net-work to use two or more configurations. This is inevitable in met-ropolitan networks, which are usually in the process of planned change from one configuration to another. Nevertheless, the MAR is considered the objective in network planning. Several tandem network configurations are capable of automatic alter-nate routing and thus lend themselves to eventual conversion to the double tandem arrangement. These include the single tan-dem sector–originating network, the single tandem sector–termi-nating network, and the central tandem system.

Single Tandem Sector–Originating Network. In the sector–originating network, shown in Figure 1–6, the metropolitan area is sectored either geographically or on a traffic basis. Traffic originating in an end office, EO_A, is routed directly to the called office, EO_B, on a direct high–usage trunk group if there is such a group. Overflow traffic is routed to home sector tandem $T1_A$. Where there is no direct high–usage group, all traffic is routed to $T1_A$. Each sector tandem has final one–way trunk groups to all end offices in the metropolitan area, including the offices in its sector.

The relative efficiencies of the trunk groups to and from the sector tandem result in a smaller number of trunks operating into the tandem office than outward. A self–regulating effect is thus provided under severe overload conditions when excess call at-tempts are blocked at the originating end offices; calls that reach the tandem office then have a reasonable chance for completion. This effect, and the fact that the sector–originating network adapts more readily and inexpensively to network management controls, make this network preferable to the sector–terminating and central tandem networks.

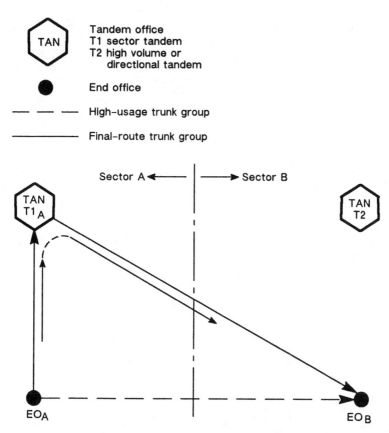

Figure 1-6.   Call routing for single tandem sector-originat-
ing network.

Single Tandem Sector-Terminating Network. In the sector-
terminating network of Figure 1–7, the tandem office switches
traffic inward to the end offices within its sector. Each sector
tandem has final one–way trunk groups from every end office in
the area. Traffic originating in EO_A for EO_B is routed via a direct
high–usage group if there is one, and the final one–way trunk
groups carry the overflow. If there is no high–usage group, all
traffic is routed to EO_B through tandem $T1_B$.

Central Tandem System. In the central tandem alternate-
routing system of Figure 1–8, a single central tandem office has
final trunk groups (usually one–way) to and from each end office

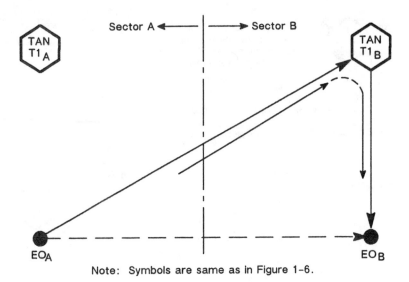

Note: Symbols are same as in Figure 1-6.

Figure 1-7.   Call routing for single tandem sector–terminating network.

in the entire metropolitan area. The area is sectored; each sector tandem has final groups to the end offices in its sector.

Traffic routing is determined by traffic volumes offered and the cost ratios involved. As shown in Figure 1–8, the first possible route is a high–usage group direct to the called office. The second and third possible routes are via high–usage groups to the subsector tandem and sector tandem, respectively. The final route is established via the central tandem.

Multialternate–Routing Network. In the MAR network, each sector tandem has final routes to and from each end office within its sector. Final intertandem trunk groups interconnect the tandem offices. High–usage groups may be established between a sector tandem and end offices outside its sector, as illustrated in Figure 1–9, by the connections from EO_A to $T1_A$ to EO_B. The four possible routings are shown in the figure.

Figure 1–10 illustrates the combined LATA access/tandem arrangement for the MAR network.

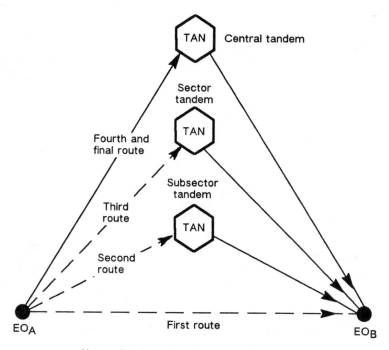

Note: Symbols are same as in Figure 1-6.

Figure 1-8. Call routing for central tandem system.

Some advantages of the MAR over the other networks are lower cost, more even distribution of traffic under distorted overload conditions, simpler application of network management, more adaptability to changes in interLATA–to–local calling patterns, more flexible routing, and superior ability to use available capacity throughout the network.

Transmission Considerations

The MAR network is a highly flexible configuration that can be used as a local trunk network or a mixed local and LATA access network. The transmission requirements, although generally consistent, vary with each configuration.

General Network Requirements. The following general requirements for satisfactory transmission performance apply to each arrangement:

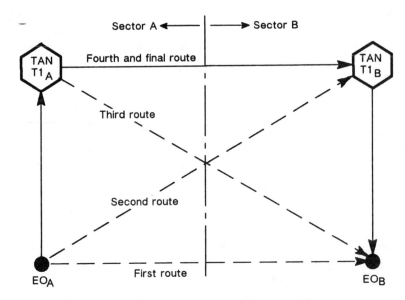

Note: Symbols are same as in Figure 1-6.

Figure 1-9. Call routing for the MAR network.

(1) There should be no more than three analog trunks in any connection between end offices (not a severe limitation in today's environment).

(2) The distance between extreme points in the metropolitan serving area should not exceed about 150 route–miles. This guideline is selected to ensure that round–trip delays in excess of 10 ms on three–link connections are rarely exceeded. Beyond these limits, echo would become a problem. The round–trip delay expected in a given net-work connection depends on the types of facilities en-countered.

To illustrate the intertandem distance that is feasible, consider a three–link connection consisting of two tan-dem trunks and one intertandem trunk, operating through a pair of digital tandem offices D and J. Each sector tandem serves a 30–mile radius on T1 span lines, reaching (for illustration) digital end offices A and Z. The tandem trunks are each routed through a digital

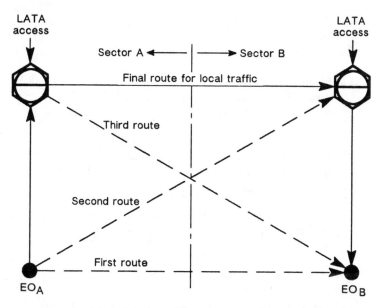

Figure 1-10. Combined LATA access/tandem arrangement for the MAR network.

cross-connect system (DCS) that adds delay of its own. If the intertandem facility is a fiber system with a round-trip delay of 0.0168 ms/mile, the interoffice echo delays are as follows:

End office A:	1.21 ms
A–D T1 spans: 30 mi × 0.0158 ms/mi	0.47
DCS at D:	1.28
Tandem D:	1.35
D–J fiber span:	x
Tandem J:	1.35
DCS at J:	1.28
J–Z T1 spans: 30 mi × 0.0158 ms/mi	0.47
End office Z:	1.21
	10.00 ms

The allowable fiber span delay x is then $(10 - 8.6) = 1.4$ ms. This equates to an allowable mileage of about 1.4 ms/0.0168 ms/mi or 80 miles.

Chapter 2, devoted to the conversion of the analog network into a digital network, discusses this consideration of increased round–trip delay in a digital environment.

(3) Combined LATA access/tandem installations, unless digital or four–wire analog, must meet through and terminal balance objectives applicable to LATA access offices (see Figure 1–11).

(4) Intertandem trunks should use four–wire facilities (in effect, carrier) and trunk terminations that meet LATA access requirements.

(5) Precision balancing networks should be used in the four–wire terminating sets of those two–wire trunks that terminate in four–wire tandem offices.

Local Networks. Figure 1–12 shows a network consisting of local trunks only; the ICL objectives are also shown. One advantage of this network is that the local tandem trunks do not require terminal balance treatment at an analog sector tandem. For digital tandem offices connected by digital facilities, the RDSN concept in Chapter 2 allows a uniform loss for all connections rather than the variable loss resulting from mixed types and numbers of trunks.

Combined Local and IntraLATA Toll Networks. It was previously pointed out that from a call–routing standpoint there is a possibility of combining local and intraLATA toll traffic on the tandem trunk groups, particularly in the larger LATAs. From a transmission standpoint, the same possibility exists. The ICL objective for analog tandem trunks is VNL + 2.5 dB or, for trunks shorter than 200 miles, 2.0 to 4.0 dB without gain or 3.0 dB with gain or carrier. These objectives fall within the range for local tandem trunks (0.0 to 4.0 dB). The same applies to zero–loss fully digital trunks.

The switching machines in these offices must be capable of switching either local or toll traffic. The tandem trunks are also

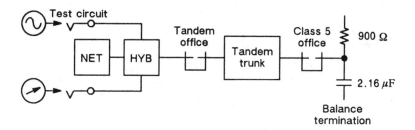

MEASUREMENTS		
	50% equal to or exceed (dB)	Minimum (dB)
Echo return loss	22	16
Singing point	15	11

(a) Test of tandem trunk

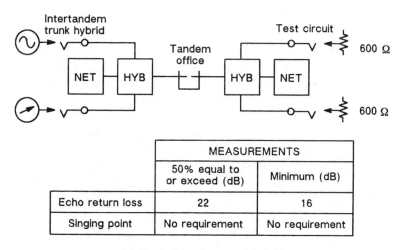

MEASUREMENTS		
	50% equal to or exceed (dB)	Minimum (dB)
Echo return loss	22	16
Singing point	No requirement	No requirement

(b) Test of tandem trunk hybrid

Figure 1-11. Balance test requirements for two-wire combined LATA access/tandem office.

used as intrasector tandem trunks. The intertandem trunks are operated at 0 dB for digital trunks between digital switches;

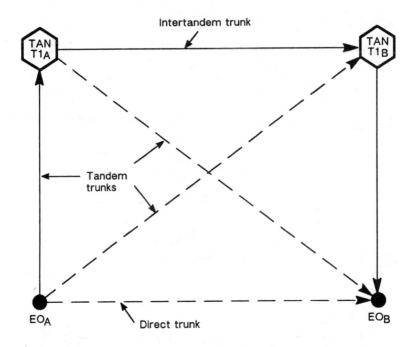

Direct trunk
- Digital switches and facilities: 0-dB trunk, 3-dB decode loss
- Other cases: ICL design, 3 dB on carrier or with gain (5.0 dB acceptable); 0-5 dB on wire without gain

Tandem trunk
- Digital switches and facilities: 0-dB trunk, 3-dB decode loss
- Other cases: ICL design, 3 dB on carrier or with gain; 0-4 dB on wire without gain

Intertandem trunk (four-wire)
- Digital switches and facilities: 0 dB
- Other cases: ICL design, 1.5 dB

Figure 1-12. Metropolitan local trunk network.

otherwise, they are operated at 0.5 dB loss, which is feasible because of adequate terminal balance of the tandem trunks. The resulting facility diagram is effectively Figure 1-12, but with reduced loss on the intertandem trunks.

For an all-digital connection, with the only digital-to-analog conversion occurring on the drop side of the digital end offices, the metropolitan network plan specifies a loss of 3 dB for inter–end–office connections. For connections involving a mixture of analog and digital transmission and switching facilities, the plan results in a typical loss range from 3 to 9 dB.

Mixed Local and LATA Access Networks. The network shown in Figure 1–13 is a combined local and LATA access network. Two types of switching are performed. Sector tandem $T1_A$ switches local traffic while the combined LATA tandem office switches both local and LATA access traffic. Trunks from the end office to the sector tandem meet tandem trunk requirements; trunks from the end office to the access tandem must meet LATA access requirements. Intertandem trunks

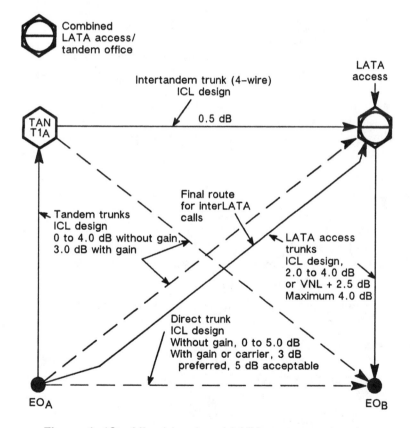

Figure 1–13. Mixed local and LATA access network.

operate at zero loss if fully digital; otherwise, they operate with an ICL of 0.5 dB.

Tandem offices may have specialized functions for special-purpose subnetworks. In the area of LATA access, an equal access tandem for Feature Group D access (covered in Chapter 18) may be separate from the tandem for Feature Group B. In the area of intraLATA routings, there may be specialized tandem functions for 976 information services, 900 mass-calling services, selective routing for Enhanced 911 service, "choke" networks for radio-station and similar group-calling situations, and special arrangements for group-address-bridging calling. Access to information-services "gateways" may require further special routing. However, transmission considerations are similar for all of these cases.

Expected Network Performance. The local network and the combined local and LATA access network have been analyzed to evaluate transmission network performance. The parameters used in the analysis were talker echo, singing-point stability, noise/volume grade of service, loss/noise grade of service, and loss contrast. On the basis of the five parameters analyzed, it was concluded that the networks shown in Figures 1-12 and 1-13 perform satisfactorily from a transmission standpoint.

References

1. Andrews, F. T., Jr. and R. W. Hatch. "National Telephone Network Transmission Planning in the American Telephone and Telegraph Company," *IEEE Transactions on Communications Technology*, Vol. COM-19 (June 1971), pp. 302-314.

2. Huntley, H. R. "Transmission Design of Intertoll Trunks," *Bell System Tech. J.*, Vol. 32 (Sept. 1953), pp. 1019-1036.

3. Wernander, M. A. "Systems Engineering for Communications Networks," *IEEE Transactions on Communications and Electronics*, Vol. 83 (Nov. 1964), pp. 603-611.

4. "Numbering Plan and Dialing Procedures," *Notes on the BOC IntraLATA Networks—1986*, Bellcore (Iss. 1, Apr. 1986), Section 3.

5. Hass, R. J. and R. B. Robrock. "Introducing the Intelligent Network," Bellcore EXCHANGE, Vol. 2, Iss. 4 (July/Aug. 1986).

6. Boese, J. O. and R. B. Robrock. "Service Control Point: The Brains Behind the Intelligent Network," Bellcore EXCHANGE, Vol. 3, Iss. 6 (Nov./Dec. 1987).

7. Duffy, F. P. et al. "Echo Performance of Toll Telephone Connections in the United States," *Bell System Tech. J.*, Vol. 54, No. 2 (Feb. 1975), pp. 209–244.

Additional Reading

Bevacqua, F., B. M. Cooney, and D. P. Coyle. "Operations Support for a Brand–New Network," Bellcore EXCHANGE, Vol. 4, Iss. 4 (July/Aug. 1988), pp. 18–23.

Chapter 2

Region Digital Switched Network Plan

Many factors have influenced the makeup of the message network in recent years: regulatory rulings, judicial decisions, standards activities, and technological progress. However, the elements of switching, transmission, and signalling are still fundamental. The unified network described in the first part of Chapter 1 has been broken into segments within and between local access and transport areas (LATAs). Exchange carriers have network plans that interconnect switching offices within their LATAs. End-to-end connections involving two LATAs require compatible transmission plans for both exchange and interexchange carriers (ICs).

A major technological influence on the message network is the deployment of digital transmission and switching. Since the conversion from analog to 100-percent digital, although proceeding vigorously, will take considerable time to complete, there will be a mix of analog and digital transmission/switching facilities for some years. The region digital switched network (RDSN) transmission plan [1] provides for the transition from (1) an analog configuration with digital portions to (2) a digital network with analog portions, and finally to (3) an all-digital design. Use of the plan conditions the message network to interoperate smoothly with the integrated services digital network (ISDN). The plan supersedes previous analog/digital transmission plans [i.e., via net loss (VNL) and fixed-loss] for the message network.

2-1 NETWORK ARCHITECTURE

The RDSN transmission plan uses a two-tier hierarchical network. The two-tier system provides for existing analog end offices (EOs) during transition, and allows consistency with the wide use of LATA and LATA-access tandems. As analog EOs are converted to digital, the network can evolve into a purely digital configuration. Therefore, within the RDSN existing transmission constraints, except for round-trip delay, no longer apply. This plan also provides for the transport of digital bit streams through the network without modification. A key value of the design is referred to as *digital signal integrity*, or simply, *bit integrity*.

Figure 2-1 is a legend of the symbols used to illustrate network configurations. Figure 2-2 shows the two-tier RDSN after full replacement of analog switches.

In the RDSN, the loss is zero between encode and decode points and the signal retains the same digital code words through all connections. The zeros next to trunk groups in Figure 2-2 emphasize their lossless design. Connection loss, a desirable feature to avoid excessive loudness on connections between analog telephone sets, is accomplished by digital-to-analog conversion at a reduced transmission level [2] to preserve a high transmission grade of service. Performance studies involving customer loops of varying lengths with both conventional and current-limited battery sources in the switches, using established network performance models, indicate that a universal decode level of -6 dB (a connection loss of 6 dB) is appropriate.* However, 0 dB on intraoffice calls or -3 dB on interlocal calls are alternate decode levels. The plan specifies end-to-end losses for all types of connection through the network during and after the transition from an analog network. This transition, accomplished in planned phases, takes advantage of the economic and functional benefits of digital technologies while maintaining the soundness of existing networks.

To ensure that network transmission goals are consistent with those of ICs, exchange carriers, and equipment manufacturers, planning effort continues to refine and evolve the RDSN transmission plan as new technology is introduced and new field data are compiled.

---

* Organizations such as the American National Standards Institute (ANSI) T1 Committee set standards; thus, their ongoing activities will describe standard loss/level values for connections of various types.

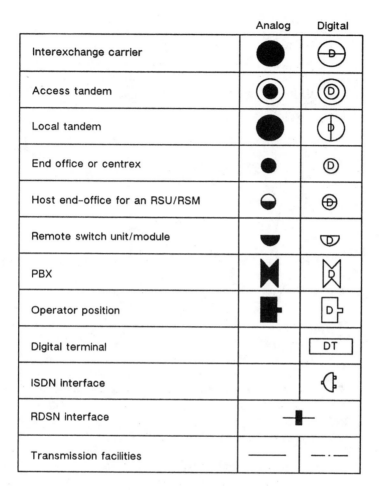

	Analog	Digital
Interexchange carrier		
Access tandem		
Local tandem		
End office or centrex		
Host end-office for an RSU/RSM		
Remote switch unit/module		
PBX		
Operator position		
Digital terminal		
ISDN interface		
RDSN interface		
Transmission facilities		

Figure 2-1. Legend of symbols used in RDSN figures.

Some of the benefits of the RDSN plan are:

(1) Improved voiceband transmission and reduced costs. Analog–to–digital and digital–to–analog processing are performed only at the originating and terminating interfaces of the network.

(2) Minimized effort in trunk design, installation, and maintenance. Elimination of VNL means that facility mileage

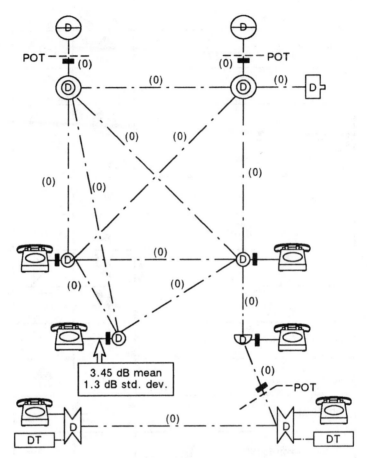

Figure 2-2. RDSN configuration.

need not be considered in design and that echo delays in digital cross-connect systems (DCSs) cannot upset the accuracy of loss calculations. Operations activities are simplified because, where RDSN trunks extend to an analog office, the decode pads in the channel banks are set to a single value.

(3) Greatly simplified pad-control tables in translations for digital switches.

(4) Improved transmission of high-speed data in the 4.8-to-19.2-kb/s range. The elimination of digital pads

slightly reduces intermodulation distortion. The loss upon decoding gives increased tolerance of echoes originating in customer–premises terminals having low return loss. Exchange carriers do not formally support a particular level of error performance through customer–premises equipment. However, the heavy and increasing usage of the network for high–speed data and facsimile traffic, and the emergence of information–services "gateways," make consistent high performance desirable.

(5) Reduced variations in transmission loss from one call to another. Because trunks are switched together on a zero-loss basis, there is no difference in loss between a direct and a tandem route. It becomes possible to use the "emergency tandem" feature of an end–office switch to press ordinary interlocal trunks into tandem use during facility or switch outages, without concern for additional losses. The transmission contrast that occurs when re-mote call forwarding is used in an analog context, caused by connecting additional analog trunks in tandem, disap-pears. Sharing of an end–office code (NXX) between of-fices becomes straightforward in terms of transmission. The reduced variations in signal level are helpful when connecting to conventional or cellular mobile systems that use analog radio channels, because consistent signal level improves the signal–to–noise performance of the ra-dio. Reduced variation is also helpful in terms of the speech volume delivered to 911 emergency operators, who often deal with soft–voiced callers. Digital intercon-nection to future digital cellular networks is similarly straightforward.

(6) Reduced maintenance and testing costs. The use of digi-tal surveillance techniques in place of routine analog per-trunk testing promises considerable savings.

(7) Elimination of test pads (TP2, TP3), with their exposure to confusion in making trunk–loss measurements.

(8) Easy reconfiguration of the network through software control instead of hardware changes.

(9) Adaptability to new services through digital transmission. Examples are easy cross–network operation with ISDN and smooth provision of digital area–wide centrex service. The RDSN provides transparent digital connection to other networks, even those that internally use special coding systems. [Examples of such unique internal codes are adaptive differential pulse code modulation (ADPCM) in private networks, 7–kHz audio in some ISDN applications, 8–kb/s coding in possible digital cellular networks, or packetized speech.]

(10) Ready interoperation with digital private networks. Interconnection is aided by the assurance of bit integrity, even though the owner of the private network has the option of internally using nonintegrity coding techniques.

Some of these values are inherent to any all–digital network; however, the RDSN concept improves even the features that are not unique to it.

2-2 REGION DIGITAL SWITCHED NETWORK FEATURES

In an RDSN, by definition, there is no analog switch or transmission facility. The digital network begins at the originating analog–to–digital conversion point with a pulse code word and ends at the terminating digital–to–analog conversion point with the same pulse code word (bit integrity). These digital–to–analog and analog–to–digital points mark the boundaries of the network in the transmission sense; network interfaces to customers or ICs may intervene in the legal sense. The RDSN is a network within a network; its boundaries are set by electrical considerations within the physical network of switches and facilities.

It is important to point out that bit integrity, in the RDSN sense, is either 56– or 64–kb/s clear–channel capability (CCC); the RDSN concept is compatible with either. With 56–kb/s CCC, the eighth or least significant coding bit is still available for signalling; it may be borrowed multiple times in successive passes through digital switches, echo cancellers, or DCSs. The seven most significant bits, however, are protected.

The RDSN plan allows increased round–trip delay within a region compared to the transmission plan for a metropolitan

network. The increase facilitates use of a hubbing facility architecture, self–healing fiber–ring facilities with their relatively long
mileage, remote switching modules (RSMs), digital switches,
DCSs, and digital loop carrier (DLC) systems that include time–
slot interchangers. All of these elements produce echo delay;
where they replace analog elements, the delay is usually increased substantially. The RDSN plan, with its use of fixed loss
on interoffice calls, provides some inherent echo control. To
improve tolerance of added echo delay, the plan takes advantage
of improvements in effective loop return loss obtained by the
progressively more effective (and more expensive) measures of
nonloaded–loaded loop segregation in digital offices, adaptive–
hybrid (echo–canceller) line circuits, and echo cancellers in selected trunk groups. Loop segregation has been modeled as
improving loop return loss from an average of 11 dB (standard
deviation of 3 dB) to an average of 13.6 dB (standard deviation
of 2.8 dB) [3]; the other two technologies are capable of return
losses of 30 dB or better.

The RDSN plan eliminates the need for traveling classmarks or
common–channel signalling packets to accommodate mixed
types of traffic on a given trunk group, i.e., to set decode levels
call by call.

When the digital network is extended to an end–user's premises, the coding and decoding are performed by the end user.
Loss insertion becomes the function of the end–user's equipment. The user gains full control of end–to–end loss for compatibility with private–network loss plans. This feature, inherent to
ISDN, is also a feature of basic DS1 interconnection of a digital
end office to a digital private branch exchange (PBX) for direct–
inward–dialing and similar trunks.

The plan provides for access for ICs and operator services.
The RDSN can provide reliable transmission connections with
enhanced capabilities. This reduces the need for unique arrangements for special–service demands and provides for integrating
other services into the RDSN transmission plan. The digital network can pass all message and data traffic without distinction and
allow for special connections without additional equipment.

The RDSN plan requires only minimal restrictions: (1) the upper tier of the network hierarchy must be digital (i.e., digital

switches interconnected with digital transmission facilities) and (2) integrated digital loop carrier (IDLC) systems must be able to operate with RDSN decode levels.

2-3 ANALOG-TO-DIGITAL TRANSITION STRATEGIES

The RDSN transmission plan can be implemented most effectively in planned phases. Although each phase contains certain defined steps, it is flexible enough to meet the needs of the local situation.

Existing analog loss and echo performance is determined by the VNL plan; for all-digital trunks, it is determined by the fixed-loss plan. Figure 2-3 depicts the classic VNL network model prior to the 1984 separation of network responsibilities.

It shows the assumed distribution of echo path delay: 10.8 ms maximum from the end (class 5) office to a toll center (class 4) in the same switching region, 35 ms to an out-of-region toll center, or 22.5 ms to the normal regional center (class 1 office). The nominal echo delay from one end office to another without echo control was a maximum of 45 ms. The inserted connection loss (VNL + 5.0 dB) corresponding to a 45-ms echo delay was about 9 dB. However, the use of VNL-based design on short regional-center-to-regional-center groups implied occasional delays beyond 45 ms without echo control (and its own user-perceived impairment of quality). The value of 10.8 ms for toll-connecting trunks would allow, in terms of today's facilities, for about 600 miles of fiber or 950 miles of digital radio. The 9.8 ms for inter-toll trunks from a switch to the next higher office (e.g., class 3 to class 2) would permit about 540 miles of fiber or 860 miles of digital radio. However, this model assumed that delays in switches and subscriber loops could be neglected, an assumption no longer valid. A survey taken just before the introduction of digital switching [4] gave a median echo path delay for transcontinental (2700-mile) calls of 45 ms.

Figure 2-4 is a nominal intraregion network of a few years ago. It assumes analog switches in use by the connecting ICs, and analog tandem switches. Trunking among these machines displays typical VNL-oriented loss values. Analog loops are shown as

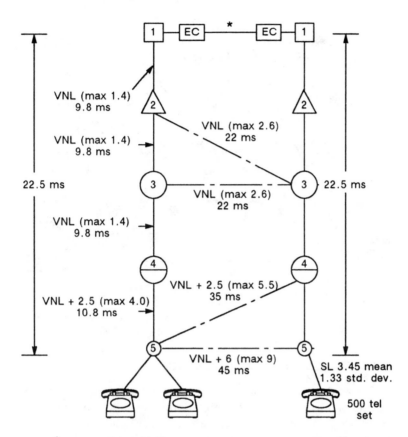

* VNL (max. 2.9) without echo control, up to 1850 miles; 0.0 dB (max. 0.5) with echo control, above 1850 miles.

Figure 2-3. Predivestiture VNL network.

having a mean loss of 3.45 dB [5] and a standard deviation of 1.3 dB, consistent with past and ongoing design criteria. Figure 2-1 gives a legend of symbols for these figures.

Table 2-1 is an example of transition phases to implement the RDSN transmission plan in the network of Figure 2-4. As transition proceeds, the domain included in the RDSN expands.

Figures 2-5, 2-6, and 2-7 illustrate the transition phases of the RDSN. Figure 2-2, shown earlier, gives the final result.

45

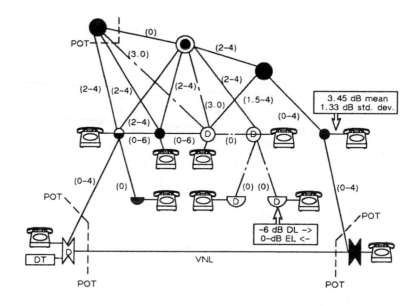

Figure 2-4. Typical intraregion network—recent past.

2-4 TRANSMISSION MAINTENANCE

The analog transmission maintenance process depends on re-curring per–trunk testing to maintain the network. This has been necessary on trunk groups between analog switches, and between an analog switch and a digital one, because there are numerous per–circuit components that can fail (e.g., switchpoints, trunk relay units, and carrier channel units) without alarms. Metallic loops receive routine line–insulation tests, while DLC channels are supervised by carrier alarms.

In the initial digital/analog environment of the RDSN, mainte-nance is supported by a network of processor–controlled network elements and an array of operations systems. However, a net-work involving digital trunk groups between digital switches has no per–channel failure mechanism and needs no routine testing, per–channel or otherwise. It can be controlled better by real–time surveillance of its digital performance on a digroup basis. Similarly, digital loop facilities (ISDN or DS1) can receive dy-namic performance monitoring instead of routine testing.

Table 2-1. Transition Phases

Phase	Planned Action	RDSN Interface	Notes	Result
1	Access tandem is digital. Remove tandem function from analog local tandem; downgrade to EO.	POT toward ICs; access tandem toward LATA.	Accommodates digital interfacing to digital IC switches.	Fig. 2-5
2	Extend digital connectivity from digital AT to EO. Use -6 dB decode level in digital switches and in channel banks at analog EOs.	Digital EO switch: line side. Analog EO switch: trunk side.	For both analog and digital EOs.	Fig. 2-5
3	Strategically locate digital EOs and RSMs. Use -6 dB decode level in IDLC terminals. As option: add ISDN features to digital EOs.	EO switch: line side (including IDLC terminals).	Permits capping of analog EO where digital connectivity is required.	Fig. 2-6
4	Provide 1.544-Mb/s connectivity between digital end office and PBXs.	Customer-premises interface.	Digital connectivity from IC through digital tandem to PBX, with bit integrity.	Fig. 2-7
5	Provide digital lines and trunks to customer-premises interface.	Customer-premises interface.	Permits customers to use a telephone, data terminal, or digital PBX to be connected to any terminal on demand.	Fig. 2-7

Performance monitoring in the mixed digital/analog early stages of the RDSN continues to depend on detecting monitored parameters (e.g., error performance, slips, reframes) that fall out

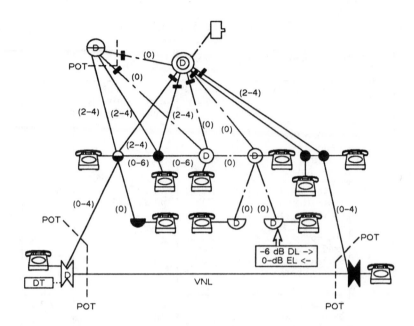

Figure 2–5. Transition phases 1 and 2—region digital switched network.

of limits. Many per–trunk measurements may be required on circuit groups to analog offices, but the trunk losses are all the same value, which simplifies the data base in the test system. Carrier group alarms (CGAs) cause trunk busy–out and reporting of transmission failure, as before. In an all–digital environment, CGAs and cyclic–redundancy check monitoring, coupled with digital switch alarms, form the surveillance method for guarding transmission continuity of the network without the need for trunk–by–trunk measurements. Because switches become effectively the digital equivalents of zero transmission level points, testing uses the digital reference signal (DRS), which eliminates the need for traditional test pads. (The DRS, pending standardization efforts that may redefine it, is the digital representation of a 1004–Hz tone at 0 dBm0.)

The bit error ratio (BER) is useful in performing out–of–service checks of facility health, sectionalizing troubles, and restoring service. However, for in–service monitoring the BER is replaced

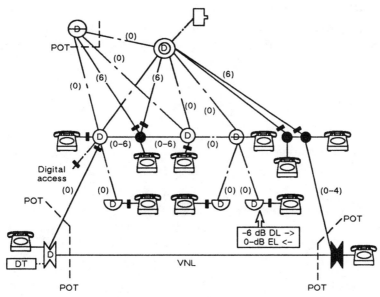

Figure 2–6.   Transition phase 3—region digital switched network.

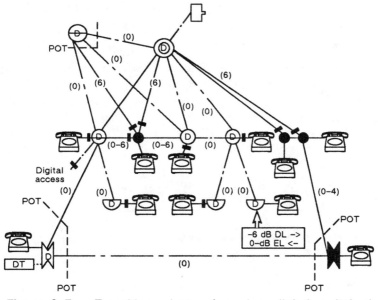

Figure 2–7.   Transition phase 4—region digital switched network.

by parameters such as errored seconds and severely errored seconds, unavailable intervals, and degraded minutes, whose automatic calculation becomes possible where the extended superframe format is implemented.

References

1. *Region Digital Switched Network Transmission Plan*, Science and Technology Series ST–NPL–000060, Bellcore (Iss. 1, Oct. 1988).

2. *RDSN Switch Generic Requirements*, Technical Advisory TA–NPL–000908, Bellcore (Iss. 1, Oct. 1988).

3. Cavanaugh, J. R., R. W. Hatch, and J. L. Sullivan. "Transmission Rating Model for Use in Planning of Telephone Networks," *Globecom 1983 Conference Record* (San Diego, CA: Nov. 1983), Vol. 2, pp. 20.2–1 to 20.2–6.

4. Duffy, F. P. et al. "Echo Performance of Toll Telephone Connections in the United States," *Bell System Tech. J.*, Vol. 54, No. 2 (Feb. 1975), pp. 209–244.

5. "Loop Loss Distribution," *Characterization of Subscriber Loops for Voice and ISDN Services*, Science and Technology Series ST–TSY–000041, Bellcore (Iss. 1, July 1987), Figure 9–64.

Chapter 3

ISDN, PPSN, and Broadband
Packet-Switched Network

The evolving integrated services digital network (ISDN) plan offers consolidation of communication circuits with improved performance for voice, data, and packet service connections between users. Presently, in addition to the usual plain old telephone service, a variety of special circuits are offered by the exchange carriers. The network implementation of special services usually involves separate engineering and craft effort encompassing signalling, transmission, and operations considerations. The latter include surveillance, maintenance, and management. The ISDN plan provides a flexible architecture to support most of these services economically with common circuits routinely assigned with little or no engineering to meet special requirements. ISDN is made possible by rapid advances in digital technology in processing and transmission areas.

Many telecommunications bodies have ISDN under consideration; it is receiving high attention by the exchange carriers. Standards for ISDN and packetized data are being addressed by groups including the American National Standards Institute (ANSI)–accredited T1 Technical Committee and the International Telegraph and Telephone Consultative Committee (CCITT).

This chapter first describes ISDN switching requirements and service transport capabilities. Then, the public packet–switched network (PPSN) and its capabilities, including ties to ISDN, are discussed. A promising approach to broadband packet switching, the Switched Multi–Megabit Data Service (SMDS), is described briefly.

3-1 ISDN NETWORK PLAN

A general representation of the network architecture for ISDN is shown in Figure 3–1. This indicates an interface for digital

51

access from the customer controller (as needed) to the ISDN switching office or node for transport over exchange and inter-exchange carrier networks to the remote ISDN office and then to the distant customer. The customer controller assembles and manages the needed telecommunications for ISDN access. Provisions can be made for customer control to change the originating or terminating offices and the channel capacities required.

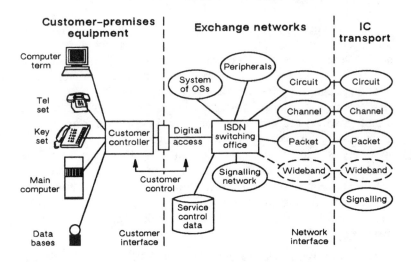

Figure 3-1. ISDN architecture.

The ISDN stored-program-control switch routes the traffic over an appropriate transport network. The network can be circuit-switched (public message), channel-switched (private-line), or packet-switched, as shown in Figure 3-2.

Basic Access Call Control

The generic call-control, switching and signalling requirements to provide ISDN capabilities, including packet service, via the basic access interface, are covered in Reference 1. Included are the switch signalling sequences for calls between two basic access ISDN lines, a non-ISDN to an ISDN line within the same ISDN switch, or between an ISDN line and an ISDN circuit- or packet-switched network.

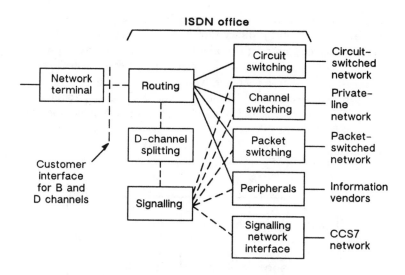

Figure 3–2. ISDN office routings.

Interoffice Call Control

The interoffice switching procedures and requirements for call control using Common–Channel Signalling System No. 7 (SS7) [2] to establish and release interoffice circuit connections for originating or terminating calls are covered in Reference 3, except for packet and some other special services.

User Equipment Call Control

Common switching and signalling requirements for ISDN circuit–switched voice services for ISDN user equipment served by a basic–rate access are covered in References 4 and 5, and for electronic key telephone service in Reference 6.

ISDN User Interfaces

The ISDN user communications interface is presently based on transporting a variety of types of digital channels listed in Table 3–1. While there may be several versions of ISDN access facilities having different digital capacities or channels, only two interfaces, basic and primary rate, are discussed in the following. These interfaces support the channel types for initial implementation. In brief, a basic-rate interface provides a customer with access to two clear 64-kb/s B voice or data channels and one 16-kb/s D channel for signalling, overhead, and packet data. The basic rate access is provided by the digital subscriber line (DSL) [7] described in Volume 2, Chapter 3.

Table 3-1. ISDN Channel Types (CCITT)

B (for bearer) channel:	64 kb/s
D (for delta) channel:	16 or 64 kb/s*
H0 channel:	384 kb/s
H11 channel:	1536 kb/s
H12 channel:	1920 kb/s
H4 channel:	†

* The D channel is 16 kb/s for basic interface and 64 kb/s for primary interface.

† Agreement on this rate has not yet been reached by the standards committees.

Each primary access interface such as for a PBX [8] may provide up to 23 B channels and 1 D channel, 24 B channels, 4 H0s, or an H11 channel over a DS1 connecting facility.

The network interface features and procedures to provide an end user with access to an ISDN end-office switch for ISDN transport is given in Reference 9. It describes the technical characteristics at the ISDN network interface of each of the lower three layers of the Open Systems Interconnection protocol that are needed for communication between the user's terminal and the ISDN switch. The reference defines access configurations and the interface requirements for establishing and maintaining the service connection to ISDN. It also describes data-link interfaces and defines the frame formats, interface procedures, and the link-access protocol-D (LAP-D) channel for a data-link connection. The messages and procedures for circuit-switched and packet-switched call control are included.

3-2 ISDN SERVICE TRANSPORT CAPABILITIES

The service capabilities planned for ISDN by bearer capability and call type are indicated in Table 3–2. These designations are needed for call–routing functions to establish a connection to the appropriate transport network. The transport network for voice-only and voiceband data capabilities may employ processing techniques such as low–bit–rate voice, time–assignment speech interpolation, digital loss, etc., and therefore bit integrity may not be preserved. It is assumed that μ –law encoded speech is present at the user–network interface for voice capability. The networks for the other capabilities are to preserve bit integrity.

Table 3–2. ISDN Bearer Capability and Call Types

Bearer Capability	Corresponding Call Type
Circuit–mode/speech	Voice (V)
Circuit–mode/3.1–kHz audio	Voiceband data (VBD)
Circuit–mode/unrestricted digital information (64 kb/s)*	Circuit–mode data (CMD)
Circuit–mode/unrestricted digital information/rate adapted 56 kb/s	Circuit–mode data
Packet–mode/unrestricted digital information (16 or 64 kb/s)	Packet–mode data (PMD)

*Including 7–kHz audio encoded into 64 kb/s.

3-3 ISDN ACCESS AND TRANSPORT SYSTEMS

ISDN basic and primary access facilities are discussed in Volume 2, Chapter 13, Part 4 for DS1 carriers. ISDN performance objectives, access interfaces, DS1 carrier access arrangements, and the DS1 bit–stream ISDN format are also covered in Part 4. Volume 2, Chapter 3 covers the DSL. Additional information on basic access system requirements and other functional topics are given in Reference 10.

Common transmission requirements for DS1 and higher–rate carriers providing for access or transport are given in Reference

11. They include availability, reliability, error performance, jit-ter, delay, formats, coding laws, etc.

Transport Networks

ISDN architecture includes three major transport networks, la-beled in Figure 3–1 as circuit–switched, channel–switched and packet–switched. A circuit–switched network provides a 56– or 64–kb/s call connection by switching through the network on a per–call basis, under control of the exchange carrier. A channel–switched network provides a dedicated 56– or 64–kb/s call con-nection (e.g., between digital cross–connections) on a time basis, often under control of the customer. A packet–switched network is where the message, usually data, is formed into prescribed packets and, under defined protocol and formats, is sent over available network paths and reformed to the original content at the terminating end. A simple comparison of circuit–switched and packet–switched operation is shown in Figure 3–3 to indicate the difference in the flow of transmission in the medium between the circuit–switched versus packet–switched path.

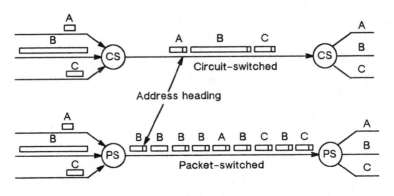

A, B, C = messages

Figure 3–3.   Example of circuit– and packet–switched connections.

Transmission connections between the ISDN switch and the circuit–switched and channel–switched networks shown in Figure 3–1 are at DS1 or higher levels. These networks are to provide

either 64-kb/s clear-channel capability with signalling provided
by Signalling System No. 7, or up to 56 kb/s with the "robbed-
bit" signalling necessary for older digital systems. ISDN connections to an exchange-carrier 56-kb/s packet-switched network is
via the digital X.75' interface supporting data rates of 9.6 and 56
kb/s. Future plans will provide for 64-kb/s and 1.544-Mb/s interface connections.

3-4 PACKET-SWITCHED NETWORK

Packet-switched networks provide for more efficient data
transport than circuit-switched or channel-switched networks
because the connections through the network are used only when
data is being transported [12,13]. In simple terms, a basic packet
switch consists of four functional blocks. Incoming packets are
stored in input buffers and their headers are decoded for routing
and other information. When the proper routing is determined,
the packets are put into output buffers via the internal switching
function and sent on to the next switch in available time slots.
Routing through the network can be done in several ways; each
packet is routed over an available path independently or a virtual
circuit is set up between the users and all packets of the same
message traveling this path. Several virtual circuits can share the
same physical path.

This section discusses a current PPSN that exchange carriers
employ for data transport. PPSN is based on 1984 and later
CCITT recommendations and on ANSI committee reports.

Public Packet-Switched Network

An example of a PPSN is shown in Figure 3-4 with transmission interfaces for many types of users and to other networks
[14]. Virtual circuit routing is employed. As noted, the connections between the packet switches are DS1 or higher level with
56-kb/s format channels (64-kb/s clear channels are not yet
generally available).

Interfaces

The interfaces are denoted X.25, X.28, X.29, X.32, X.75
(CCITT) and X.75' (exchange carrier). These are well defined

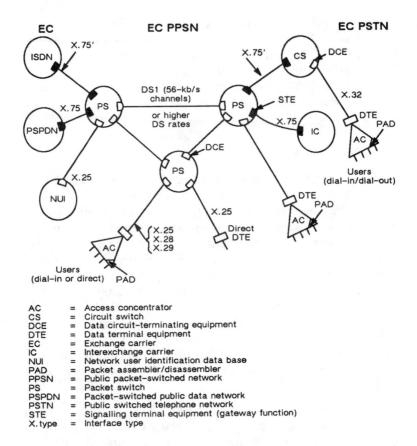

AC	=	Access concentrator
CS	=	Circuit switch
DCE	=	Data circuit–terminating equipment
DTE	=	Data terminal equipment
EC	=	Exchange carrier
IC	=	Interexchange carrier
NUI	=	Network user identification data base
PAD	=	Packet assembler/disassembler
PPSN	=	Public packet–switched network
PS	=	Packet switch
PSPDN	=	Packet-switched public data network
PSTN	=	Public switched telephone network
STE	=	Signalling terminal equipment (gateway function)
X.type	=	Interface type

Figure 3–4.   Example of a current public packet–switched network.

as to bit rate, format, protocol, etc. Other interface types are being considered. There are three hierarchical layers of protocol for these interfaces, as shown in Figure 3–5 for X.25 [15]. The first layer is the "physical level," which defines the electrical and mechanical connections, voltage levels, pulse rates, etc., between the data terminal equipment (DTE) and the X.25 modem. The second layer is the "link level," which defines the protocol between the DTE and the data circuit–terminating equipment (DCE) for flow control, link connection, error control, etc. The third layer is the "packet level," which defines the procedures to set up, maintain, and clear virtual calls (same network path for a complete message), and to maintain permanent virtual circuits (same network path for a certain time period). The X.25 DTE/ DCE interface supports synchronous data signalling rates of 2.4, 4.8, 9.6, 19.2 and 56 kb/s. The X.28 and X.29 DTE/DCE interfaces support asynchronous data signalling with additional rates of 300 and 1200 b/s for dialed and direct access.

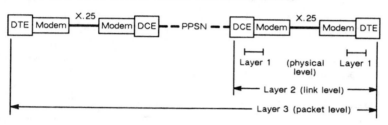

DCE = Data circuit–terminating equipment *
DTE = Data terminal equipment *

   * May include modem.

Figure 3–5. Interface levels for X.25.

The X.32 interface [16] provides users with dial–in/dial–out access to the PPSN via a circuit–switched network such as the public switched telephone network. This interface supports 1.2–, 2.4–, 4.8–, and 9.6–kb/s data signalling rates. Provision is made for nonidentified, identified, and customized services. The latter two services use automatic number identification or an exchange carrier calling–card number to authenticate the user for billing and for allowed services.

The X.75 and X.75′ STE/STE (signalling terminal equipment) interfaces between various packet networks currently support synchronous rates of 9.6 and 56 kb/s. The access concentrators and packet assemblers/disassemblers noted in Figure 3–4 are needed to accept and convert the users' various data rates and protocols into the standard interface links noted above.

PPSN Performance

The performance objectives for PPSN accuracy, delay, and dependability are listed in Table 3–3. The delay objectives pertain to the reference connection of Figure 3–6.

Table 3-3. PPSN Accuracy, Delay, and Dependability Objectives

ACCURACY[1]		
Access		
Misrouted calls	1 in 10^4 attempts	
Information transfer		
Errored packet rate	1 in 10^8 packets sent	
Misdelivered packet rate	5 in 10^8 packets sent	
Lost packet rate	5 in 10^8 packets sent	
Duplicated packet rate	1 in 10^9 data packets sent	
Out-of-sequence packet rate	1 in 10^9 data packets sent	
DELAY	*end-to-end* [5]	*per node* [5]
Access [2,3]		
Call setup	0.80 sec	0.145 sec
Information transfer [2,3]		
Data transfer (128 octets)	0.35	0.055
Data transfer (256 octets)	0.40	0.055
Virtual circuit data	95% of 9.6 kb/s or	
transfer rate [4]	throughput class if lower	
DEPENDABILITY		
Access-blocking[2]	1 in 100 attempts	
Information transfer [2,3]		
Reset rate		
Virtual call	2 in 10^8 data packets sent	
Permanent virtual call	2 in 10^6 per second	
Premature disconnect	2 in 10^6 per second	

Notes:
1. Long-term average.
2. Applies to the nth high-day busy hour.
3. For virtual calls only.
4. Applies simultaneously to each direction.
5. Average times.

3-5 BROADBAND PACKET-SWITCHED NETWORK

An evolutionary step toward broadband ISDN is the proposed Switched Multi-Megabit Data Service [17,18,19], a public packet-switching service in a wide area network employing

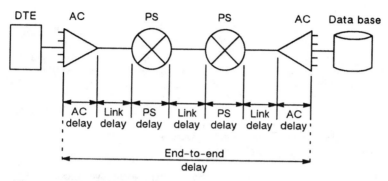

AC = Access concentrator
DTE = Data terminal equipment
PS = Packet switch

Figure 3-6. PPSN reference connection for delay objectives.

fiber-optic transmission facilities (Figures 3-7 and 3-8). It would extend local area network data transport capabilities across a wide area. SMDS would provide for the exchange of variable-length data units up to a maximum of 8191 octets per unit.

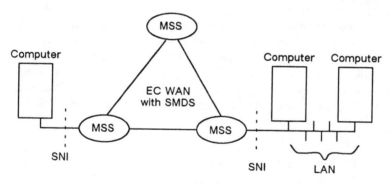

[Connections to operations systems
are not shown.]

LAN = Local area network
MSS = MAN switching system with SMDS
SNI = Subscriber network interface
SMDS = Switched Multi-Megabit Data Service
WAN = Wide area network

Figure 3-7.   Proposed   Switched   Multi-Megabit   Data
Service in a wide area network.

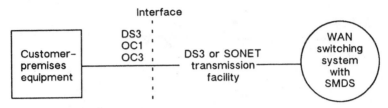

SMDS   = Switched Multi-Megabit Data Service
SONET   = Synchronous optical network
WAN   = Wide area network

Figure 3-8. Access to Switched Multi-Megabit Data Service.

Initially, the proposed SMDS could provide a customer with high-quality DS3 (45-Mb/s service), and in the future with OC1 or OC3 (50 or 150 Mb/s) throughput with a data message unit transfer delay objective of less than 20 ms average, and less than 30 ms in 95 percent of the cases.

The initial criteria for providing SMDS are contained in Reference 17. It includes a service description, discussions on access and network transmission needs, and a treatment of interface protocols. Reference 18 contains the generic transmission, operational, and performance requirements for a DS3-based local access system. Topics that are covered in this reference include physical and electrical interface requirements, synchronization, performance monitoring, failure detection, alarms, etc. DS1 and synchronous optical network local access (when available) will be covered in future issues.

Broadband Packet Switching

With optical fibers providing multi-megabit transmission facilities, advances are being made in broadband packet switching to use the high bit rate that can be presented. Research is being conducted on a promising experimental packet switch employing a new switching technology [20], the Batcher-banyan switch, which can provide self-routing capability with each packet sent over an available path through the switching fabric. This avoids putting heavy routing demands on the switching processor. A simple packet switch, based on this technology and using a custom very–large–scale integrated chip for a 32 x 32 matrix of input/output lines, each operating at 140 Mb/s, has been demonstrated in the laboratory. With such speed, the time delay (an important characteristic) through a metro packet–switched network is anticipated to be low, in the range 5 to 287 ms, depending on propagation, switching, and processing delays. Delay–sensitive services may require priority assignment to avoid excessive delay. However, considerable development will be required to put this technique into practice.

References

1. *ISDN Basic Access Call Control, Switching, and Signalling Requirements*, Technical Reference TR–TSY–000268, Bellcore (Iss. 2, Nov. 1988).

2. *Bell Communications Research, Inc. Specification of Signalling System No. 7*, Technical Reference TR–NPL–000246, Bellcore (Iss. 1, Rev. 2, June 1987).

3. *Switching System Requirements Supporting ISDN Access Using the ISDN User Part*, Technical Advisory TA–TSY–000444, Bellcore (Iss. 2, Mar. 1988).

4. *ISDN Features—Common Switching and Signalling Requirements*, Technical Advisory TA–TSY–000847, Bellcore (Iss. 1, Apr. 1988).

5. *ISDN Basic Access Call Offering*, Technical Advisory TA–TSY–000795, Bellcore (Iss. 1, Dec. 1987).

6. *ISDN Electronic Key Telephone Service*, Technical Advisory TA–TSY–000205, Bellcore (Iss. 2, Apr. 1988).

7. *ISDN Basic Access Digital Subscriber Lines*, Technical Reference TR–TSY–000393, Bellcore (Iss. 1, May 1988).

8. *ISDN Primary Rate Interface for SPCS–to–PBX Signalling*, Technical Advisory TA–TSY–000035, Bellcore (Iss. 1, Mar. 1985).

9. *Network Interface Description for ISDN Customer Access*, Technical Advisory TA–NPL–000776, Bellcore (Iss. 1, Nov. 1988).

10. *ISDN Basic Access Transport System Requirements*, Technical Advisory TA–TSY–000397, Bellcore (Iss. 2, May 1987).

11. *Transport Systems Generic Requirements (TSGR): Common Requirements*, Technical Reference TR–TSY–000499, Bellcore (Iss. 1, Dec. 1987).

12. Bellamy, J. C. *Digital Telephony* (New York: John Wiley and Sons, Inc., 1982), Chapter 8.2.

13. Chen, T. M. and D. G. Messerschmitt. "Integrated Voice/ Data Switching," *IEEE Communications Magazine*, Vol. 26, No. 6 (June 1988).

14. *PPSNGR, Public Packet–Switched Network Generic Requirements*, Technical Reference TR–TSY–000301, Bellcore (Iss. 1, Sept. 1986) and Proposed Revisions, Technical Advisory TA–TSY–000301 (Iss. 2, May 1988).

15. Geery, W. B. "Integrating X.25 into Data Networks," *Telecommunications Products & Technology*, Vol. 4, No. 11 (Nov. 1986).

16. *PPSN X.32 Interface Requirements*, Technical Advisory TA–TSY–000926, Bellcore (Iss. 1, Nov. 1988).

17. *Metropolitan Area Network Generic Framework System Requirements In Support of Switched Multi–Megabit Data Service*, Technical Advisory TA–TSY–000772, Bellcore (Iss. 1, Feb. 1988).

18. *Local Access System Generic Requirements, Objectives, and Interfaces in Support of Switched Multi–Megabit Data Service*, Technical Advisory TA–TSY–000773, Bellcore (Iss. 1, Dec. 1988).

19. Hemrick, C. F., R. W. Klessig, and J. M. McRoberts. "Switched Multi–Megabit Data Service and Early Availability via MAN Technology," *IEEE Communications Magazine*, Vol. 26, No. 4. (Apr. 1988).

20. Sincoskie, W. D. "Frontiers in Switching Technology, Part Two: Broadband Packet Switching," Bellcore EXCHANGE, Vol. 3, Iss. 6 (Nov./Dec. 1987).

Telecommunications
Transmission
Engineering

Section 2

Loops

Loops connect customer locations with central offices. They form the end links for the message network and for most special services. The loop is important in every connection because such transmission impairments as noise or loss affect every call and because failure of a loop isolates the station served by it.

The loop was originally a simple pair of wires in a cable, often extended with open wire. Now most loops are in fine–gauge cable, with transmission often aided by electronic range extenders, while most growth of loop plant uses facilities provided by digital loop carrier. Much of the digital span capacity to support those carrier systems employs optical fiber. At the same time, trials are underway of fiber facilities extending directly to the customer's premises.

The shift to digital pair–gain systems using the carrier serving area (CSA) concept, and to copper plant ready for digital operation, is a major trend. The declining costs of electronics compared to wire plant and the growing demand for digital services have spurred this shift. The arrival of the integrated services digital network (ISDN) makes loop transmission at total speeds of 160 or 1544 kb/s almost a routine matter.

Chapter 4 covers the physical and electrical characteristics of loops as found in the latest nationwide survey. It presents results relevant to regular message service, to special services, and to ISDN applications.

Chapter 5 describes the planning and engineering of loop facilities in terms of the process and tools that support planning. It gives the technical principles underlying practical engineering (the revised resistance design, modified long–route design, and CSA concepts). It covers the application of loops to special services and the digital subscriber line for ISDN basic–rate access. Fiber is covered as both the base capacity for multiplexing and as a potential loop facility in its own right.

Chapter 4

Loop Plant Characteristics

Since loops are integral portions of every connection, whether through the message network or as part of a private–line channel, a knowledge of loop plant characteristics is required to establish criteria for other network or channel components. Periodic surveys are taken to determine the nature of the loop plant and to evaluate important trends in the physical and electrical characteristics that will influence technology choices and administrative planning.

Characterization resulting from a loop survey is broad in nature and is not necessarily valid for small segments of the total loop plant. The topography of one wire–center serving area may differ sharply from another and from the average: some center–city wire centers have no loop longer than one mile; some rural wire centers serve customers 20 miles away. Without electronic treatment, some wire–center distributions in rural areas would not provide the desired grade of loop transmission. However, each design method presently in use provides the flexibility to accommodate topographical differences and to achieve a distribution of transmission parameters that meets loss and noise requirements.

The evolution of design and construction practices, covered in Volume 2, also influences the actual distributions found in surveys, since most wire–center serving areas contain some plant placed according to earlier design methods or installed before a new wire–center area was established. These loops may tend to skew the overall loss distribution in the low or high direction. These situations are gradually corrected by applying more modern design methods as feeder routes are extended or enlarged, or as older cables are replaced with digital loop carrier (DLC). However, in areas that have experienced little growth, some of the earlier plant remains.

4-1 PHYSICAL CHARACTERISTICS OF LOOPS

Comprehensive surveys of loop facilities are used to character-
ize the loop plant in costs and performance. A typical loop survey includes detailed information on loop lengths, percent of
loop carrier, bridged–tap lengths, wire gauges, type of construction, and type of service provided. There has been a gradual
trend toward longer loops, which is economically significant.
Costs for long loops are proportionately much higher than costs
for short loops where equivalent transmission, signalling, and supervisory performance are provided. The historical development
of long–route design concepts involving loop electronics and of
DLC systems was stimulated heavily by economic considerations.

Three major surveys of exchange carrier loops have been conducted in the last 25 years. The first survey was in 1964 [1], the
second in 1973 [2], and the third in 1983 [3]. These were supplemented, and are being supplemented, by specialized studies of
long routes and of DLC applications. Results obtained from these
surveys are of value for distribution–network planning and management, as well as for judging the broad potential of loop facilities to support basic–rate integrated services digital network
(ISDN) access or repeaterless 800–kb/s digital subscriber lines
(DSLs). Recent technological advances and the increasing types
of service, coupled with the wide use of digital carrier technology
in the loop plant, have a significant impact on the assumptions
that distribution–network planners and engineers use to evolve
the network.

A description of the principal results of the 1983 loop survey
follows. It provides a statistical profile of the distribution network.

Definitions and Terminology

The following terms are used in the survey results. Figure 4–1
illustrates terms describing the distribution plant.

Total length: Total length of the loop is the sum of all cable
segment lengths including all of the bridged taps.

Working length: Working length of the loop is the sum of all
cable segment lengths from the office to the network interface

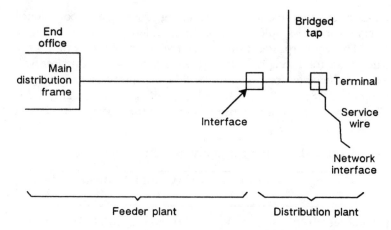

Figure 4-1. Representation of loop.

(NI) at the customer's premises. Working length is less than or equal to the total length of the loop.

Service: Service refers to the type of service provided by the sampled pair (i.e., either a business service or a residence service) and to whether the residence or business service is a special service.

Drop length: Drop length is the total drop wire length from an outside terminal location to the NI.

Survey Results

Composite. Table 4-1 contains the summary statistics of lengths for all of the 2290 sampled working pairs. An average total length for a sampled pair is 12,113 feet, a mean working length is 10,787 feet, an average bridged tap length is 1299 feet, a mean airline distance is 7692 feet, an average drop length is 73 feet, and a mean planned ultimate route length is 29,850 feet.

Table 4-1 also contains the standard errors in the estimation of means for each of these statistics. For example, to calculate a 90-percent confidence interval for the sample of total lengths, one can multiply 196, the standard error of mean (SDM) for total length from Table 4-1, by 1.645, the 90-percent confidence coefficient. The resulting 90–percent confidence interval for the sample mean of the total length is 12,113 ± 322 feet. This is interpreted to mean that with a probability of 0.9 the true mean total length of the working pairs lies in this interval. The coefficients for various confidence intervals are: for 99 percent, 2.58; for 95 percent, 1.96; and for 80 percent, 1.28. These coefficients can be used to modify the survey data to fit a desired confidence interval.

Table 4–1. Length Statistics—1983 Loop Survey

	Minimum (ft)	Maximum (ft)	Mean (ft)	SDM (ft)
Total length	250	114,838	12,113	196
Working length	186	114,103	10,787	188
Total bridged tap	0	18,374	1,299	34

Figures 4–2, 4–3, and 4–4 present the distributions of total, working, and bridged–tap lengths as determined by the 1983 survey.

Table 4–2 shows a comparison among the loop lengths of working pairs from the 1964, 1973, and 1983 loop surveys. Residence pairs from the 1983 survey are compared to the total sample results from the earlier surveys. This comparison is made because earlier surveys emphasized residential main stations. Working loop lengths show an increasing trend while average bridged–tap length decreases over these years. The latter factor presumably reflects the phasing out of multipled plant in favor of interfaced designs.

Figure 4–5 shows the distribution of cable gauges at various distances from the serving office, not including bridged tap. This distribution was derived by determining the gauge of each working pair sampled at 500–foot intervals from the end office. The figure shows that as one moves away from the office the gauge becomes coarser. For example, at a distance of 10 kft the approximate cable mix is 30–percent 26–gauge (ga), 51–percent 24–ga, 18–percent 22–ga, and only 1–percent 19–ga. To illustrate the historical shift toward finer gauges: in 1964, the distribution at the 10–kft point was only 17–percent 26–ga, 50–percent 24–ga, 27–percent 22–ga, and as much as 6–percent 19–ga.

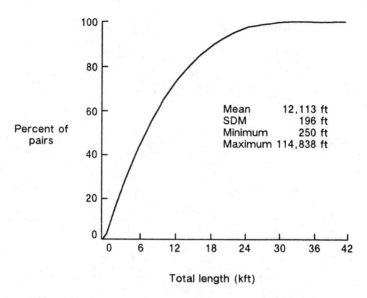

Figure 4-2. Total-length distribution—1983 loop survey.

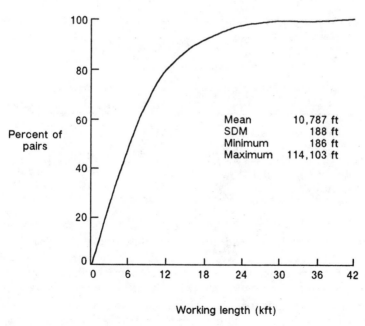

Working length (kft)

Figure 4-3. Working-length distribution—1983 loop survey.

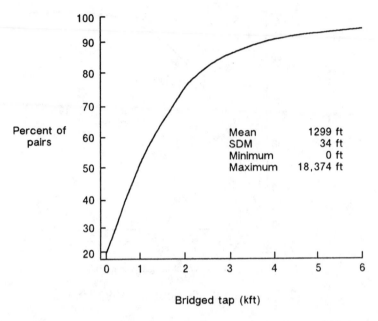

Figure 4-4.   Distribution of bridged-tap lengths—1983 loop survey.

Table 4-2. Loop Surveys

	Total	Total	Residence
Year of survey	1964	1973	1983
Average working length (ft)	10,613	11,413	11,723
Total bridged tap (ft)	2478	1821	1490
Airline distance (ft)	7758	8410	8387

Figure 4-6 shows the cable-structure distribution as a function of distance from the end office, excluding bridged taps. This information was also derived by determining the structure type at 500-foot intervals from the end office. Aerial cable is mounted on poles, underground cable is in ducts, and buried cable is directly buried. This figure shows that over 85 percent of the cable structure is underground close to the office. As one moves away from the office, buried and aerial structures begin to surpass the

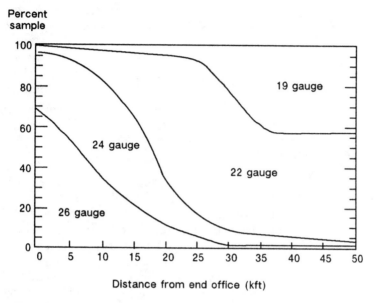

Figure 4-5. Cable–gauge distribution—1983 loop survey.

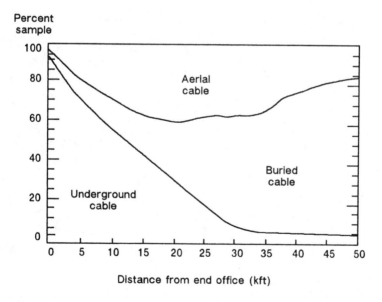

Figure 4-6. Cable-structure distribution—1983 loop survey.

underground structure. For example, at a distance of 10 kft from
the office, about 53 percent of the structure is underground, 16
percent is buried, and 31 percent is aerial. Showing the trend
toward out-of-sight plant: in 1964, the distribution at 10 kft was
only 46 percent underground, only 11 percent buried, and 43
percent aerial, the latter being a combination of cable, rural/ur-
ban wire, and open wire. The virtual disappearance of open wire
is evident: in 1964, at 50 kft (9.5 mi) from the central office, 27
percent of the structure was open wire, whereas the later surveys
showed no appreciable open wire at all.

Table 4–3 contains summary statistics of lengths of sampled
residence and business working pairs. Sampled residence pairs
have an average total length of 13,190 feet and an average working length of 11,723 feet. The average bridged-tap length is 1490
feet. Business pairs have an average total length of 9840 feet, a
mean working length of 8816 feet, and an average bridged-tap
length of 894 feet. The average working loop length for a business service is about 30 percent shorter than for a residence service. Figures 4–7, 4–8, and 4–9 present cumulative distribution
plots for these statistics.

Table 4–3. Length Statistics—1983 Loop Survey

	Minimum (ft)	Maximum (ft)	Mean (ft)	SDM (ft)
Residence				
Total length	495	114,838	13,190	245
Working length	186	114,103	11,723	236
Total bridged tap	0	18,374	1,490	44
Business				
Total length	250	100,613	9,840	302
Working length	200	99,569	8,816	296
Total bridged tap	0	11,333	894	47
Special Service				
Working length	391	63,348	9,059	635
Total bridged tap	0	8,685	736	99

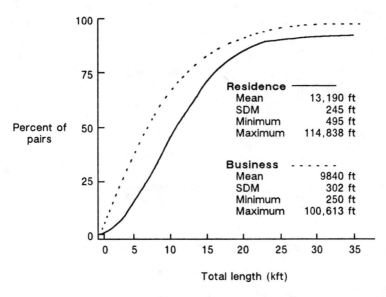

Figure 4-7.   Total-length distributions, business versus residence—1983 loop survey.

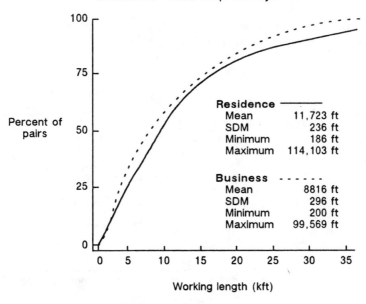

Figure 4-8.   Working-length distributions, business versus residence—1983 loop survey.

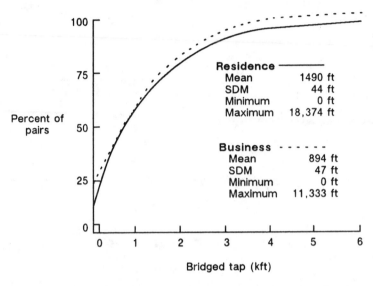

Figure 4-9. Distributions of bridged-tap lengths, business versus residence—1983 loop survey.

Of perhaps greatest interest for transmission purposes, Figures 4-10 and 4-11 give the distribution of dc resistance and ac

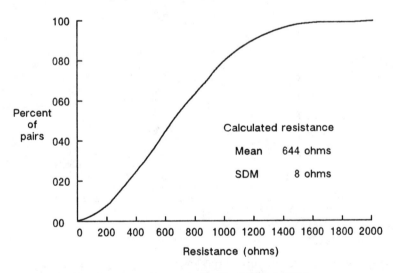

Figure 4-10. Loop resistance distribution—1983 loop survey.

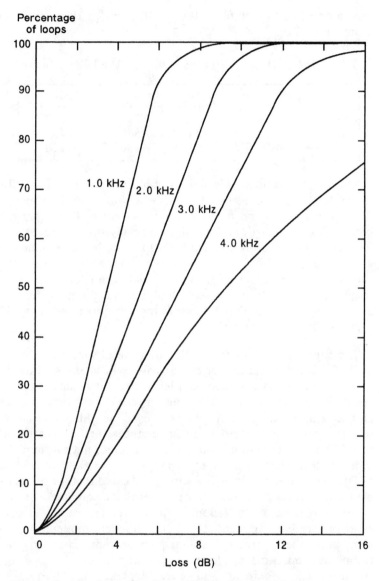

Figure 4-11. Insertion loss distribution—1983 loop survey.

insertion loss for the whole universe of sampled loops. Table 4–4 summarizes the distributions of resistance. The mean resistance was sampled as 574 ohms in 1964, 646 ohms in 1973, and 644 ohms in 1983.

Table 4–4. DLC Resistances (Ohms)—1983 Loop Survey

	All	Business	Residence
Mean	644	544	692
Median	690	540	760
90th percentile	1190	1130	1220

The survey included calculations of loop parameters for DSLs for ISDN basic–rate access. Figure 4–12 shows the distributions of loop loss at 40 and 80 kHz for the subset of loops that fall below 18 kft in length and are thus normally nonloaded. About 98 percent of the loops sampled have 40 dB or less attenuation at 40 kHz, exclusive of customer–premises cable. Included in the survey are distributions of image impedance for loops at ISDN frequencies. As seen at the central–office end, the mean impedance at 40 kHz is $134 - j69$ ohms; at the subscriber end, it is $105 - j65$ ohms.

The survey provides considerably more results than can be included here: distributions of dc resistance, business versus residence, feeder–versus–distribution length, tentative small–sample results for length from DLC remote terminal to customer, losses at 1 kHz and multiple frequencies relevant to ISDN operation, scatter plots of calculated input impedances, etc. It was found that 77 percent of working loops are nonloaded, not substantially different from the 1973 figure of 76 percent, but likely to decline as DLC continues to grow. The survey indicated that, of the sampled pairs, the connected customer terminal equipment broke down as follows: voice station, 78 percent; key system, 11 percent; private branch exchange, 4 percent; data modem, 1 percent; and other (private lines, etc.), 6 percent. The signalling techniques on working loops were loop–start, 78 percent; ground–start, 10 percent; and other (duplex, single–frequency, loop reverse–battery, etc.), 12 percent.

It is necessary to point out one characteristic of the later loop surveys: the electrical characteristics that they report are calculated, by modern reliable means, from physical makeups of

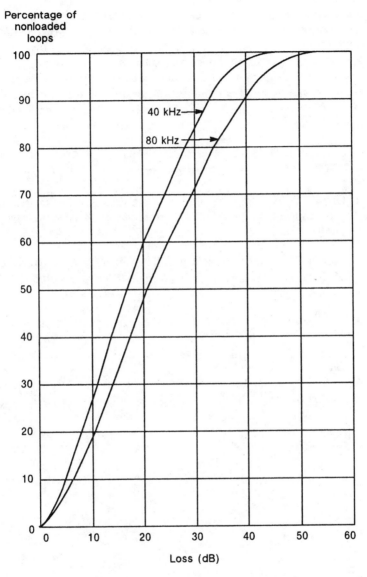

Figure 4-12. Losses—ISDN basic-rate access, 40 and 80 kHz.

the facilities. Compared to making several thousand field measurements, this greatly reduces the cost of making the study.

Unfortunately, it does not capture those electrical parameters that cannot be calculated reliably. Good examples are crosstalk and noise. The 1964 loop survey contained measured distributions of crosstalk coupling for loops. There are numerous reasons to believe that the crosstalk performance of cable plant has improved since then, but the later surveys did not include this physical measurement. Special surveys [4] have been used to sample special parameters like loop noise.

References

1. Gresh, P. A. "Physical and Transmission Characteristics of Customer Loop Plant," *Bell System Tech. J.*, Vol. 48 (Dec. 1969), pp. 3337–3385.

2. Manhire, L. M. "Physical and Transmission Characteristics of Customer Loop Plant," *Bell System Tech. J.*, Vol. 57, No. 1 (Jan. 1978).

3. *Characterization of Subscriber Loops for Voice and ISDN Services*, Science and Technology Series ST–TSY–000041, Bellcore (Iss. 1, July 1987).

4. Batorsky, D. V. and M. E. Burke. "1980 Bell System Noise Survey of the Loop Plant," *AT&T Bell Laboratories Technical Journal*, Vol. 63, No. 3 (May/June 1984).

Chapter 5

Loop Plant—Planning and Engineering

Since every connection between end users includes a loop at each end, the transmission performance of the loop plant significantly affects overall connection performance. The large number of loops, variety of lengths, varying density of customers along routes within a wire—center area, and high inward and outward movement of customers would make transmission design of individual loops prohibitively expensive and operationally infeasible. Therefore, the design of loop plant is treated statistically. The gauging of feeder and distribution routes is planned so that certain maximum resistance values are not exceeded in any distribution area, conforming to general prescription transmission and signalling designs. These designs vary with resistance range, taking economic advantage of fine—gauge cable with electronic supplements and of digital loop carrier (DLC). Since a central office is generally located near the population center of its serving area, a distribution of transmission losses results so that most losses are less than the loss corresponding to the limiting resistance. This approach to the design of loop facilities is economically sound and, when the entire local—plant universe is considered, produces transmission that meets objectives.

5-1 OUTSIDE PLANT ENGINEERING

The provision of outside plant facilities requires the application of engineering skills to the efficient layout and utilization of cables and carrier systems. At the same time, these facilities must be capable of rendering customer satisfaction for a wide variety of telecommunications services in terms of signalling, supervision, and transmission quality.

As demands for service have increased and expenditures have risen to satisfy these demands, new concepts and administrative procedures have been introduced for designing the plant. The serving area concept (SAC) has been coupled with DLC to form the carrier serving area (CSA). Used with T1 and fiber span facilities, these approaches help solve today's loop engineering problems of increasing demand, digital orientation, and pressure to reduce costs.

Changing Usage Patterns

Population growth, the changing distribution of population, and the general increase in the standard of living have brought an unprecedented demand for telecommunication services and a concomitant need for facilities. The growth in service is notable, but even more remarkable is the fact that for a net gain of one working service, about ten must be installed; i.e., to accommodate residential or business moves. Providing loop facilities in such a situation presents difficult engineering problems.

In addition to population growth and mobility, there has been an increase in the percentage of households that order service and a near–disappearance of party–line service. There is a growing demand for extra lines for general use, computer modems, and facsimile machines. These factors bring the need for additional loop facilities.

In suburban areas there is a gradual trend toward longer loops. This is one of the forces shaping the demand for DLC facilities.

Evolution of Designs

The outside plant provides a signalling and voice transmission path between a central office and the end user. This is usually accomplished over pairs of wire bundled in a cable, by themselves or as extensions of DLC–derived pairs. Distribution cables provide facilities in local residential and business areas. Cable sizes are chosen on an "ultimate" basis, to provide for the maximum service expected to evolve within the area under expected land–usage and zoning plans. Distribution cable pairs are connected to the central office through larger branch feeder and main feeder cables, or through DLC–derived feeder pairs, which are sized on an economic basis.

The geographic arrangement of streets, the ease of providing for growth, the availability of property easements, and the need for protection from construction activity all influence the layout of feeder routes. The most economical route is not necessarily the shortest, especially where DLC is used; it may be determined by studies that involve the determination of equal–cost boundaries within a fairly large area or, preferably, by specific studies using the tools (the LEIS™ system, etc.) described below. The problem is to meet transmission requirements with the most economical facilities.

Carrier Serving Area Concept. While the serving area concept enhanced the provision of an all–metallic distribution network, it also led to the CSA concept. This plan ensures digital–service capability in the loop network. The CSA concept is to sectionalize the route into discrete geographic units, or CSAs, so that when DLC is deployed along the route, every termination within the CSA has access to DS0–level digital service. The DLC terminal, or a serving area interface (SAI) wired to it, forms the nucleus of the CSA. In this sense, it differs from the terminals distributed along the route commonly used with multichannel analog loop carrier.

Administration of the Local Cable Network. Since the introduction of the SAC, procedures have evolved to provide for the administration of local cable networks and the application of various design methods. Each design has administrative methods associated with it.

When multipled plant design is used, feeder and distribution cable pairs must be reassigned as changes are required. In addition, pairs may be reassigned 30 to 90 days, depending on local policy, after release from a previously assigned address. These two features are called *reassignable* and *connect–through* methods of administration, respectively.

The *permanently connected plant* method of administration is used predominantly with dedicated outside plant design. In this method, a cable pair is permanently connected from the central office to each residential unit.

LEIS is a trademark of Bellcore.

An SAI may be administered by reassignable, connect–through, or permanently connected plant methods. As previously mentioned, the SAI is an integral part of the serving–area and CSA concepts. It provides a portion of the savings achieved because of its inherent flexibility. The SAC is attractive for cable extensions and conversion of existing multipled plant.

Outside Plant Engineering Functions

With changes in customer usage, many new combinations of design and administration have been introduced. Other factors have also produced changes in outside plant engineering. Technological changes in the construction industry have shortened the time available to provide facilities for new buildings. The use of computers has increased the amount and accessibility of data used to optimize expenditures.

Planning of the outside plant requires that an *outside plant plan* be prepared and kept up–to–date and that the construction budget be prepared regularly. The outside plant plan is completely restructured only when the validity of the existing plan is seriously in doubt. Many computer programs are available to aid the analysis and development of an outside plant plan and an optimum budget. The expansion and improvement of the plant involve the processes of feeder administration, rehabilitation studies (the facility analysis plan), and detailed distribution–area planning.

Additions to facilities must conform with the outside plant plan and should agree with company objectives for the provision of digitally oriented facilities, underground plant, etc. Coordination is necessary to ensure scheduling compatibility with major undertakings, such as central–office cutovers and area transfers.

5-2 THE OUTSIDE PLANT PLAN

The outside plant plan is designed to minimize the long–term cost of engineering, constructing, administering, and maintaining the local network. The plan is intended to appraise the outside plant requirements of a wire center through an evaluation of the available alternatives. When it is fully implemented, loop transmission characteristics may be expected to stabilize and to be much more predictable, allowing ready assignment of loops for integrated services digital network (ISDN) and other digital applications.

Once developed, the plan should be reviewed periodically (every three to five years) and modified to reflect the most current data. The outside plant plan is the fundamental plan for the area; it must be consistent with long–range road and land–use plans and with zoning changes. Switch modernization plans, interoffice facility plans, and the outside plant must be consistent with each other. A developed and documented plan is the basis for orderly expansion of outside plant facilities in line with overall objectives and optimal costs.

Procedures are also available for establishing an economical outside plant improvement program. These procedures are covered by a *facility analysis plan* [1] that acts as a monitoring and managing system on the loop network. With this plan, a wire-center area is studied for areas of high repair and rearrangement costs. These are used as the basis for collecting facility and maintenance activity data and for planning changes to feeder and distribution plant. The plan provides cost analyses, problem diagnosis, analysis of proposed plant improvements, and objectives for reductions in maintenance effort.

Development of the Outside Plant Plan

In order to provide a systematic approach with periodic review and modernization, certain basic steps are followed in the development of an outside plant plan.

Feeder and Branch Feeder Cable Locations. The economical layout of the local cable network is closely related to its physical arrangement. Branch feeder cables intersect the main feeder route and provide facilities to the feeder–route boundary. This configuration is commonly referred to as *pine–tree geometry*. Figure 5–1 shows pine–tree geometry and, for comparison, another configuration called *bush geometry*. Studies have indicated that the savings of the pine–tree over the bush geometry range from 5 to 30 percent of present worth of expenditures.

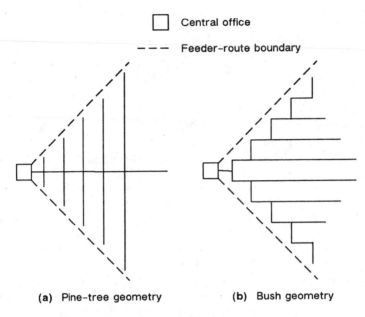

□ Central office

‐ ‐ ‐ Feeder-route boundary

(a) Pine-tree geometry   **(b) Bush geometry**

Figure 5-1. Loop cable layout geometry.

The ideal central–office area, having outside plant cabling in a perfect pine–tree configuration, never exists. The configuration in a given area must be determined by computerized economic analysis of alternative configurations. All plans under study provide enough cable pairs of the proper gauge, or pairs derived by loop electronics, to serve the demand. In order to ensure that each plan provides the necessary facilities, the area is subdivided into allocation areas.

Allocation Areas. An allocation area is a geographic area with defined boundaries that is the primary unit of plant for feeder-pair administration. It has one connection point to the feeder cable plant and consists of one or more distribution areas. An allocation area, in the absence of DLC, is served by only one transmission design; i.e., the same cable gauge or pair of gauges is used throughout. In developing an outside plant plan, each allocation area should be delineated to minimize variations in loop length.

Distribution Areas. Feeder cable pairs are committed to serve distribution areas, one or more of which may make up an allocation area. Complements of feeders are first allocated to an allocation area and then committed (physically spliced) to individual distribution areas. The distribution area is called a serving area if it is fed from an SAI.

Growth Forecast. The outside plant forecast provides the basis for the timing and sizing of additions to the outside plant network. Predictions of the types, magnitudes, and locations of future demands for service are used for the development of an outside plant plan and an optimal construction program. After the forecast has been made, the expected growth is spread in the applicable distribution areas as logically as possible on the basis of available information. The manner in which the expected growth is spread in these relatively small areas is a critical step in the planning process. If it is done properly, the resulting outside plant plan and construction program are nearly optimum [2]. After the forecast is spread, the existing and alternative configurations may be evaluated. Special needs—DS1 facilities for high-capacity services, ISDN access lines, special fiber-to-the-premises requirements—must be included.

A well-developed outside plant plan considers all feasible alternatives that may provide economy. Alternatives to be considered include physical design choices, transmission design options, and the application of electronics and fiber optics.

Boundary Changes. The boundaries of feeder routes, the areas served by different wire centers, and allocation and distribution areas are well defined, but there are occasions when changes are needed. The economic advantages of changing a piece of territory, such as the wire–center area, should be evaluated completely since savings are often possible through the coordinated efforts of several planning and engineering organizations. The wire center of an exchange area (i.e., the "center of mass" or centroid of all customer terminations) is subject to movement over time as growth occurs. This, coupled with the exhaustion of an existing switch, may induce splitting of a wire–center area.

Out-of-Sight Plant. The normal choice for new urban and some suburban/rural construction, and for some modernization of existing facilities, is out–of–sight plant. The economic factors involved in providing below-ground facilities are frequently evaluated via the LEIS system and action is taken on the basis of company policy. The decision is not necessarily one of least cost but rather one of planning a program to meet ecological objectives with the least penalty.

Resistance Design. The concept of resistance design originally referred, in an environment predating carrier and electronic range extension, to establishment of a common maximum–resistance limit for an office as a fundamental design method for loops of almost any length. The limit was applied to the longest forecasted loop in each distribution area served, even to distances that can now be served more economically via loop electronics. As modernized to fit a loop–electronics/pair–gain situation, it refers to designing loops within 24 kft of the central office to a particular resistance; it is still used widely.

Long–Route Design. This transmission design is applicable to those routes that serve customers relatively distant from the central office where forecasted customer density is low. The design achieves economies by using fine–gauge cables and adding electronic equipment to maintain transmission, signalling, and supervision performance comparable to that achieved under the resistance–design plan. While this technique has been overshadowed by DLC, it is still applicable in low–growth areas.

Pair–Gain Systems. In the preparation of an outside plant plan, a major alternative is the use of a DLC system or a remote switching module (RSM) attached to a suitable "host" office. The application of pair–gain systems often saves cable and structural additions.

Subscriber loop concentrator switching systems provide pair gain by concentrating a number of subscriber lines onto a smaller number of trunks between the central office and remote terminals (RTs). They have been largely replaced for new construction by DLC and RSMs.

A digital loop carrier system is the usual economical solution to the provision of new service in rural areas, industrial parks, and subdivisions. It includes an RT from which service is extended by loop distribution facilities to customer premises. In addition to being a pair–gain system, it brings about the CSA concept for the provision of DS0–level digital and ISDN service. The economics of DLC (avoidance of conduit reinforcement, compatibility with high–capacity services, etc.) have made it the first choice for facility growth at decreasing distances from the central office. The integration of DLC with a digital switch eliminates the central–office terminal, reducing investment further. The trend, then, is toward digital loop facilities away from the central office.

Evaluation of the Alternatives

A family of minicomputer decision–support programs is available to assist in the evaluation of outside plant design alternatives, based on a series of predefined study algorithms. Without these programs, such studies would be impractical. Termed the LEIS system [3], this software is a loop engineering information system that assists the planning, engineering, administration, and rehabilitation of outside plant. It thus supports fundamental or strategic planning, current or tactical planning, construction budgeting, and design engineering. It generates a series of standard reports and, via a query language, a variety of user–requested special outputs.

The LEIS computer programs are grouped into eight modules that can be accessed from the LEIS system. Each module is briefly described below.

Loop Engineering Assignment Data. Using a relational data base manager, the loop engineering assignment data (LEAD) module produces detailed information about cable, pair, terminal, crossbox, and living–unit usage, giving assignments and fills in the feeder and distribution plant. Data from LEAD are derived from and updated periodically by extracts of assignment data from the Loop Facility Assignment and Control System (LFACS) or other assignment systems. LEAD provides fill data to other LEIS modules (CARL, PLAN, LEIM), and the Loop Activity Tracking Information System (LATIS), which resides outside the LEIS computer. It also generates a variety of reports for direct use by the engineer.

Loop Electronics Inventory Module. The loop electronics inventory module (LEIM) provides an inventory of loop–electronics equipment and facilities, aiding in the engineering of equipment and facilities for digital transmission in the loop network [4]. LEIM replaces paper records and assists in better utilization of spare plug–ins and available equipment slots. Mechanized assistance to the design function is provided by the LEIM Catalog and a set of application programs that provide an inventory of existing loop–electronics equipment, with its locations and interconnections. It covers both span lines (repeater slots, etc.) and terminals in considerable detail.

Current–Planning Functions. The current–planning functions (PLAN) module uses route and forecast data to monitor the route and determine where and when to place cable, DLC, multiplexers, and fiber cable to relieve shortages, as well as what size facility to place. This is the main loop–technology planning tool of the LEIS system. PLAN uses data from the economic–study module to allow the engineer to determine the most cost–effective alternative for relieving or rearranging a route. It uses widely accepted optimization algorithms [5] to derive a least–cost plan. PLAN also provides loop administration tools to aid in the "allocation" and "commitment" processes referred to earlier. Fill data can be obtained mechanically from LEAD or input manually. PLAN is designed to let the user refine a plan by making changes, quickly rerunning, and making further modifications as needed.

Economic Study Module. The economic study module (ESM) performs economic evaluation of outside plant investment alternatives. Given general economic parameters, item–specific data, and study–specific information input on both a local and an area basis, ESM will evaluate the set of plans supplied to it. ESM is a general–purpose, economics tool based on the Capital Utilization Criteria (CUCRIT) program. Although it gives a number of economic evaluators, such as internal rate of return (IROR), net present value (NPV), discounted payback period (DPP), and long–term economic evaluator (LTEE), it ranks plans by present worth of expenditures (PWE). Plans developed by PLAN can be automatically transferred to ESM for detailed economic evaluation.

Computerized Administrative Route Layouts. Administrative route layouts are tools to assist in making loop–technology planning decisions. The computerized administrative route layouts (CARL) module produces concise, graphical representations of an outside plant feeder route by subroutes, feeder sections, manholes, poles, or pedestals. The CARL module is based on data that users enter into a file using plant location records through an editor. It also obtains data from LEAD or LATIS. Data outputs from CARL are also fed to LEAD for use in PLAN.

Loop Activity Data. The loop activity data (LAD) module allows engineers to access loop activity data collected by LATIS. Summary data by tracking unit and month are passed periodically by a tape interface to LAD. The LAD environment and a small set of LAD applications allow the engineer to generate flexible ranking reports, trouble and activity data by tracking unit, or other reports as designed by the engineer. With LAD, the engineer can generate custom reports to solve specific problems such as locating areas of high maintenance cost or checking the accuracy of growth forecasts.

Distribution Area Rehabilitation Tool. The outside plant planner/engineer is responsible for improving profitability by planning and designing quality distribution rehabilitation projects. The LAD module provide the planner with the tools needed to pinpoint the geography and type of problem to be solved. The planner must then define rehabilitation alternatives to solve problems in high-cost areas and evaluate their economics.

Making explicit predictions of operations activities manually is an inaccurate and labor-intensive effort. The distribution area rehabilitation tool (DART) module includes an activity predictor, as well as an economic model, user interface, and interfaces with other LEIS system modules for data extraction.

Digital Line Engineering Program. The digital line engineering program (DILEP II) module provides a tool for designing T1 digital lines, taking into account the necessary transmission parameters, DS1 and primary-rate access demand, and route characteristics. Mechanized layout aid is particularly useful for spans in the loop area, where the cable layouts are much more complex and variable than in interoffice routes. It helps choose repeater locations and calculates dc power-loop resistance. The output is a span-line design giving the proper repeater spacings for cables

with gauge changes, route junctions, and, where necessary, customer end sections.

Figure 5–2 depicts the modules of the LEIS system, the data
flows between them, and their relationship to outside software
systems.

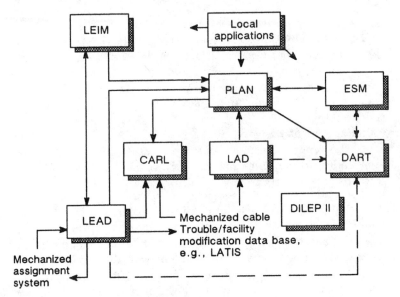

Figure 5–2. The LEIS system.

Other Planning Considerations

While considerations of design configuration, administration,
and transmission performance are the backbone of a properly
developed outside plant plan, additional items are involved in the
development. The items covered here represent those most frequently encountered.

Special Services. The effects of special requirements on the
outside plant plan must be taken into account. Generally, there
are three alternatives: (1) the use of electronics such as repeaters, (2) the application of derived–local–channel or data–over–
voice carrier systems, and (3) the use of digital loop carrier. The
use of DLC with special low–loss channel units usually satisfies

the transmission requirements of these services while providing installation and maintenance savings. For particularly large special–services demands, DLC remote terminals coupled to a digital cross–connect system (DCS) in the serving central office give a configuration termed integrated network access (INA). For digital high–capacity services and ISDN primary–rate access, the available choices are (1) T1 spans, central office to premises, (2) fiber facilities from the central office to a "fiber hub," with T1 span extensions from there to the customer's premises, and (3) dedicated fiber facilities, central office to premises.

Centrex. The normal procedure for provision of centrex service requires basic steps if service is to be provided economically while meeting transmission objectives. For example, periodic marketing reviews of anticipated service requirements are conducted for each large centrex location. Engineering and traffic studies must be made to determine the best serving arrangement for each case. These studies must consider the user location, present and expected service requirements, any expected need for conversion to a digital centrex, and outside plant considerations. Sales of medium–size centrex installations may require extra facilities to locations not expected in the forecast.

Once determined, the potential centrex requirements can be included in the outside–plant and central–office planning processes. Centrex customer sites located well outside the central office may require special loop facilities, or use of RSMs, to meet transmission requirements.

ISDN. ISDN users require either primary–rate access, which involves construction of DS1 spans to the customer's premises, or basic–rate access, which involves either the digital subscriber line (DSL) on nonloaded pairs or a CSA equivalent of the DSL.

ISDN has several planning impacts. DSL applications like ISDN centrex obtain two four–wire "B" channels from a single DSL pair, suggesting some reduction in requirements for pairs. Where the DSLs also use DLC facilities, there is the potential for some increase in carrier requirements because a 2B+D basic–rate access line occupies 160 kb/s of span–line capacity. Where primary–rate access is involved, one DS1 primary–rate access line

(23B+D) replaces 23 voice–frequency (VF) pairs, more if full–duplex services occupy the B channels.

5–3 RANGE LIMITS

Range considerations for the various types of central–office switching equipment have had a significant effect on the nature of the loop plant. For older switches, the length and gauge of cable are limited by the signalling or supervisory range rather than transmission performance. The limitations that determine loop ranges must be considered in designing loop plant. With the latest generation of switches, the switch range is long enough that the limitation on loop resistance derives from the need to supply at least 20 mA of loop current to the customer equipment, rather than the sensitivity of the switch itself. Likewise, the newer switches have expanded ranges for coin control on coin lines, typically 2400 ohms or more.

5–4 REVISED RESISTANCE DESIGN

Revised resistance design is a method for designing loops up to 24 kft in length based on the resistance limit for an office without electronic treatment. This limit is applied to the longest forecasted loop (far point) in each distribution area contained within a perimeter called the *resistance–design boundary*. In most urban and some suburban areas, the resistance–design boundary coincides with the wire–center boundary. In rural areas, carrier or a long–route design are more suitable economically for serving distribution areas on the extremities of a route. By applying rules for controlling transmission loss and meeting signalling range, use of the revised resistance design method produces a distribution of loop losses with the majority well below 8 dB. It is the basis on which most loop plant is installed.

In addition to the resistance–design boundary, certain other terms must be defined:

(1) The *resistance–design limit* is the maximum outside plant loop resistance to which the resistance design method is applicable. Office signalling permitting, this limit is dual:

1300 ohms for nonloaded loops up to 18 kft and 1500 ohms for loaded loops between 18 and 26 kft [6]. These limits serve to control transmission loss.

(2) The *resistance–design area* is the area enclosed within the resistance–design boundary.

(3) The *office supervisory limit* is the conductor loop resistance beyond which the operation of central–office supervisory equipment is uncertain.

(4) The *office–design limit* is the maximum resistance to which loops should be designed for a particular office. This is the supervisory limit for those electromechanical offices, now rare, with supervisory limits less than 1300 ohms; otherwise, the resistance limit of 1300 ohms controls.

(5) The *design loop* is the loop under study in a given distribution area to which the office–design limit is applied to determine the conductor gauge. It is normally the longest loop expected during the period of fill of the cable involved.

(6) The *theoretical design* is the cable makeup consisting of the two finest consecutive gauges permissible in the design loop to meet the office–design limit. Theoretical design does not take into consideration any economic advantage of reusing existing coarser–gauge cable pairs.

Basic Procedures

The application of resistance design begins with three basic steps or procedures. These are (1) the determination of the resistance–design boundary, (2) the determination of the design loop, and (3) the selection of the cable gauge(s) to meet the design objective.

Determination of the Resistance–Design Boundary. The resistance–design method is applied to the bulk of loops close to the central office, in areas where density and growth potential are moderate to heavy. For more sparsely settled areas, particularly those beyond 24 kft, carrier or long–route design are applicable.

Determination of the Design Loop. The length of the design loop is based on local service requirements, marketing forecasts, and other relevant data.

Some cables contain pairs that extend outside the resistance-design area. These pairs do not control the gauging of the cable but are are used for long–route design. Also, if a major branching point occurs before a gauge–change point, each branch may have a different gauge requirement; hence, it may have a separate design loop and is designed on a separate basis.

Selection of Gauge. The theoretical design is used to determine the gauge(s) required for any loop. When more than one gauge is required, the most economical design, if existing plant is temporarily neglected, results from use of the two finest consecutive standard gauges that meet 1300 or 1500 ohms. It is usual practice to place the finer–gauge cable close to the office where a larger cross–section is normally required. Since the design loop length is known, and the resistance per kilofoot for each gauge may be determined from such sources as Table 5–1, the theoretical design can be obtained from the solution of two simultaneous equations. For example, if the design loop is to be 18 kft, it can be seen from Table 5–1 that 18 kft of 26–gauge cable would exceed 1300 ohms and 18 kft of 24–gauge would give much less than 1300 ohms; therefore, a combination of 26– and 24–gauge is required.

One of the simultaneous equations may be written

$$x + y = 18 \qquad\qquad (5–1)$$

where x is the length of 26–gauge cable and y is the length of 24–gauge cable making up the 18–kft design loop. Since the total resistance should equal 1300 ohms, the second equation may be written

$$83.3x + 51.9y = 1300 \text{ ohms} \qquad (5–2)$$

where 83.3 and 51.9 are the resistances in ohms per kilofoot at 68°F of 26–gauge and 24–gauge wire pairs, respectively. Equations 5–1 and 5–2 can now be solved simultaneously to yield

$$x = 11.7 \text{ kft of } 26\text{-gauge cable}$$

and

$$y = 18 - x = 6.3 \text{ kft of } 24\text{-gauge cable}.$$

Table 5-1. Loop Resistances of Common Facilities

Type of Conductor	Ohms/kft		
	at 68°F	at 100°F	at 140°F
Cable			
19 ga	16.1	17.2	18.6
22 ga	32.4	34.6	37.4
24 ga	51.9	55.5	60.0
26 ga	83.3	89.1	96.2
Open wire (for comparison)			
109 high-strength steel	12.4	13.3	14.3
104 copper-steel*	4.7	5.1	5.5
Rural wire			
1 pr 0.064 CS † (14 ga)	17.0	18.2	19.6
6 or 12 pair 19 ga	16.4	17.6	19.0
Underground wire			
1 pair 19 ga	16.0	NA	NA
Service wire			
1 pair 18-1/2 ga CS†	43.0	46.0	49.6
2 pair 24 ga	52.0	55.6	60.1
2 pair 22 ga	34.0	36.3	39.2
Loading coils (88 mH)			
9 ohms each	—	—	—

*Conductivity is 40 percent that of copper.
†Conductivity is 30 percent that of copper.

Other resistance values from Table 5-1 might be used if local temperatures were significantly different from 68°F, as with long aerial routes.

While solution of simultaneous equations yields the correct design and is used in computer applications, a simpler and more flexible method for manual use is a graphical solution using a resistance–design worksheet such as that shown in Figure 5–3. This sheet has preplotted slopes corresponding to the resistance

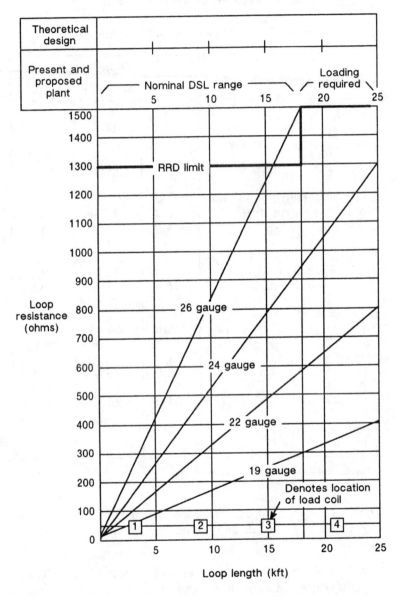

Figure 5–3. Worksheet for revised resistance design.

per kilofoot at 68°F of 19–, 22–, 24–, and 26–gauge nonloaded cable (19–gauge cable is considered primarily in terms of reuse of existing facilities.) As shown, the resistance–design limit is a constant 1300 ohms out to 18 kft. The original resistance–design concept allowed for 1300–ohm loops, with coarse–gauge cable and H88 loading, far beyond 18 kft.* With relatively long office signalling range, however, the resistance can be increased. The revised resistance design concept covers the zone from 18 to 24 kft by allowing a 1500–ohm cable resistance, office supervision permitting. If the office–design limit is different from 1500 ohms, the line representing this limit must be drawn in parallel to the resistance–design limit line and used in establishing the theoretical design. For nonloaded loops, the limit of 18 kft includes bridged tap; a maximum of 6 kft of bridged tap is allowed. Load-coil designations, based on H88 loading, are shown near the bottom of the worksheet. Ranges of recommended and permissible end–section lengths are also given on standard worksheets and the recommended positions of load coils are shown. Final locations of coils may vary somewhat, depending on the actual length of office end–section and physical constraints such as available manhole locations. In addition, rules for load–spacing deviations are followed.

When the worksheet is used to establish the theoretical design, a horizontal line equivalent in length to the proposed design loop, or ultimate–design loop if longer than the proposed loop, is drawn to scale in the box labeled "Theoretical Design." Next, the point of equivalent length is located on the resistance–design limit line. Through this point, a line is drawn parallel to the preplotted slope for the gauge to the immediate right, and its intersection with the slope for the next finer gauge is determined. The length to this intersection represents the gauge–change point for the theoretical design. The resultant length of each gauge is marked on the line in the theoretical design box.

Once the theoretical design is determined, the makeup of the existing plant is drawn in the upper portion of the box labeled

---

* While H88 is by far the most common loading plan, there was at one time some use of H44 loading of loops by the Bell companies. The Rural Electrification Administration was active in promoting D66 loading (66–mH coils at 4500–ft intervals) among its loan recipients.

"Present and Proposed Plant." A second, heavier line drawn below the existing plant makeup is then entered for the proposed
plant; a dashed–line extension to the ultimate length is used, if
required. The proposed plant makeup may be gauged identically
to the theoretical design, or existing plant conditions may be
taken into account to establish the actual gauge–change point.
This point, when a two–gauge makeup with the finer gauge at the
office end is assumed, must be moved closer to the office if the
point falls in the middle of a conduit section and may be changed
if there is a major branching point within two or three sections of
the theoretical gauge change. To meet the resistance limits, it
cannot be moved farther from the office. When duct space is at a
premium (under rivers, highways, etc.), it is permissible to devi-
ate from the two–gauge plan by utilizing finer–gauge cable in
such a conduit section, provided that coarser–gauge cable is
added in another part of the loop so that the overall office–design limit is met. Deviation from the two–gauge plan may also be
desirable: where only part of a route is being reinforced, where
the gauging of existing portions does not correspond to the new
theoretical design, to avoid ordering and splicing a short length of
small–pair–count fine–gauge cable, to avoid a splice at a danger-
ous street intersection, or to reuse existing plant. The resistance-
design worksheet is quite flexible as a tool for handling these
situations.

Transmission Considerations

Control of total resistance does not ensure a satisfactory transmission–loss distribution unless some additional rules are followed. These include loading all loops longer than 18 kft.

Loading. The maximum number of 88–mH load coils consistent with H spacing (6000 ft) should be placed on loops longer
than 18 kft. In general, the load–coil spacing should meet an
objective of 6000 $\pm$ 120 ft after the half–section leaving the central office. Occasionally, deviations greater than $\pm$120 ft may be
allowed for economic reasons, provided the transmission short-
coming are analyzed and weighed against the service provided in
the route, especially when data and special–services objectives
are considered. It is also desirable to take deviations greater than

120 ft on the short side so that correction may be applied by adding building–out capacitors or lattices.

The central–office end section for each office is determined by the expected length of frame jumpers so that the combination is equivalent to 3000 ft of cable. As far as spacing is concerned, the first coil is the most critical in achieving acceptable return loss and is placed as close to the recommended location as is physically and economically possible. The range for customer end sections is between 3 and 12 kft, *including all bridged taps.*

Bridged Tap. A bridged tap is considered to be any branch or extension of a cable pair in which no direct current flows when a station set is connected to the pair in use. It is thus nonessential to operation of that station. The cumulative bridged–tap limit for nonloaded loops is 6 kft, no matter where the customer interface is to be connected. An example of how this limit applies is shown in Figure 5–4. If the working station set were connected at points C or D, the cumulative bridged tap would be 6 or 5 kft, respectively, apparently within the 6–kft limit. However, if a working station set were located at points A or B, the cumulative bridged tap would be 13 or 7 kft, respectively. Therefore, such a layout violates the 6–kft limit.

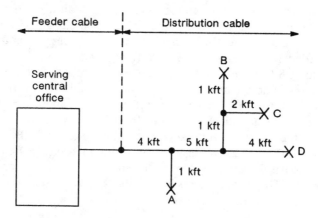

Figure 5–4. Application of bridged–tap limit.

For loaded loops, no bridged tap is permitted between load coils and no loaded bridged tap is permitted under any circumstance. [Loaded bridged tap causes frequency–response dips that are intolerable to dual–tone multifrequency (DTMF) signalling or data/facsimile transmission.] In some cases, bridge lifters may be employed to eliminate the effect of bridged tap, e.g., for party–line services, although central–office bridging is now more common.

Loop concentrators are used under resistance–design rules without significant transmission impairment. The rules are applied to the complete connection from the central office to the customer interface. However, different types of concentrator have specific signalling, control, and supervisory limits that must be taken into account.

Since customer locations served by more than one cable route have some chance of transmission contrast, an attempt should be made to select the gauges for each route so that similar losses are obtained.

Transmission Losses. Figure 5–5 shows the computed 1–kHz insertion loss versus loop length for theoretical revised resistance design loops when a temperature of 68°F and no bridged tap are assumed. Variations from the ideal design in the loaded range (18 to 24 kft), i.e., variations in customer end sections and load spacing variations, would change the locations and, to some extent, the magnitudes of the peaks from those shown. However, the figure shows that the highest theoretical losses occur in the nonloaded range between 13 and 18 kft, with a maximum approaching 8.0 dB at 18 kft. If the effects of bridged taps are considered, additional loss may be encountered. With a bridged tap of 2.5 kft on a 15.5 kft loop (18 kft total), about 0.6 dB of loss would be added. Hence, the insertion loss for nonloaded loops in the 14–to–18–kft range can slightly exceed 8 dB, which is taken as the maximum desirable insertion loss so that the mean and standard deviation of loss distributions in loop plant are not excessive. (The survey results in Chapter 4 indicate that about 98 percent of loops have losses below 8 dB).

5-5   MODIFIED LONG-ROUTE DESIGN

Application of revised resistance design to cable plant serving scattered end users at great distances from the central office

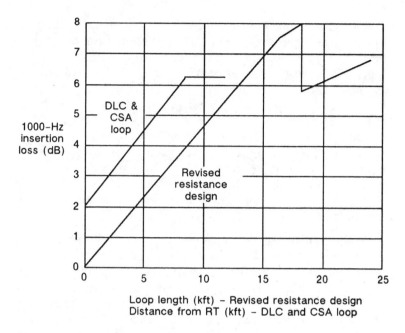

Figure 5–5. Expected insertion loss—revised resistance design and DLC/CSA concepts.

would involve high per–station costs due to the large amount of coarse–gauge cable required. Moreover, many stations are so distant that resistance design cannot be used at all. Therefore, alternative designs have been developed. Multichannel loop carrier and long–route design offer increased economic flexibility and still provide a satisfactory distribution of transmission losses.

Long–Route Design

Conceptually, three zones that correspond to ranges of resistance between 1500 and 2800 ohms are established by the modified long–route design (MLRD) procedure. It provides for a specific combination of electronic range extension and fixed gain devices to be applied to all loops falling within each range so that the maximum insertion loss is limited to 8 dB (8.5 dB with office loss included). The overall distribution of losses obtained by the use of long–route design thus provides a grade of service not significantly poorer than that generally received by customers [7].

Basic Zones and Transmission Layout. Figure 5–6 illustrates a long–route design worksheet. The vertical scale on the left is in terms of loop resistance and the scale on the right shows the upper and lower boundaries of zones designated 16, 18, and 28 for their resistances in hundreds of ohms. Zone 16 requires no gain device but does require a range extender in many offices. The range extender is required primarily so that the ringing signal can be tripped. Since most electronic and digital offices have a ringing–trip range of 1600 ohms or more, most companies administer zone 16 on an individual–office basis and provide the range extender only as required. Beyond zone 16, both range extension and gain equipment are required. The gain is applied uniformly to all loops within a given zone.

Gauging. In Figure 5–6, lines corresponding to the smoothed loop resistance versus length for each gauge of loaded cable are drawn for both 68°F (lower boundary) and 100°F (upper boundary). These lines are used to lay out prospective gauging plans and plot the corresponding zone boundaries. They automatically include the resistances of the loading coils. In contrast to revised resistance design, however, there is no set rule concerning selection of the gauge or gauges that would yield the most economical theoretical design; the cost of the office electronics must be included. In theory, there are an infinite number of gauge combinations and, in reality, many of these alternatives are possible. Generally, forecasts of end–user densities in each section of the route must first be obtained; then several of the most viable gauging alternatives are laid out on the worksheets upon which the zone boundaries are plotted. Next, the quantities and associated costs of range–extension equipment are determined and added to the costs of the cable in each plan; the plans are then compared economically on the basis of present worth of expenditures. As previously mentioned, computer programs are available to aid in analysis for long–route design or carrier alternatives.

Gain Options. A range extender with VF gain (REG) is available from a variety of makers for zones 18 and 28 [8]. As detailed in Chapter 3 of Volume 2, the range extender operates in two basic modes. In the signalling mode, sensitive balanced–bridge circuits detect the off–hook condition and the presence of dial pulses. A shunt resistance is then placed across the line to

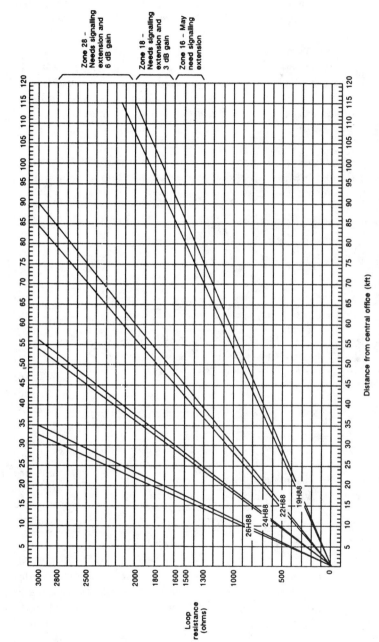

Figure 5-6. Typical long-route design worksheet.

increase the current that operates the central–office equipment. In the transmission mode, a negative–impedance repeater is switched into the circuit to provide for DTMF dialing and voice transmission. A compromise line build–out network that adequately matches an H88 loaded cable pair of any gauge combination is designed so that no detailed measurement is required during installation. An auxiliary power unit provides an additional −30 volts dc boost to the −48 volts provided by the normal central–office battery. The boost is applied to the loop through the REG output circuit to ensure adequate loop current. Gain is automatically switched from 3 to 6 dB if the loop resistance exceeds 2000 ohms. With these features, no manual selection of options is required.

Other Transmission Considerations. In addition to the above considerations regarding transmission layout for each zone, other transmission requirements must be satisfied.

Load Spacing and End Sections. Permissible load–spacing deviations are the same as for resistance design. Customer end–section plus bridged–tap length is similarly limited to a range of 3 to 12 kft. However, for cables with deviations that must be built out, theoretical return loss performance should be computed since capacitance build–out alone may not provide adequate return loss.

Noise. Noise on MLRD loops has the general objective of 20 dBrnc or less at the station end. However, due to the greater exposure of long loops and the occurrence of multiparty services using grounded ringers, it is economically impossible to bring each loop to within this limit. Therefore, although the objective remains 20 dBrnc, the maintenance limit for noise on long–route design loops is 30 dBrnc. Loops having noise in excess of 30 dBrnc are corrected by special treatment as required. This results in most multiparty loops being under the 30–dBrnc limit, thus maintaining a satisfactory noise distribution in the long–route universe. In particular, a 1980 analysis of loops of all lengths [9] (made before widespread use of DLC with its exposure–reducing properties) gave a 90th–percentile noise of 13 dBrnc, and a 99th–percentile noise of 31 dBrnc, measured at the customer interface.

Loss. Early studies showed that the best balance between plant cost and grade of service is achieved by establishing the maximum insertion loss for long–route designed loops at 8 dB (excluding central–office loss) [7]. Costs of loop plant and associated range extension and gain equipment required to meet each of the insertion–loss objectives have been analyzed on a present worth of annual charges (PWAC) basis. Figure 5–7 shows a comparison of the percent change in poor–or–worse (PoW) grade of service for customers on the long route for each insertion–loss objective with the percent change in PWAC.

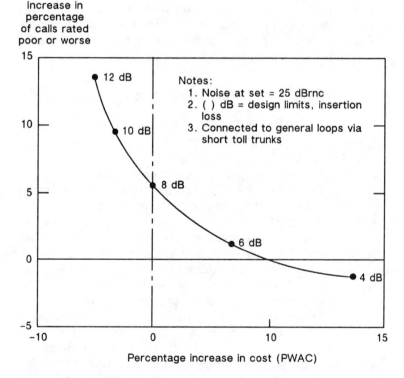

Figure 5–7.   Ratings of calls versus costs for various design limits.

The relationship between cost and grade of service is nearly linear as the insertion–loss objective is reduced from 12 to 8 dB. With a further reduction in the objective, the figure shows that the cost increases faster than the grade of service. Thus, the

8–dB objective appears to be an appropriate compromise between cost and grade of service. Later design plans like the CSA concept, while driven by the need for provision of digital service, are consistent with this approach.

Another factor supporting the 8–dB objective for long routes is that it is compatible with the 8–dB maximum insertion–loss objective for revised–resistance–design loops. It would not appear reasonable to make the long–route maximum objective more stringent, although a higher proportion of long routes may be near the limiting loss. The 5–percent impairment (Figure 5–7) in PoW grade of service for long routes compared with the general loop universe appears consistent with the economic factors governing long routes, especially since long routes are a small portion of the total loop population.

5-6 DIGITAL LOOP CARRIER SYSTEMS

An appealing alternative to providing long–route (and long resistance–design) loops by VF facilities is to use a digital loop carrier facility. In addition to being more economical in many situations, DLC systems can provide improved transmission–loss distributions. The 1–kHz insertion loss of a typical DLC system is only 2.0 dB between the central office and remote terminal, regardless of the length of the digital facility. Additional losses are incurred in the distribution cables between the RTs and the station interfaces. The low insertion loss of the digital portion of such loops allows substantial additional loss in the fine–gauge distribution cables. The cable loss is actually controlled by losses for 56– or 64–kb/s digital transmission, but results in worst–case VF loss, station–to–central–office, of the order of 6.5 dB as shown in Figure 5–5. In certain circumstances, it is possible to reduce the carrier system insertion loss to 0 dB, or even to provide gain transfer (negative loss) and thereby to extend further the range of the connected plant. Idle noise on the digital portion of the loop is less than 20 dBrnc so that loop noise objectives are rarely stressed. The relative shortness of the VF cable greatly reduces exposure to 60–Hz power induction. Quantization noise is low and found to be satisfactory over the entire normal dynamic speech input range; when the DLC is integrated into a digital switch, no step of coding or decoding is added in any event.

Carrier Serving Area

A CSA is defined as an area in which every customer has access to DS0–level digital services to include the capability of providing locally–switched voice–grade exchange service, special services, and the ISDN without special circuit design.

The CSA concept is to sectionalize a route into discrete geographic units, or CSAs, along that route. An RT site is chosen so that the DLC–served end user farthest from the RT site will be within the 64–kb/s serving limits of a properly designed cable. Figure 5–8 depicts a series of CSAs.

Typically, cable design for CSA transmission has the following limitations:

(1) Nonloaded cable only

(2) 26–gauge cable is not to exceed a total of 9 kft, including bridged tap

(3) Total length, including bridged tap, for single or multi-gauge cable of 19, 22, or 24 gauge, is not to exceed 12 kft

(4) Total bridged tap is 2.5 kft or less; a single bridged tap is not to exceed 2 kft

(5) Multigauge cable is restricted to two gauges (excluding short links for stubbing and fusing)

(6) Total length of a multigauge design that contains 26–gauge cable is not to exceed $12 - 3[(L_{26})/(9 - BT)]$ kft where L_{26} = total length of 26–gauge cable in kft and BT = total length of bridged tap in kft (any gauge).

Cable designs that surround a central office and meet the above restrictions are not included in a CSA as such, but are also capable of supporting the services afforded by a DS0–level signal. These cables qualify for a CSA designation for administration purposes.

Some companies have found it attractive to devise an "extended CSA" in rural areas where only telephone service and voiceband special services are envisioned. In this case, the use of loaded cable is allowed. The extended CSA can be quite large in size, as controlled by the supervisory limit of the DLC terminal, the minimum loop current of 20 mA, and a maximum cable loss of about 6 dB.

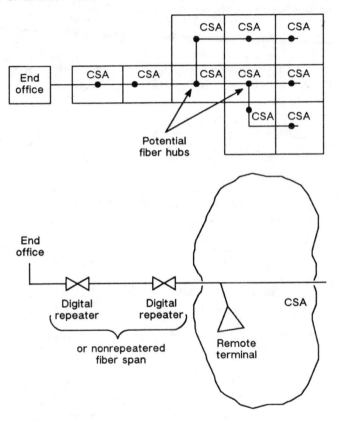

Figure 5–8. Carrier serving areas.

5–7 SPECIAL–SERVICES ASPECTS

Services with specified slope or envelope–delay requirements often require special engineering to supplement basic loop design. For example, a slope equalizer may be required; sometimes long bridged taps must be reduced even though they are within the normal resistance–design limits. DS1 span lines are utterly intolerant of bridged taps, even below 100 feet. Even loops that are loaded under resistance–design rules may require gain to reduce losses in some special–services applications.

Special–services requirements are generally accommodated on an individual–case basis rather than as a modification of global design procedures. However, it is important that special–services requirements in a given cable route be forecast accurately. In many cases, economies can be realized when the size, gauging, loading or nonloading, and bridged–tap content of certain cross sections are planned initially to meet special–services requirements, or when a metallic design is changed to carrier. For example, an industrial park 19 kft from the office is likely to involve numerous four–wire digital services, so either a DLC system or a special complement of nonloaded pairs will be required.

Since centrex users generally require special features such as conferencing and add–on, the transmission objectives for centrex station lines are more stringent than those for ordinary loops. For centrex station lines, the supervisory limit remains at the resistance–design limit. Centrex station lines must meet a maximum expected measured loss objective of 5.5 dB, including the assumed 0.5–dB test access loss to the local milliwatt supply. Hence, the 1–kHz loss of the loop facility should not exceed 5 dB. It can be seen from Figure 5–5 that resistance–design loops longer than 11 kft may require additional transmission treatment, such as loading or gain, to meet this objective when used as centrex station lines.

Specially Conditioned Data Loops

The envelope–delay and slope limits for data loops may precipitate supplemental loop design procedures, including equalization in the central office. Envelope–delay equalization for long loaded loops generally involves converting to four–wire operation with a hybrid repeater, delay–equalizing in the four–wire path, then reconverting to two–wire operation, all within the central office. Slope equalization by itself requires only a 2–2 type repeater.

Message–circuit noise on loops is not critical for data use. On the other hand, electromechanical switching equipment is often the source of impulse noise; hence, proper design of loop facilities does not guarantee meeting impulse noise limits. Impulse-noise mitigation techniques are employed if these limits are exceeded.

5-8 DIGITAL SUBSCRIBER LINE

The standard DSL for ISDN basic–rate access, as described in Chapter 3 of Volume 2, is intended for use on ordinary non-loaded loops up to 18 kft in length. The corresponding loss limit is 42 dB at 40 kHz. This is roughly comparable with the limits of 32 to 38 dB at 80 kHz for typical "prestandard" basic–rate access loops. The survey results in Chapter 4 indicate that about 97.5 percent of all nonloaded loops sampled meet this criteria. The 2B1Q coding method and other features of the DSL are intended to give margin to allow for customer building cable, cable crosstalk, office wiring, changes in cable gauge, and bridged taps up to 6 kft. For DLC systems that provide DSL transmission within the CSA, the CSA loop rules are thus conservative. However, loaded plant is wholly unsuited to DSL use.

This broad applicability of the DSL reduces the need for pre-qualification of loops for ISDN applications; however, interference coordination remains necessary, particularly with regard to the potential for interference to analog systems operating in the same or adjacent binder groups. Considering joint use of the DSL with other systems in the same binder group, the operating range of the other system depends on the specific type. It may be unaffected (Digital Data System, T1 spans, some data–above-voice multiplexers), reduced by 2 to 33 percent (other data-above–voice systems), or reduced to zero (yet another data system).

5-9 FIBER IN THE LOOP PLANT

The use of fiber in the interoffice network and the feeder portion of the loop network is now routine. The application of fiber in the distribution portion of loop plant is not far behind. The design limits are presently controlled by the terminal requirements and fiber capabilities. As developments in these areas progress, so do the loop design criteria. One design trend involves use of a "fiber hub" in which some DS3 signals on a 405– or 565–Mb/s repeaterless feeder are converted to VF (or low–speed digital) to serve users in the local CSA, while other DS3s are converted to DS1 (or passed through) to serve surrounding CSAs. This drop–and–add application is well suited to use of synchronous optical network (SONET) terminals incorporating synchronous transmission (SYNTRAN) coding. This approach minimizes the need to construct T1 loop spans to provide high–capacity services. For major customer sites, one or more DS3s may be directed to the customer's premises for on–site demultiplexing.

The use of fiber for the DLC in CSA applications has been the initial thrust. However, the use of light–to–analog terminal units for installation at the customer's location provides transmission advantages. One benefit of an all–fiber loop is the elimination of inductive interference from power–line sources. More importantly, a great variety of services can be offered. The use of fiber in the loop plant paves the way for information–age services: video on demand, Switched Multi–Megabit Data Service, "dark fiber" leased to customers, cable television trunks, etc. Many of these applications are suited to relatively inexpensive multimode fiber. At present, several trials are under way; consistent design techniques for premises fiber will emerge in the next few years.

References

1. Aughenbaugh, G. W. and H. T. Stump. "The Facility Analysis Plan: New Methodology for Improving Loop Plant Operations," *Bell System Tech. J.*, Vol. 57, No. 4 (Apr. 1978), pp. 999–1024.

2. Long, N. G. "Loop Plant Modeling—Overview," *Bell System Tech. J.*, Vol. 57, No. 4 (Apr. 1978), pp. 797–804.

3. Webb, E. F. "The LEIS System," *Proceedings*, National Communications Forum (1988), pp. 225–229.

4. "LEIM Enhancements," Bellcore EXCHANGE, Vol. 4, Iss. 1 (Jan./Feb. 1988), p. 36.

5. Bulchav, B. et al. "Feeder Planning Methods for Digital Loop Carrier," *Bell System Tech. J.*, Vol. 61, No. 9 (Nov. 1982), pp. 2129–2142.

6. *Notes on the BOC IntraLATA Networks—1986*, Technical Reference TR–NPL–000275, Bellcore (Iss. 1, Apr. 1986), Sections 7 and 12.

7. Howe, C. D. "Transmission Design, Long Subscriber Routes," *Conference Record*, IEEE International Conference on Communications, Cat. No. 71C28–COM (June 1971), pp. 30–7 to 30–11.

8. Burgiel, J. C. and J. L. Henry. "REG Circuits Extend Central Office Service Areas," *Bell Laboratories Record*, Vol. 50 (Sept. 1972), pp. 243–247.

9. Batorsky, D. V. and M. E. Burke. "1980 Bell System Noise Survey of the Loop Plant," *AT&T Bell Laboratories Technical Journal*, Vol. 63, No. 3 (May/June 1984), p. 808.

Telecommunications Transmission Engineering

Section 3

Trunks

This section consists of five chapters that provide transmission engineering information on trunks supporting the switched networks. Many types of message trunk are needed to provide the communication paths between switching machines. The functional and transmission characteristics of these trunks have been established mostly by consideration of the wide range of services that are offered, the supporting trunk network arrangements, and the estimated end-user perception of the quality of service provided. This section defines and identifies the principal trunk types. It also describes the elements of traffic engineering for routing and load capacity, and provides the major trunk requirements such as loss, noise, and return loss. Trunks supporting operator services networks are discussed separately.

Chapter 6 covers the definitions of the current principal trunk categories and the COMMON LANGUAGE® code identifications used for traffic, transmission, and administrative purposes.

Chapter 7 gives some engineering concepts for traffic distribution and routing through networks. This is needed so that trunk groups can be provisioned and administered to carry a wide range of traffic loads efficiently.

In Chapter 8, the local access and transport area (LATA) switched message network is shown, for convenience, to consist of three parts: access, intraLATA, and metropolitan subnetworks. The trunks involved in these networks are defined and the principal trunk transmission requirements for intraLATA and metropolitan networks are given.

Chapter 9 covers the return losses (through and terminal balance) for LATA access trunks. This includes direct and tandem

---

COMMON LANGUAGE is a registered trademark and CLEI, CLLI, CLFI, and CLCI are trademarks of Bellcore.

routing. Balance requirements are given for a two-wire electronic tandem office and for the two-wire ports of a digital end office. Cases covered include direct and tandem routing.

Chapter 10 deals with the present digital operator services systems (OSSs), the older analog OSSs, and auxiliary services. The trunking and major transmission requirements for the digital OSS are covered, with emphasis on operator circuit and headset characteristics. The calculated grade-of-service performances are shown for the operator and for the customer, based on a system model. The older systems are briefly covered, followed by a short section on auxiliary services in which conferencing services and centralized automatic message accounting transfer are discussed.

Chapter 6

Trunk Types and Uses

There are two general kinds of trunk: message and special-services. Message trunks can be defined generally as the transmission facility and associated terminal equipment to provide a message channel (nominal 4–kHz bandwidth for analog or nominal 64–kb/s bit rate for digital) between switching entities or between a switching entity and operator position equipment. Trunks are classified broadly for transmission and specifically for traffic. For transmission, trunks are defined by the connection, e.g., as a tandem–connecting trunk for an end–office–to–tandem connection, with specified transmission parameters (loss, noise, etc.). For traffic purposes, the trunks are operationally categorized as to class and usage. Special–services trunks are discussed in Chapter 11.

There are many traffic trunk groups. For traffic identification, the groups have been assigned various class codes and use codes (the traffic–use codes are based on transmission categories) and combined in a COMMON LANGUAGE format. The identification is used by exchange carriers for circuit administration, provisioning, design, forecasting, and servicing purposes.

This chapter illustrates the message and operator trunk network for local access and transport area (LATA) and exchange access and the associated trunk names commonly used in transmission parlance. For traffic purposes, this chapter also describes the COMMON LANGUAGE message–circuit identification codes, formats, and types of signalling used.

6–1 TRANSMISSION NOMENCLATURE

For message transmission purposes, the three LATA networks and their associated trunks as shown in Figure 6–1 are:

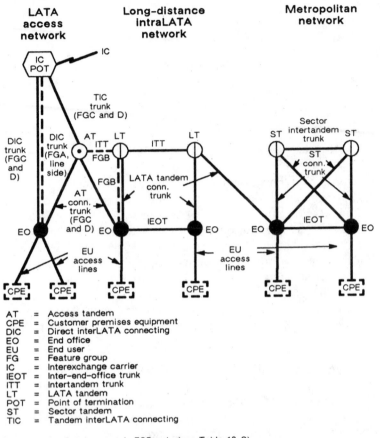

AT   =   Access tandem
CPE =   Customer premises equipment
DIC =   Direct interLATA connecting
EO   =   End office
EU   =   End user
FG   =   Feature group
IC   =   Interexchange carrier
IEOT =   Inter–end–office trunk
ITT =   Intertandem trunk
LT   =   LATA tandem
POT =   Point of termination
ST   =   Sector tandem
TIC =   Tandem interLATA connecting

Note: Equal access is FGD only (see Table 18–2).

Figure 6–1. LATA networks.

(1) The LATA access network, containing the direct inter-
    LATA connecting, tandem interLATA connecting, and ac-
    cess tandem connecting trunks

(2) The intraLATA tandem network, containing the intertand-
    em, LATA tandem connecting, and inter–end–office
    trunks

(3) The metropolitan tandem network, containing the sector
    intertandem, sector tandem connecting, and inter–end–of-
    fice trunks.

118

It should be noted that in many LATAs there is only a single tandem switch providing access, intraLATA, and metro tandem functions. The trunking pattern becomes simpler accordingly.

Detailed discussions on the transmission parameters and requirements for the intraLATA and metropolitan network trunks are given in Chapter 26 of Volume 1 and Chapter 8 of this volume. Similar information on LATA access trunks is given in Chapter 18 of this volume.

For operator services networks, the older main trunk networks are shown in Figure 6-2 for the analog No. 5 Automatic Call Distributor (ACD) System and the No. 1 Automatic Intercept System (AIS). The network architecture for the newer digital operator services is shown in Figure 6-3 using inter- and intraLATA-transmission-type trunks. Transmission considerations for operator services are discussed in Chapter 10.

6-2 COMMON LANGUAGE FORMAT AND CODES — MESSAGE TRUNKS

The COMMON LANGUAGE code message format is composed of 41 character positions in eight sets, needed for circuit identification, as shown in Table 6-1. The first set consists of positions 1 to 4 for the *trunk number* assigned to a particular circuit within a trunk group. Positions 5 and 6, the second set, are for *traffic class* codes, which identify whether the group is designed for blocking grade-of-service criteria or has an alternate route. Descriptions of the major traffic class group codes are given in Table 6-2.

Positions 7 and 8, the third set, are for one of five *switching level* codes for offices A and B respectively. An end office (including centrex) is level code 5. A tandem serving an end office is level code 4. A tandem serving one or more code 4 tandems is level code 3, and so on, although these higher hierarchy levels are little-used today.

Positions 9 and 10, the fourth set, are for *traffic use* codes denoting the kind of traffic to be carried, such as trunk groups for: (1) inter-end-office, tandem-connecting, intertandem,

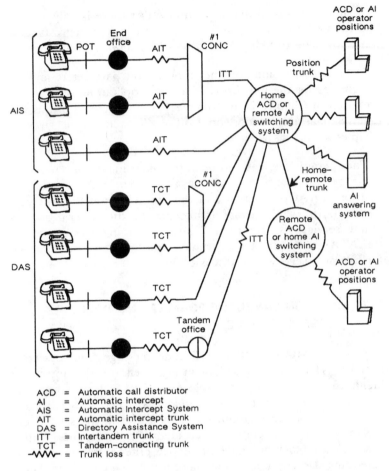

Figure 6-2.   No. 5 Automatic Call Distributor System and
No. 1 Automatic Intercept System trunks.

interLATA connecting, and operator-connecting services, (2)
auxiliary services (directory assistance, repair, etc.), and (3)
administrative services (operator, coin, etc.). Descriptions of the
most important traffic use codes for the groups in (1) above are
listed in Table 6-3, for auxiliary groups in Table 6-4, and for
administrative groups in Table 6-5.

Positions 11 through 17, the fifth set, indicate the coding used
for *trunk-type modifiers* of various character lengths to identify a

120

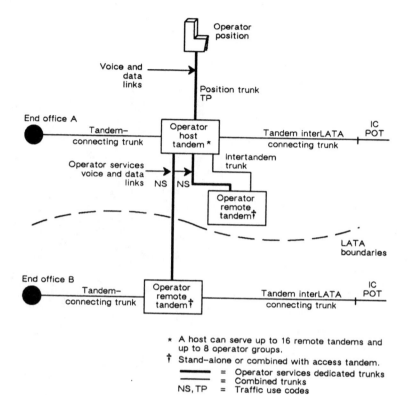

Figure 6-3. Operator services digital architecture.

trunk group uniquely. For example, a traffic use code, ET, for an
end–office–to–tandem connection has 33 possible modifier codes
covering various uses such as many operator, coin, and noncoin
types, choked–call routing, hotel–motel, party number, bit–rate
indicators, etc. Except for the bit–rate indicators shown in Table
6–6, the modifier codes are too numerous to list here. While the
bit–rate codes cover a range from 2.4 kb/s to 6.312 Mb/s, only
the ones currently used are listed.

Positions 18 through 28 and 31 through 41, the sixth and
eighth sets, cover *location identification* codes for offices A and
Z, respectively, in the form of city (four characters), state (two
characters), office (two characters), and switching entity (three
characters).

Table 6-1. Code Format for Identification of Message Trunks

Character	Trunk Number	Trunk Type				Location Identification (Office A)	Type And Direction Of Pulsing	Location Identification (Office Z)
		Traffic Class	Office Class	Traffic Use	Trunk-Type Modifier			
Positions	1-4	5-6	7-8	9-10	11-17	18-28	29-30	31-41
Sets	NNNN	AA	XX	AA	XXXXXXX	AAAAAAXXXXX	XX	AAAAAAXXXXX

Legend: A = letter
          N = number
          X = letter, number, or hyphen

Example: 1042 DF5SIE - - - - - - LNTNWCDS1- -M- ZNTHWCCG1- - is the 42nd trunk in the lxxx trunk group. This group is direct-final, class-5-to-class-5, inter-end-office trunks from the Linton, WC office, digital switch 1 (DS1), to the Zenith, WC office, control group 1 (CG1). The M- indicates a one-way trunk using multifrequency pulsing in the A-to-Z (Linton-Zenith) direction. The blank spaces in this example are for illustrative purposes.

Positions 29 and 30, the seventh set, encode the *type and direction of pulsing* of address information between switching offices, as shown in Table 6–7. The first position is outgoing from office A; the second is outgoing from office Z.

Table 6–2. Description of Traffic Class Trunk Group Codes

CODE	DESCRIPTION
AF	**FINAL** *Alternate–route final*: Provided as the last–resort path in the final–route chain, this group carries direct traffic and switched overflow from high–usage trunk groups. It may also carry calls that have not been routed over a high–usage group of any type and that instead are first–routed over the final group.
DF	**NONALTERNATE ROUTE** *Direct final*: Commonly referred to as a nonalternate-route trunk group, this group does not receive overflow and is provided as the only route between two offices for the items of traffic it carries.
DN	*Nonhierarchical routing*: A group used in such a network.
FG	**FULL GROUP** *Full group*: This group would be high–usage in the basic routing pattern, but for an expedient reason (service advantage or equipment limitations) it is engineered for low incidence of blocking and is not provided with an alternate route.
IF	**HIGH USAGE** *Parallel protective high usage*: A group that parallels a route–advance final group and overflows to it. It receives first–routed traffic and switched–overflow traffic, but not route–advance traffic.
IH	*Intermediate high usage*: This group is provided to carry a combination of overflow traffic and first–routed traffic between any two offices when the combined volume of first–routed and overflow load makes direct routing economical. The group is designed to pass a predetermined amount of offered–load overflow to an alternate route during the busy hour.

Table 6–2.   Description of Traffic Class Trunk Group Codes (Continued)

CODE	DESCRIPTION
PH	*Primary high usage*: A group provided to carry only first–routed or primary traffic between any two offices when the volume of traffic makes direct routing economical. It is designed to pass a predetermined amount of offered–load overflow to an alternate route during the busy hour.
	OTHER
MI	*Miscellaneous*: This group is provided for traffic administration or operations maintenance, and for trunks that do not fall into one of the other categories.

Table 6–3. Traffic Use Codes—Message Services

CODE	DESCRIPTION
CA	*Tandem access—CAMA*: Carries customer–dialed calls to centralized automatic message accounting (CAMA) equipment, where recording and timing of calls are done.
CT	*Tandem–connecting—CAMA*: A two–way trunk group with combined functions of tandem access—CAMA and tandem–completing—toll.
DD	*Tandem access—LAMA*: Carries customer–dialed message calls (ordinary or mobile) from an end office, using local automatic message accounting (LAMA), to a tandem for completion of toll traffic or combined toll and local traffic over the message network.
DT	*Tandem–connecting—LAMA*: A two–way trunk group with combined functions of tandem access—LAMA and tandem–completing—toll.
EB	*Direct interLATA connecting—7–digit carrier access code for Feature Group B*: Interconnects an inter-exchange carrier's (IC's) point of termination (POT) with an exchange carrier's end–office switching system.

Table 6-3. Traffic Use Codes—Message Services (Continued)

CODE	DESCRIPTION
ED	*Direct interLATA connecting—1— or 5—digit carrier access code for Feature Group D:* Interconnects an IC's POT with an exchange carrier's end—office switching system.
EE	*Direct interLATA connecting—radio:* Interconnects an IC's POT with an exchange carrier's end—office switching system for domestic mobile and paging services.
EF	*Direct interLATA connecting—radio:* Connects a cellular mobile radio carrier's (CMC's) POT to one exchange carrier's end office for the completion of local or interLATA calls associated with that particular end office.
ET	*End—office tandem:* Interconnects an end office and a tandem that is equipped to bring in an operator to assist in the completion of interLATA or intraLATA calls.
IA	*Intra—end—office:* Carries calls between subscribers served by the same end office.
IE	*Inter—end—office—local:* Carries local or multi—message—unit calls between end offices in the same or different buildings.
IM	*Inter—end—office—marker:* Interconnects two No. 5 Crossbar marker groups in the same building by inter—marker—group operation.
IT	*Intertandem—toll:* Interconnects tandems for the completion of toll traffic or combined local and toll traffic, but not local only.
JT	*Intra—end—office—junctor:* Used to provide such special features as coin control, billing supervision, etc.
MB	*Mixed Feature Group B:* Carries FGB, FGC, intraLATA, local, and toll traffic between an end office and a tandem.

Table 6-3. Traffic Use Codes—Message Services (Continued)

CODE	DESCRIPTION
MD	*Mixed Feature Group D*: A one- or two-way trunk group from an equal-access end office to an access tandem, or from an access tandem to an end office that may carry local and interLATA traffic, including FGD.
MT	*Intertandem—local*: Interconnects tandems for the completion of local, extended area service (EAS), and multi-message-unit traffic. No toll traffic is routed over this group.
NS	*Network services*: A one- or two-way trunk group that carries operator services system (OSS) traffic between host and remote tandems or that interconnects a stored-program-control end office and a network teleconferencing system.
OD	*Operator Feature Group D*: Interconnects an equal-access end office and a tandem that is equipped to bring in an operator to assist in the completion of an interLATA call for ICs subscribing to exchange carrier operator services.
OG	*Tandem-connecting—local*: Interconnects a tandem and an end office to handle local traffic exclusively.
OR	*Optical remote*: A trunk group that carries a combination of voice, control data, and administrative signals over an optical link between a host and a remote switch.
PK	*Packet*: A trunk group carrying packet traffic.
RM	*Remote switch channel—voice*: A trunk group provided between a host switching system and a remote module, between two clustered remote modules, or between a master remote module and a slave remote module to carry customer traffic.
SP	*Tandem access—service position*: Carries customer-dialed traffic from an end office to a tandem and is equipped to bring in a position-system operator, if required, to assist with the completion of the call. SP is also used to identify position-system base-remote trunks.

Table 6-3. Traffic Use Codes—Message Services (Continued)

CODE	DESCRIPTION
TB	*Tandem interLATA connecting—7-digit carrier access code for Feature Group B*: Interconnects an IC's POT with an exchange carrier's access tandem switching system.
TC	*Tandem—completing—toll*: A trunk group from a tandem to an end office for completion of toll traffic or combined toll and local traffic, but not local only.
TD	*Tandem interLATA connecting—1- or 5-digit carrier access code for Feature Group D*: A trunk group that interconnects an IC's POT with an exchange carrier's access tandem switching system.
TE	*Inter—end—office—toll*: A trunk group that interconnects end offices and is provided to carry toll traffic. The group may carry some local, multi—message—unit, or EAS traffic.
TF	*Cellular mobile tandem—connecting*: Connects a CMC's POT with an exchange carrier's tandem switching system for access to local and interLATA call completion.
TG	*Tandem—completing—local*: A one—way trunk group from a tandem to an end office to handle local traffic exclusively.
TO	*Tandem access—local*: A trunk group from an end office to a tandem that carries local traffic exclusively.
TT	*Tandem access—operator*: A trunk group between a service position and a tandem that provides message—network access to operators.

Table 6-4. Traffic Use Codes—Auxiliary Services

CODE	DESCRIPTION
DA	*Directory assistance*: Provides the means for customers or operators to obtain listed directory numbers and newly connected numbers.
ES	*Emergency service*: Provides dedicated routing between central offices for E911 traffic.

Table 6-4. Traffic Use Codes—Auxiliary Services (Continued)

CODE	DESCRIPTION
IR	*Intercept*: Provides callers the means of obtaining information concerning called line numbers.
OF	*Official*: Carries official business traffic as part of an official company network that may cross LATA and state boundaries. Carries customer-dialed traffic to a business office.
PA	*Public announcement*: Provides customers with announcements, such as time, weather, lottery, or sports results, etc., by the destination code dialed.
RI	*Requested information*: Provides the means for customers to obtain specialized information other than that offered through directory assistance or rate-and-route.
RR	*Rate-and-route*: Provided for operators to obtain charge and routing information.
RS	*Repair service*: Used by customers to report telephone lines in need of repair.
TI	*Time*: Provides callers with time-of-day information.
VS	*Voice storage*: Used to provide access to a voice storage/announcement system. The VS system may be used by customers to enter a message, retrieve messages left by callers, remove stored messages, etc. Callers who receive messages from the VS system may also leave word for the customer to call back.
WE	*Weather*: Provides callers with weather information.

Table 6-5. Traffic Use Codes—Administrative

CODE	DESCRIPTION
AN	*Announcement*: Provides customers and operators with information when a call cannot be completed.
AP	*Operator service AP*: Connects an ACD or AIS control unit to operator positions.
CP	*Operator service CP*: Used to obtain billing information (such as the calling number) manually for the completion of CAMA traffic.

Table 6-5. Traffic Use Codes—Administrative (Continued)

CODE	DESCRIPTION
MI	*Miscellaneous*: A category for trunk groups that are not otherwise defined in this table.
SC	*Service code*: Permits the concentrated routing of service codes and their subsequent distribution to the appropriate terminus.
SL	*Signalling link*: One or more data links between processor-equipped switching systems and signal transfer points, used solely for support of the operation of the message network. These data links carry information for error and network control, as well as information to control trunking connections, monitoring, and disconnection. Some examples are common-channel signalling (CCS), remote trunk arrangement, and ACD.
TP	*Operator service TP*: Connects an OSS's base unit to OSS positions.
VC	*Vacant code*: Notifies a caller that a vacant code was dialed for calls that are attempted with invalid end-office (NXX) or numbering plan area (NPA) codes (live-operator intervention only).
VR	*Verification*: Used to verify subscriber line conditions (i.e., busy, out-of-order, etc.).

Table 6-6.   Bit-Rate Indicators Used with Signalling Links (Traffic Use Code SL)

Modifier	Data Speed (kb/s)
KA	2.4
KB	4.8
KC	9.6
KL	19.2
KJ	32.0
KN	50.0
KD	56.0
KE	64.0

Table 6-7.   Type and Direction of Pulsing (Other Than Supervisory Signals)

CODE	DESCRIPTION
D	*Dial*: A system of pulsing in which the digits are transmitted to the called end. The number of pulses, one to ten, corresponds to the digits dialed, one to zero.
J	*Tone dialing (12-button)*: A signalling system that uses combinations of dual-frequency signals originating in a 12-button tone set or equivalent sender.
M	*Multifrequency*: A system of pulsing where the identity of digits is determined by sending a combination of two out of five frequencies. A sixth frequency is used with one of the other five to provide priming and start signals.
N	*No pulsing*: A code that signifies no pulsing of address information.
6	*CCITT No. 6*: A common-channel interoffice signalling (CCIS) system that adheres to the CCITT No. 6 recommendation.
7	*Common-channel signalling*: A CCS system that adheres to the American National Standards Institute (ANSI) Signalling System No. 7 (derived from the CCITT No. 7 recommendation).
–	*No operation*: A hyphen is entered in character position 29 and/or 30, whichever is appropriate, when no pulsing is done at location A and/or location Z. (A one-way trunk group will always have a hyphen [–] in one of these positions.)

Additional Reading

Wright, A. J. "Where Things Are—In COMMON LANGUAGE," *Bell Laboratories Record*, Vol. 55 (July/Aug. 1977), pp. 192–197.

Chapter 7

Traffic Engineering

A communications network serving traffic from a large area of users is designed with common–usage links for economic and other reasons. The traffic capability required for these links depends on the flow of traffic and the quality–of–service objectives for the busy hour, when traffic is heaviest. Traffic engineering is used for evaluation of link performance (servicing) and for economic network configuration and trunk–group sizing. It is concerned with providing common transmission facilities (such as interoffice trunks and shared lines from digital loop carrier terminals to the serving office), common operator positions for service, and common–usage switch hardware such as signalling receivers.

This chapter gives busy–hour engineering terms and service criteria for modern local switching systems and briefly discusses two techniques that have been used to determine trunk capacity to meet traffic demands. One is the Neal–Wilkinson (N–W) method that has been used heavily in the past, supplemented by a newer technique, the extreme value engineering (EVE) technique. Brief discussions of trunk network configurations with alternate routing and preplanned and real–time dynamic routing are included.

7-1 QUALITY OF SERVICE

A measure of the quality of service for defined traffic conditions in a modern local switching system is generally stated in terms of percent busy–hour blocking for various switch connections and circuits, and in terms of time for dial–tone delays, attachment of signalling receivers, etc. These criteria [1, Section 11] are shown in Table 7–1 with N–W and EVE headings defined below. The criteria defined under the first three headings are for the N–W technique; they have evolved over the years. They represent the quality of service that customers are currently experiencing and are considered de facto standards [2]. The next two headings are for EVE criteria that have been calculated from equations (noted later) based on data of the older technique. The percent blocking given in Table 7–1 as one percent, for example, is usually denoted as B.01.

Table 7–1. Busy–Hour Blocking Criteria

		Neal-Wilkinson			EVE	
	Criterion Type	ABSBH	THDBH	HDBH	ABBH	EHD
Network Service						
Line–to–trunk	Blocking (ML)	1%	–	2%	1%	2%
Trunk–to–line	Blocking (FFM)	2%	–	–	2%	–
Line–to–line	Blocking (ML)	2%	–	–	2%	–
Trunk–to–trunk	Blocking (FFM)	2%	–	–	2%	–
Trunk–to–trunk	Blocking (ML)	0.5%	–	2%	0.5%	2%
Call Origination						
Dial–tone delay	Delay (t >3 sec)	1.5%	8%	20%	1.5%	20%
Dial–tone delay	Average (sec)	0.6	–	–	0.6	–
Service Circuits						
Customer digit receivers	Delay (t >0 sec)	–	–	5%	–	5%
Interoffice receivers (IR)	Delay (t >0 sec)	1%	–	–	1%	–
Interoffice receivers	Delay (t >3 sec)	–	–	0.1%	–	0.1%
IR attachment delay	Delay (t >3 sec)	1.5%	8%	20%	1.5%	20%
Interoffice transmitters	Blocking	–	1%	–	–	–
Ringing circuits	Blocking	–	–	0.1%	–	0.1%
Coin circuits						
Coin control	Blocking	–	–	0.1%	–	0.1%
Overtime announcement	Delay (t >0 sec)	1%	–	–	1%	–
Announcement circuits	Blocking	1%	–	–	1%	–
Tone circuits	Blocking	1%	–	–	1%	–
Reorder tone circuits	Blocking	–	–	0.1%	–	0.1%
Conference circuits	Blocking	0.1%	–	–	0.1%	–
Billing	Error	–	–	2%	–	2%

Notes:
    ML   = average final–failure–to–match loss.
    FFM = average first–failure–to–match loss.

Time-Consistent Engineering Terms

The N-W technique for determining trunk group sizes uses three terms that are based on time-consistent (TC) data; i.e., the traffic data are to be taken during the same hour over a number of days. The terms are as follows.

Average Busy-Season Busy Hour (ABSBH). The three months, not necessarily consecutive, that have the highest average traffic in the busy hour are termed the "busy season." The busy-hour traffic level averaged across the busy season is termed the ABSBH load.

Ten High-Day Busy Hour (THDBH). To calculate the THDBH, traffic data for the TC busy hour are processed all year to identify the ten highest traffic days of the year. The ten-day average traffic level for this TC busy hour is the THDBH load.

High-Day Busy Hour (HDBH). The one day among the same ten days that has the highest traffic during the busy hour is designated the (annually recurring) "high day." The traffic level in the busy hour of the high day is termed the HDBH load. There may be some other hour of the high day or another day of the year with a higher traffic level, but normally it would not be used in the engineering data base.

Busy-season data exclude Mother's Day, Christmas, or extremely high traffic days that can be attributed to unusually severe weather or catastrophic events, and are not reasonably expected to recur from year to year. Generally (but not always) the busy-season data exclude weekends.

Extreme Value Engineering Terms

The newer EVE method also uses three definitions. These follow (with the same exclusions as noted above).

Average Bouncing Busy Hour (ABBH). The ABBH load is the average of the individual peak hourly loads (whatever the hour) measured each day during the busy season (not the same hour as for the ABSBH noted above).

Once-A-Month (OAM). The OAM load is the hourly load that is expected to be met or exceeded once each month during the busy season. Since traffic data are generally tracked during about 20 business days a month, OAM is taken to be the load that is expected to be met or exceeded once every 20 days.

Expected High Day (EHD). The EHD load is the EVE–derived load that estimates the above THDBH load. Selection of a hundred–day measurement period is recommended since new criteria can be established to be the same as the old TC HDBH level.

7-2 TRAFFIC-CAPACITY ENGINEERING

Generally, traffic–capacity engineering is based on complex probability distributions that are modified to fit observations. Mechanized programs [3] are usually used (in place of a book of tables) to provide the estimated trunk capacity to meet blocking criteria. The N–W technique [4] was developed over the years and is heavily used in practice. It was based on several important considerations and assumptions including: (1) traffic load is measured at the same hour over a period of days, (2) a Poisson statistical distribution of traffic load is assumed, (3) blocked calls are cleared rather than held, (4) offered traffic is from independent and statistically infinite sources, and (5) adjustments are required for nonrandomness of traffic (hour–by–hour "peakedness") for day–by–day variations, etc. The programs relate the offered load, the number and kinds of trunks provided, and the probability of blocking for low, medium, and high levels of day–to–day load variation. The N–W technique monitors traffic data to determine the values for three types of busy–hour terms noted above and for other data. Table 7–2 is an example of a trunk–capacity table that can be provided by this technique.

The EVE technique monitors traffic data for 24 hours each day (instead of a certain hour as above for N–W) to provide a more robust estimate of extreme load values for peak–sensitive service. The EVE technique accounts automatically for the effects of ABBH and day–to–day variation that, for the N–W technique, require a judgement modification of the calculations. Use of the EVE technique for precutover and postcutover analysis for a local digital switch has been presented by Reference 5. The precutover results were based on converting existing

Table 7–2. Example of a Neal–Wilkinson B.01 Trunk–Capacity Table—Final or Only–Route Groups—Low Day–By–Day Variation

Number of Trunks	Average Offered Load (CCS)				
	Peakness Factors				
	1.0	1.2	1.4	1.6	2.4
1	2	0	0	0	0
2	8	0	0	0	0
4	34	25	18	0	0
6	68	59	49	39	0
8	109	97	87	76	0
10	153	140	128	116	72
12	201	186	172	160	111
14	250	234	219	205	154
16	300	283	267	252	199
18	353	334	317	301	245
20	406	386	368	351	292
22	460	439	420	402	340
24	515	493	473	454	389
42	1035	1005	976	950	861
44	1095	1064	1035	1008	916
46	1155	1123	1093	1066	971
48	1215	1182	1152	1124	1027
50	1276	1242	1211	1182	1083

Notes:

    B.01 = one–percent blocking
    CCS = hundred call seconds (per hour)

time–consistent data to equivalent EVE criteria that would provide, on the average, the same traffic service. The postcutover analysis was based on EVE measurements. It was found that estimates of EVE loads based on TC data are:

$$OAM = ABBH + 1.74\sigma \text{ and}$$
$$EHD = ABBH + 2.6\sigma$$

where σ is the standard deviation.

The amount of traffic to be served by a range of switch capacities is defined by traffic demand information given in Section 17 of the LSSGR [1]. This reference indicates the overall load distributions resulting from surveys for major traffic variables such as the originating and incoming rate, the originating and terminating usage, and other data. Figure 7–1 indicates an example for the expected customer call flow for average busy–season conditions. Several significant tables of data are given for: (1) local switches of fewer than 1000, 1000 to 10,000, and more than 10,000 lines (Table 7–3) and (2) stored–program–control (SPC) switches of several ranges over 10,000 to 40,000 lines (Table 7–4). Tables are given in Reference 1 for morning, afternoon, and evening customer–line calling rates and traffic usage.

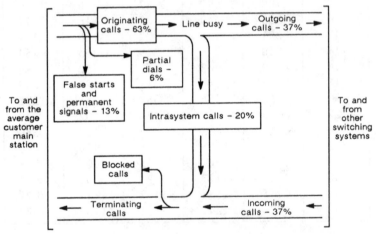

Gross average percentages are illustrative for ABS conditions and SPC switch

Figure 7–1.   Conceptual customer call flow through a local switch.

7–3   TRUNK NETWORK DESIGN

The present design of the trunk network has developed over many years, partially on the basis of an evolving body of traffic theory and partially as a result of advancing technology. Problems of trunk–group efficiency and size, service criteria (such as the probability of blocking), alternate routing, and load allocation are all involved in modern design.

Table 7–3. Traffic Variable Distributions Among Local Switches—Survey Results

Traffic Variable	Switch Size × 1000 Lines	Percentiles				
		10	25	50	75	90
O+I calling rate:	<1	0.6	0.8	1.1	1.4	1.7
ABS calls/MS	1–10	0.8	1.0	1.3	1.6	2.1
	>10	1.1	1.3	1.6	2.0	2.6
O+T usage:	<1	1.0	1.8	2.2	2.7	3.9
ABS CCS/MS	1–10	1.6	2.4	2.6	3.1	4.2
	>10	2.4	2.7	3.0	3.5	4.5
MS/line ratio	1	1.02	1.04	1.11	1.21	1.35
(Party lines)	10	1.0	1.0	1.01	1.03	1.07
	>20	1.0	1.0	1.01	1.01	1.02
Line/trunk ratio	1	6	8	12	18	24
	5	6	9	14	23	34
	>10	6	8	10	16	25

Notes:
Intrasystem call distributions are not shown.
O+I   =   originating and incoming
ABS   =   average busy season
MS   =   main station
O+T   =   originating and terminating
CCS   =   hundred call seconds (per hour)

Example: For O+I calls from the top row, 10 percent of switches having fewer than 1000 lines generate fewer than 0.6 ABS calls per MS, 50 percent generate fewer than 1.1, and 90 percent generate fewer than 1.7.

Trunk Group Efficiency and Size

For any trunk group to which access is provided in a particular sequence, the first trunk in the sequence carries the highest load. This is true because the same trunk is always selected first and is reselected when idle. Succeeding trunks in the access sequence carry decreasing amounts of load; the trunk that is selected last carries the least load because it is chosen only when all other

Table 7–4.   Traffic Variable Distributions Among Switches—
SPC System Statistics

Traffic Variable	Switch Size × 1000 Lines	Percentiles				
		10	25	50	75	90
O+I calling rate:						
ABS calls/MS	10–25	1.3	1.6	2.1	3.0	3.6
	25–40	1.2	1.5	1.8	2.3	2.7
HD calls/MS	10–20	1.6	2.0	2.8	3.8	4.7
	20–40	1.5	1.8	2.3	3.2	3.8
HD/ABS ratio	10–40	1.1	1.15	1.20	1.35	1.55
O+T usage:						
ABS CCS/MS	10–25	2.5	2.9	3.4	4.4	5.2
	25–40	2.5	2.8	3.1	3.6	4.4
HD/ABS ratio	10–40	1.08	1.10	1.15	1.20	1.37

Notes:
    Intrasystem call distributions are not shown.
    HD = high day.

Example: For O+I calls from the top row, 10 percent of switches having
10,000 to 25,000 lines generate fewer than 1.3 ABS calls per
MS, 50 percent generate fewer than 2.1, and 90 percent generate fewer than 3.6.

trunks are busy. The trunk selected last, therefore, is the least
efficient trunk and is commonly referred to as the *last trunk*. It
can be seen that if the access sequence is low–to–high, the last
trunk will be the highest numbered trunk; the reverse occurs if
the sequence is high–to–low. In those switching systems where
trunk access is random, rather than ordered, it is not possible to
identify a particular trunk as the last trunk and the load is evenly
distributed across the trunks in the group. However, the total
capacity of the group is the same regardless of access sequence.
Rather than identify the last trunk with a particular trunk
number, as is often done to visualize the effect, it is better to
determine the load carried by n trunks (where n is the number of
trunks in the group), subtract the load carried by $n - 1$ trunks,
and call the remainder the capacity of the nth or last trunk. With

the same offered load, the probability of blocking is different for n trunks than for $n - 1$ trunks.

This concept of diminishing returns as trunks are added to a group becomes important in the economic sizing of high–usage groups where the cost of an added trunk is weighed against the cost of an alternate route in selecting the proper route for the traffic.

The amount of load that can be carried per trunk in a group is a function of both offered load and group size. As the load offered to a group increases, the load carried per trunk increases until it approaches its full capacity (36 CCS). Large groups are more efficient than small ones, but the increase in efficiency levels off as the group becomes larger. The greatest increase in capacity per trunk occurs when small trunk groups are made larger.

Trunk efficiency is usually expressed in terms of percent occupancy and is defined as the ratio of carried load to total full capacity. Typical efficiency (occupancy) for groups of 1 to 100 trunks at two values of blocking probability are shown graphically in Figure 7–2. It is evident from the figure that as the group size increases, the total available capacity can be used more efficiently without degrading service. Also evident is the fact that the higher the efficiency, the smaller the margin that remains for traffic peaks caused by surges of traffic. A practical example of this occurs in large metropolitan areas where, on days of severe storms, the percent overflow on large groups runs far in excess of the percent overflow on smaller groups. Particular caution is necessary where large groups engineered for one–percent blocking (final and only–route) may be subjected to heavy surges of traffic.

Alternate Routing

In a multioffice network that must be capable of full interconnection, direct interconnection would require many trunk groups, most of which would carry extremely low volumes of traffic. This would clearly be economically unacceptable. Therefore, other methods of organizing and grouping the flow of traffic have evolved. These methods use such concepts as a multilevel switching hierarchy and alternate routing.

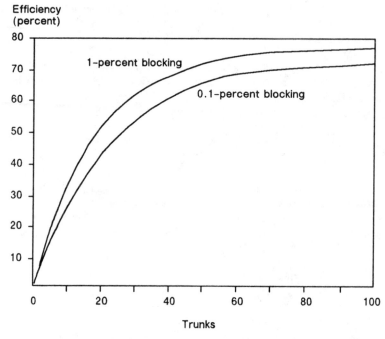

Figure 7-2. Efficiency versus number of trunks in a group.

The alternate-routing concept and the way it improves the traffic handling capabilities of a network are illustrated in Figure 7-3. Direct interconnection of the six end offices shown would require 15 trunk groups. By routing all traffic through a tandem, only six trunk groups are needed. These six groups would operate much more efficiently than the original 15 would. However, total network cost per CCS carried per hour would not necessarily be minimized by this configuration.

If the volume of traffic between offices 2 and 3 is high, it may prove economical to install a trunk group between those two offices, as indicated by the dashed line. This possibility suggests the concept of alternate routing. The direct route between offices 2 and 3 may be designed to carry only a portion of the traffic, with the overflow carried over the alternate trunk groups through the tandem. It can be seen that this arrangement can generally be operated more efficiently, i.e., less cost per CCS carried, than a network made up entirely of directly interconnected offices.

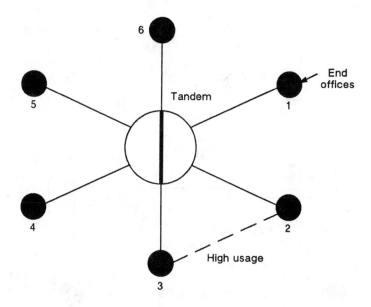

Figure 7-3. Routing of traffic through a tandem.

The concept of a multilevel hierarchy of switching and trunking may be expanded as in Figure 7-4. Here, direct interconnection of several tandems results in a three-level network. As can be seen, the alternate-route possibilities increase as the network expands.

Network design and configurations (nodes, routing, blocking, etc.) are covered in Reference 6.

In the concept of multilevel network configurations, every office has a single office of higher class to which it is connected and on which it is to "home." The offices of highest class are completely interconnected. There is a logical progression of traffic in the hierarchy such that there is a set of backbone or final routes available to call from any one point in the hierarchy to any other. High-usage trunk groups may be placed between any pair of offices with enough traffic to justify them economically.

Load Allocation

The process of determining the portion of a given load to be carried by a high-usage group and the portion that should be

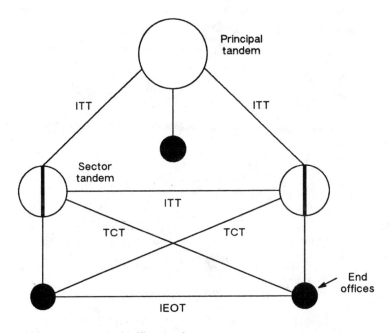

IEOT = Inter-end-office trunk
ITT = Intertandem trunk
TCT = Tandem-connecting trunk

Figure 7-4. Routing via tandems.

offered to an alternate route is called *load allocation*. The objective is to provide trunks in such numbers in both the direct and
alternate routes that the traffic between a given pair of offices or
switching entities can be handled at the least possible cost per
CCS carried consistent with service requirements.

Load allocation requires a knowledge of the offered load between terminals (during the hours of interest), the cost of a direct
path between terminals, and the cost of an alternate route (including intermediate switching) between terminals. To select the
most economical trunking arrangement, it is necessary to relate
costs and trunk efficiencies of the direct and alternate routes for
a stated busy–hour offered load. Such a procedure is essentially
cost balancing in which the cost of carrying a unit of traffic on
the high–usage group is balanced against the cost of carrying it on
the more expensive but more efficient alternate–route group. A

practical consideration in digital switches, and in analog switches equipped with digital carrier trunk (DCT) frames, is that trunks are most conveniently added or removed in DS1 modules of 24. This tends to quantize the sizes of trunk groups, thus affecting the economics of routing.

7-4   TRAFFIC MEASUREMENTS

In SPC switches, traffic data are programmed to be taken on a specified time schedule. They consist of measurements of peg counts (the occurrence of calls) and usage counts (the duration of a specified condition such as a call in progress) [1, Section 8.2]. These data may be collected by the Engineering and Administrative Data Acquisition System and are usually monitored by the circuit administration center in the exchange carrier. The processed data, however, are used for many functions by many other areas, such as network engineering, management, etc.

Data measurements are taken on the traffic demand for originating, incoming, and terminating services, and the system's ability to meet the demand. Some of the peg–count measurements are on originating–call line seizures and call attempts, on incoming–call trunk seizures and attempts, and for intrasystem and tandem calls. Some of the usage data are origination, incoming intrasystem, outgoing, terminating, and tandem call usage. Measurements are also made to show ineffective attempts on all of these types of calls caused by blockage, both inside and outside the system (such as false starts). The time schedule for the data period depends on the intended use and may cover 30–second, 5–, 15–, 30–minute, and total–day intervals. For the usual trunk traffic engineering purposes, hourly data are taken to determine the busy–hour terms defined earlier for the time–consistent or EVE traffic loading methods.

7-5   TRUNK ADMINISTRATION

Trunk administration is the continuing evaluation of changes in traffic flow from week to week and season to season, and the determination of the number of trunks required to meet the service objective. The trunk administration program must be based on an adequate traffic measurement schedule since the current traffic load and number of trunks in service must be used as a base for projecting future needs. Consequently, inadequate trunk administration could result in trunk estimates of dubious accuracy, yielding poor service due to trunk shortages or unnecessary expense due to trunk surpluses.

Effective trunk administration depends on timely availability of trunk terminations and interoffice facilities, and assigning trunks to switching systems in a way that maintains an optimum level of usage on each trunk frame or switch module. The ability to achieve the latter is largely determined at the time a traffic order is prepared. The traffic order designates the number and types of trunk to be located on each frame. In many instances, a layout that permits optimum loading may require a substantial number of trunk transfers from existing frames to new frames.

7-6 DYNAMIC ROUTING

Modern communication networks provide for common usage links and alternate routing for increased efficiency to provide service economically to a large body of users. The predivestiture five-level toll network was originally based on hierarchical routing with fixed alternate routing paths and rigid routing rules. The "final" route was through all levels of the hierarchy. The capacity of this route was engineered on the basis that only one percent of the calls would be blocked during the average busy season.

Alternate routing called "high-usage routes" between hierarchy levels was established mostly on economic considerations (i.e., the capacity of high-usage routes depended on the relative costs of carrying traffic to a higher level versus using the alternate route).

The fixed-routing rules result in two main disadvantages: quick response to unpredicted congestion is not possible, nor is it possible to take advantage of the spare capacity that becomes available in off-peak traffic periods.

It has been found in networks having common-channel signalling capability that dynamic routing [7] can make a more efficient network by having traffic bypass congested areas and by using spare capacity. There are several dynamic routing methods for large networks. These have been based on a time dependency, to take advantage of noncoincident busy periods around the network, or a state dependency, where congested links are sensed and traffic routed to other links. There are different ways of sensing and reporting congestion; the frequency of updating this information is an important consideration.

Dynamic Nonhierarchical Routing

Dynamic nonhierarchical routing (DNHR) [7] has been employed in the AT&T intercity network since 1984 with a reported network saving of 15 percent. In brief, DNHR routes calls via minimum–cost paths from the originating toll center either directly to the terminating center or via one of several tandems as a needed path becomes available. Common–channel signalling is used for network control, routing, and management purposes. DNHR takes advantage of different time–zone trunk usage by dividing the day into ten time periods and using a different set of preplanned routes for each period. The routing plans are evaluated over weekly and semiweekly intervals. In addition to preplanned routing, DNHR automatically switches reserve capacity for direct connections and thus provides some real–time adaptive capability to meet changing traffic patterns.

A proposal [8] has been made to enhance DNHR with real–time routing decisions by employing a real–time trunk status map (TSM) to implement the dynamic routing strategy. TSM provides a frequent and timely update of the number of idle trunks at each office in the DNHR network, and accordingly, the routing patterns can be adjusted frequently to minimize traffic blocking.

References

1. *LSSGR, LATA Switching Systems Generic Requirements*, Technical Reference TR–TSY–000064, Bellcore (Iss. 2, July 1987), Sections 8.2–8.6, 11, and 17.

2. Kappel, J. G. "Service Standards: Evolution and Needs," *ITC–11* (Japan: International Teletraffic Congress, 1985).

145

3. Jagerman, D. L. "Methods in Traffic Calculations," *AT&T Bell Laboratories Technical Journal*, Vol. 63, No. 7 (Sept. 1984).

4. *Trunk Traffic Engineering Concepts and Applications*, Special Report SR–EOP–000191, Bellcore (Iss. 1, Apr. 1985).

5. Friedman, K. A. "Precutover Extreme Value Engineering of a Local Digital Switch," *ITC–10* (Montreal: International Teletraffic Congress, 1983).

6. *Notes on the BOC IntraLATA Networks—1986*, Technical Reference TR–NPL–000275, Bellcore (Iss. 1, Apr. 1986), Section 4.

7. Hurley, B. R., C. J. R. Seidl, and W. F. Sewell. "A Survey of Dynamic Routing Methods for Circuit–Switched Traffic," *IEEE Communications Magazine*, Vol. 25, No. 9 (Sept. 1987).

8. Ash, G. "Use of a Trunk Status Map for Real–Time DNHR," *ITC–11* (Japan: International Teletraffic Congress, 1985).

Chapter 8

Trunk Design

Message service is provided to users in the local access and transport area (LATA) by the three general LATA networks indicated in Figure 8-1. The networks encompass several kinds of trunk and switching offices differing in transmission and signalling capabilities as needed to establish connections between users in the same LATA or via an interexchange carrier (IC) if in different LATAs.

This chapter discusses the present transmission design of the analog, combination, and digital trunks involved in the networks. The topics covered are definitions of loop and trunk, office wiring loss, analog and digital loss, and trunk transmission levels and requirements. Simplified illustrations of various trunks are included to indicate the test levels to be used to adjust the trunk loss to meet requirements. Operator and other auxiliary services are covered in Chapter 10.

Trunk signalling is discussed in Volume 1, Chapter 13, Part 2. Discussions on signalling for digital systems using the "robbed-bit" DS1 format and using common–channel signalling are included in Volume 2, Chapter 13, Part 1.

8-1 LATA NETWORKS AND TRUNKS

The three networks and their associated trunks are:

(1) *LATA access network:* Direct interLATA connecting trunk—Feature Group A (FGA), FGB, FGC, FGD; tandem interLATA connecting trunk—FGB, FGC, FGD; access tandem–connecting trunk—FGB, FGC, FGD; intertandem trunk (access tandem – LATA tandem)—FGB

(2) *Long–distance intraLATA network:* Intertandem trunk, LATA tandem–connecting trunk, inter–end–office trunk

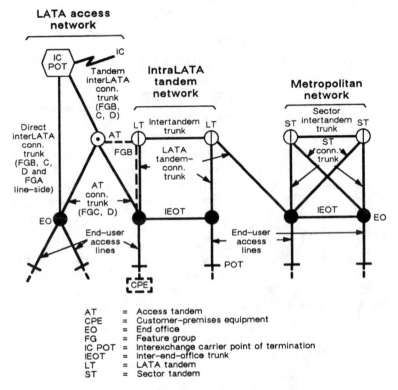

LATA access network

AT   = Access tandem
CPE   = Customer-premises equipment
EO   = End office
FG   = Feature group
IC POT = Interexchange carrier point of termination
IEOT   = Inter-end-office trunk
LT   = LATA tandem
ST   = Sector tandem

Note: Equal access is provided only by FGD (see Table 18–2 for FG comparisons).

Figure 8-1. LATA networks for public switched network services.

(3) *Metropolitan network:* Sector intertandem trunk, sector tandem–connecting trunk, inter–end–office trunk.

In many LATAs, of course, the networks are small enough and simple enough to use a single tandem for all three functions.

The detailed transmission parameter requirements (loss deviation, slope, return loss, noise, interface, etc.) for the LATA access network are given in Chapter 18, which indicates requirements for the four feature groups and the three transmission types that can be specified for voice–grade switched access service. The feature groups were established at divestiture to cover connections from the interexchange carrier point of termination (IC POT) to LATA networks.

American National Standards Institute Telecommunications Standard

The performance of switched access networks is of considerable importance since two access networks (originating and terminating) are involved in every interexchange connection. The American National Standards Institute (ANSI)–accredited standards committee, T1–Telecommunications, has prepared *ANSI T1.506–1989* to define performance of these networks. The standard will apply to direct or via–access–tandem connections and has been modeled after the capabilities offered by FGC and FGD.

Feature Groups

The feature groups and transmission types are described briefly in the following paragraphs. FGA is used by an IC or other network provider for foreign–exchange service that is terminated on the line side of an end office (telephone number assigned), or for general access purposes. FGB is used by an IC or other network provider (including private–network users) for call termination, or for "10xxx" dial access. FGC is the AT&T connection that was usually in place prior to divestiture. FGD is the equal–access trunk connection that provides full postdivestiture features and will eventually fully replace FGC.

The three transmission types, designated A, B, and C, are:

(A) Four–wire interface at the IC POT with digital and selected analog facilities

(B) Same as (A) for message and C–notched noise but somewhat less stringent requirements on slope and return loss

(C) Two–wire interface; same as (B) but has wider slope and return loss requirements.

8-2 RELATIONSHIPS OF LOOPS AND TRUNKS

A call may be routed over several possible paths between the originating switching office and the terminating switching office.

The end–user–to–end–user connection shown in Figure 8–2 is one possible routing of a call; the connection includes two loops and two trunks in tandem and illustrates two–wire switching. Many differences in detail are found in the telephone plant. The figure shows switching equipment at both ends of each trunk with transmission facilities between. A trunk circuit having relay equipment to signal and supervise the connection is associated with the switching machines at both ends of each trunk. Between the trunk circuits, the transmission facilities may include transmission equipment such as repeaters, four–wire terminating sets, etc., together with an associated cable pair or an analog or digital carrier channel.

Signalling equipment, also an important trunk component, does not appear in Figure 8–2 because its location is difficult to generalize. In some cases, it may be a separate entity on the line side of the trunk circuit. In other cases, it is built around the four–wire terminating set. In T–type carrier systems it may be a part of the channel equipment or included in a digroup termination on the switch. With common–channel signalling there is no per–trunk signalling equipment at all.

Definitions

The transmission characteristic of an overall connection is the sum of the characteristics of two loops, any trunks, and switching paths. As shown, the switching path in the originating office is not included in either the loop or the trunk.

Loop. By definition, a loop for POTS (plain old telephone service) extends from the end–user network interface (NI) to the line side of the end–office switch. Private branch exchange (PBX) lines providing basic service features also connect to the line side. However, for direct outward/inward dialing features, PBXs connect to the trunk side of the switch.

Trunks. Trunks are defined as analog, combination, or digital types. An analog trunk connects to analog switches at each end, using analog or digital facilities. A combination trunk connects to a digital switch at one end and an analog switch at the other end, employing analog or digital facilities. A digital trunk connects digital switches via digital facilities.

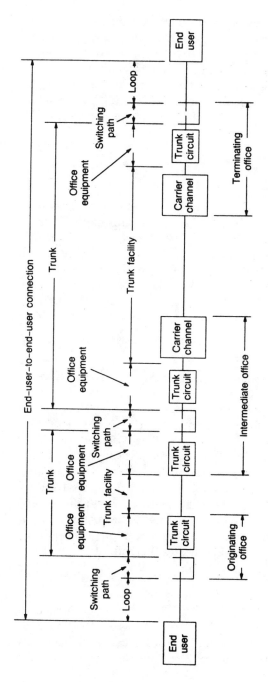

Figure 8-2. Relationships of loops and trunks in a connection.

Notes:

1. Switching can be two-wire or four-wire.

2. Signalling equipment, not shown, may be associated with the trunk circuit, the carrier channel unit, or the switch.

Generic trunk definitions for through transmission are illustrated in Figure 8-3, which indicates the points for establishing the trunk inserted connection loss (ICL). It is noted that a digital trunk encompasses the path from an analog or digital interface through the digital switch, digital facility, or the remote digital switch to an analog or digital interface. Switching system interfaces [1] are illustrated by Figure 8-4 for analog systems and by Figure 8-5 for digital systems.

Office Wiring Losses

Office wiring that increases the voice-frequency (VF) channel losses is noted on each figure. Table 8-1 estimates the average office loss for analog and digital end offices [1]. Wiring loss is accounted for by a small adjustment of the associated repeaters or digital channel units when adjusting for the required trunk loss.

Loss Adjustment

In analog systems and facilities, the loss of each trunk can be adjusted by setting attenuators (pads) at appropriate points in the circuit. In digital systems, after encoding the analog signal into the digital realm at the originating office, the intent is to obtain the loss at the terminating office only immediately before or after decoding to analog. At the originating office, encoding a 0-dBm signal at the zero encode level point (0 ELP) produces the digital reference signal (DRS). At the terminating end office, the DRS decoded at a zero decode level point (0 DLP) produces a 0-dBm signal. Analog loss, as the preferred method, is used to set the trunk or connection loss to zero or any other required value. Digital loss, described in the following paragraph, may be used but has limitations. Table 8-2 indicates the current loss required for various trunks terminating at an end office [1].

Digital Loss

At an end-office digital switch, the trunk loss can be adjusted by code conversion as noted in Figure 8-6. Code conversion is a

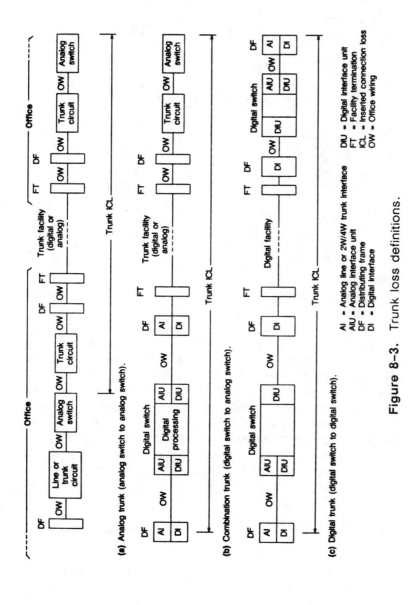

Figure 8–3. Trunk loss definitions.

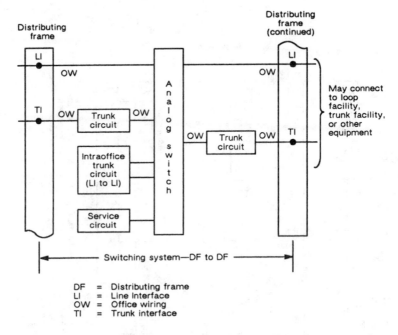

Figure 8–4. Generic transmission interfaces—analog switching system.

process where all of the received pulse code modulation (PCM) code words are converted to new code words representing the signal after a loss of x dB. This process produces digital loss [2], causes slight impairments similar to another encode–decode step and, further, does not preserve bit integrity (needed for digital data transport). Digital loss reduces the signal–to–distortion ratio characteristic of a digital–to–analog decoder as shown in Figure 8–7 over a range of input VF signals. Further, the value of the digital loss is a function of the signal level and will vary somewhat over the dynamic range of operation.

8-3 ECHO CANCELLERS

Echo cancellers [2,3,4] have replaced echo suppressors to control echo on connections with excessive transmission delay. In addition, cancellers [5,6] may be used to meet the office balance requirements of Chapter 9 without the costly return loss

154

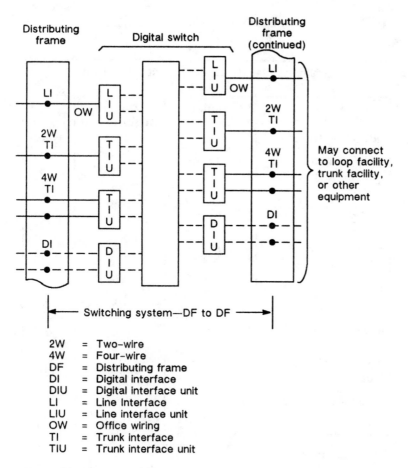

2W   = Two-wire
4W   = Four-wire
DF   = Distributing frame
DI   = Digital interface
DIU = Digital interface unit
LI   = Line Interface
LIU = Line interface unit
OW   = Office wiring
TI   = Trunk interface
TIU = Trunk interface unit

Note: DI and DIU are digital; the other interfaces are analog.

Figure 8–5.   Generic   transmission   interfaces—digital
switching system.

balance testing and adjustment routine that is normally carried
out.

General

The canceller is based on integrated circuits using very–large–
scale integration (VLSI) techniques. It employs digital processing

Table 8-1. Average Office Loss for Current Analog and Digital End Offices

Average 1004-Hz Loss—Distributing Frame to Distributing Frame			
Connection Path Interface-Interface	Analog Requirement (dB)	Digital Requirement (dB)	Notes
LI or remote LI to LI or remote LI	1.0	1.0	1,2
LI or remote LI to 2WTI	0.75	0.75	1,2,3
LI or remote LI to 4WTI	NA	0.75	1,2,3
LI or remote LI to DI	NA	0.5	2
2WTI to 2WTI	0.75	0.75	1,3
4WTI to 2WTI or 4WTI	NA	0.75	1,3
DI to 2WTI or 4WTI	NA	0.4	3

LI = Line interface
TI = Trunk interface

Notes:

1. For these connection paths, the average loss in most offices is 0.3 to 0.5 dB. Margins are allowed in recognition of the long wiring runs occurring in some large offices.

2. Values given are for a 0-dB loss remote-to-host link.

3. Interoffice facility loss design may compensate with the addition of gain for part of the wiring loss included in the requirements.

to analyze the echo path of the near-end echo-producing circuit continuously. In the process, estimates are made (using a tapped delay line) of the echo amplitudes and delays of the near-end

Table 8-2. Current Digital End-Office Loss

DRS to 2W Loss (dB)	Trunk Type
0	Intraoffice
3	Short metro inter-end-office
5	Analog LT (-2 TL)
6	Long inter-end-office and digital LT (-3 TL)

DRS = Digital reference signal
LT = LATA tandem
TL = Transmission level

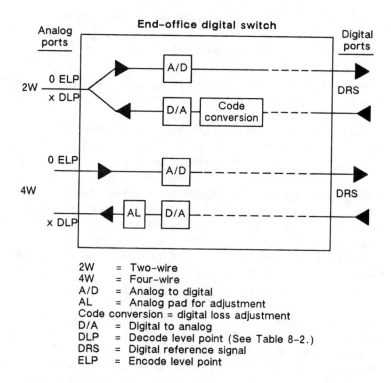

2W   = Two-wire
4W   = Four-wire
A/D = Analog to digital
AL   = Analog pad for adjustment
Code conversion = digital loss adjustment
D/A = Digital to analog
DLP = Decode level point (See Table 8-2.)
DRS = Digital reference signal
ELP = Encode level point

Figure 8-6. End-office digital loss by code conversion.

echo path. These estimates are subtracted from the near-end transmitted signal of speech and echo, ideally leaving just the

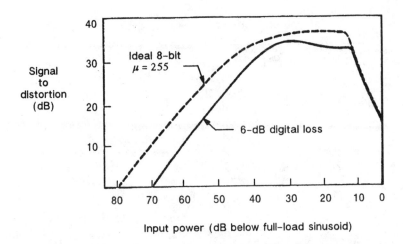

Figure 8-7. Effect of digital loss on signal-to-distortion.

near-end speech. A near-end speech detector is needed to inhibit the processing during the times that this speech is present.

A canceller is used at each end of a circuit to control echo in both directions of transmission. For operation, some cancellers require that the echo path loss be at least 6 dB; all require the echo path delay between the location of the canceller and the echo-producing point (impedance mismatch) be within the operating delay range of the canceller. Thus, the important delay is not that to the remote end, but only that to the near end, commonly called tail-circuit delay. The delays of some typical equipment and facilities are given in Volume 2, Chapter 13, Part 7. When used for improving end-office four-wire/two-wire balance, only one canceller of short delay range is needed.

An example of the performance of an early echo canceller in a simulated circuit using a Gaussian noise signal (with residual echo present) is given in Figure 8-8. Tests with a −10 dBm0 Gaussian noise signal on a single 16-ms canceller and two of these units in cascade for a 32-ms canceller have been reported with a performance of 30 to 33 dB of echo return loss enhancement (not including the hybrid echo return loss). Proposed CCITT requirements are: (1) for a −10 dBm0 Gaussian signal, the return loss enhancement is to be 32 dB or greater with residual echo and (2) the returned echo level with the residual echo suppressed is to be −65 dBm0 or less.

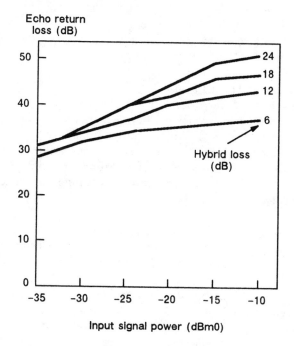

Figure 8-8. Example of echo canceller performance.

The canceller is to be transparent to transmission of voiceband data. As digital bit integrity is not preserved, digital data (56 or 64 kb/s) or test data may be impaired, so the canceller is to be disabled. CCITT recommends a canceller tone disabler that operates on periodic phase reversals of a 2100–Hz signal. Upon disablement, bit and sequence integrity are to be preserved.

Application

The application of echo control for some connections in the predivestiture network and the allocated delay allowance are given in the following to serve as a guide for application to similar connections today. In the original via net loss (VNL) network plan for distance dialing, echoes on connections in which the echo path delay exceeded 45 ms were controlled by echo suppressors or cancellers. Intraregional connections did not require echo control devices because the maximum round–trip delay did not exceed the 45–ms limit. However, interregional trunks were equipped with these devices when the round–trip delay on interregional calls using these trunks was likely to exceed the limit.

Selection of Trunks Requiring Echo Cancellers. There were two categories of interregional trunk groups: regional–center–to–regional–center final groups and high–usage interregional groups. The latter category included interregional grade–of–service groups. Usually, trunks between regional centers required echo control and interregional high–usage trunks did not. However, there were exceptions in both cases depending on the loss and delay involved.

Regional–Center–to–Regional–Center Trunks. The round–trip echo delay between any toll office and the most distant end office in the final–routing chain was called end delay for that toll office. The end delay for a regional center could be as great as 22.5 ms. Thus, a call routed from one regional center through another could encounter a 22.5–ms end delay in each region or a combined end delay of 45 ms without allowing for any delay in the interregional trunk. For that reason, most interregional trunks were equipped with echo control. Exceptions were interregional trunks with short delays between regions that were compact geographically (e.g., Wayne, PA to White Plains, NY).

Interregional High–Usage Trunks. The end delay from lower–class toll offices was less than that from regional centers. Thus, in determining the need for echo control on high–usage trunks, it was common practice to assume 10 ms as the maximum end delay since at least one end of each high–usage trunk terminated below the regional center in the hierarchy. Thus, echo control was to be used on all high–usage trunks having round–trip delays of 25 ms or more. This delay corresponded to a VNL of 2.9 dB.

In term of miles, echo control has been employed on trunks derived over radio facilities of 1850 route–miles or greater. Optical fiber facilities have more delay and digital transmission is quieter; echo control may be needed for such facilities greater than about 1000 route–miles. With the application of echo control, the above trunk losses were set to zero.

Fixed–loss network plans not employing VNL for echo reduction may require echo control at lower mileage unless the loop return loss is increased. Echo will continue to be a significant problem; the main source is the impedance mismatch of the two–wire loop at the end office. The match can be improved by loop segregation methods requiring several fixed balance networks applied appropriately (see Volume 1, Chapter 26, Part 1) or by adaptive two–wire–to–four–wire conversion circuits. Such circuits use echo canceller technology to increase the return loss greatly, virtually eliminating echo impairment. The evolving region digital switched network (RDSN) transmission plan in Chapter 2 addresses this problem.

8-4 CURRENT LOSS RELATIONSHIPS

If performance is to be satisfactory on multitrunk connections, the overall loss must be held to values that represent a compromise between transmission performance (adequate received volume, singing margin, echo, contrast, and noise) and circuit costs. Presently, the loss is allocated to the individual trunks in the overall connection as described in previous chapters. Inserted connection loss, expected measured loss (EML), and actual measured loss (AML) are used to ensure that individual trunks are designed, installed, and maintained within allowable loss tolerances.

In the future, with an all–digital network, the multitrunk connection loss will be established at the terminating office after decoding to analog. The digital trunks will be aligned to the DRS and the encoded information bit stream will flow through the network unaltered. Hence, the individual trunk losses are zero in analog terms, up to the terminating office. Such a digital network will be the RDSN, which is discussed in Part 11 of this chapter and in Chapter 2.

Inserted Connection Loss

The ICL of a trunk is the net 1004–Hz loss that is inserted into the connection by an operating trunk.

Expected Measured Loss

The EML of a trunk is the 1004–Hz loss that is expected to be measured under specified test conditions. This loss is calculated by summing all gains and losses in the specified measuring configuration; it is provided as a reference for comparison with actual measurements.

Actual Measured Loss

The AML is the 1004–Hz loss measured by the proper test equipment with the proper measuring configuration. Upon installation, it must be compared to the EML to ascertain that the trunk meets the loss objective. Minor deviations exist between the AML and the EML due to discrete step sizes of pads used in the trunk, differences between average and actual central–office cabling losses, differences between average and actual test access losses, and the unpredicted impedance interactions of the various parts of the circuit. If the AML does not fall within tolerable limits, the trunk is not placed in service. Similarly, the subsequent periodic AML measurements made for maintenance purposes are compared to the EML; differences that do not fall within maintenance limits require corrective action. These actions may include immediate removal of the trunk from service.

Office Transmission Levels

In the analog world, the outgoing direction of a two–wire toll switchboard or tandem switch was historically considered to be a zero transmission level point (0 TLP). However, a transmission upgrade program involving the removal of 2–dB switched pads resulted in tandems being redefined as −2 TL points; it was easier to redesignate the switch and add test pads than to realign all the affected trunks.

The assignment of office levels became embroiled in analog trunk design, testing, and maintenance procedures. Transmit and receive test pads were used in applying 0–dBm test signals in aligning and measuring trunk loss. This was carried over to the digital 4ESS™ office, which was designated a −3 TL because it originally operated in an analog network. In some applications, this office still retains a −3 TL designation. Table 8−3 provides the current office transmission levels for analog networks.

Table 8−3. Office Transmission Levels for Analog Networks

OFFICE	TL
Analog LATA or sector tandem	−2 *
Digital LATA or sector tandem	−3 †
Analog access tandem	−2
Digital access tandem	−3 †
Analog end office	0
Digital end office	0 †
Operator services system	Same as associated tandem office (0, −2, or −3) †
Operator services system in local DA−charge recording application	−2
No. 1ESS CAMA in local DA−charge recording application	0 or −2
Phase II No. 5 Crossbar ACD	−2
Automatic intercept system	−2

* Some existing switches are operated as 0 TL.

† This value is designated to facilitate trunk design and testing. It does not represent a true TL value at any portion of the switch where digital signals are present.

In future digital networks [7], all digital offices will be 0 TL and test levels will be set to produce the DRS in the encoded stream. The connection loss will be established at the decoder at

_____

ESS is a trademark of AT&T.

the network terminating end. This will eliminate the test pads mentioned above. In addition, testing and maintenance criteria probably will be based on surveillance of the quality of the digital signals without converting to analog for analog testing.

8-5   LOSS

Loss is one of the important parameters to be considered in trunk design since it affects such channel characteristics as received volume, echo, stability, noise, and crosstalk.

Loss Requirements

The current message trunk losses (ICLs) for the three LATA networks are given in Table 8-4 for analog, combination, or digital trunks [1]. For LATA access network trunks, only trunk Feature Groups C and D are given here. Complete transmission parameters for all access trunks are given in Chapter 18.

Loss Design

Several trunk loss design examples are given in Figure 8-9 for inter-end-office trunks, tandem-connecting trunks, and inter-tandem trunks [8]. These illustrate the relationships between test-tone levels, encode-decode levels and the office transmission levels involved in establishing the required losses for the various trunks. There are other arrangements (not shown) that provide the same insertion loss but different test-tone loss, such as a digital tandem without test pads.

Combined Function Trunks

A switch can be designed to function as several combined offices, such as an end-office/LATA tandem, an end-office/access tandem, an end-office/LATA tandem/access tandem, etc. The end-office requirements are given in Reference 1; the LATA tandem and access tandem requirements are given in Reference 9. Such combined offices may interface trunks serving combined

Table 8-4. LATA Networks—Message Trunk Losses (ICLs)

TRUNK	ANALOG TRUNK W/O GAIN (dB)	ANALOG TRUNK W/O GAIN (dB)	COMBIN. [1,3] TRUNK (dB)	DIG. [2,3] TRUNK (dB)
ACCESS NETWORK				
● Direct interLATA connecting				
Feature Group C (Note 4)	2.0–4.0	VNL + 2.5	3.0	3.0
Feature Group D	2.0–4.0	3.0	3.0	3.0
● Tandem interLATA connecting				
Feature Group C (Note 4)	–	VNL	1.0	0.0
Feature Group D	–	0.0	0.0	0.0
● Access tandem–connecting				
Feature Group C (Note 4)				
a) ≤200 miles	2.0–4.0	3.0	3.0	3.0
b) >200 miles	–	VNL + 2.5	3.0	3.0
Feature Group D				
a) ≤ 200 miles	2.0–4.0	3.0	3.0	3.0
b) >200 miles	–	3.0	3.0	3.0
INTRALATA AND METRO NETWORKS				
● Inter–end–office				
a) ≤ 200 miles	0.0–5.0	3.0	3.0	3.0
b) >200 miles	–	VNL + 6.0	6.0	6.0
● LATA tandem–connecting				
a) ≤ 200 miles (Note 5)	2.0–4.0	3.0	3.0	3.0
b) >200 miles	–	VNL + 2.5	3.0	3.0
● Intertandem	–	VNL	1.0	0.0
● Sector tandem–connecting	0.0–4.0	3.0	3.0	3.0
● Sector intertandem (Note 6)	–	1.5	1.0	0.0

Notes:

1. A combination trunk uses digital or analog facilities and connects analog and digital switching offices.

2. A digital trunk uses digital facilities and connects digital switching offices.

3. Trunk losses are based on a fixed-loss plan, i.e., 6 dB for an end office to end office via a sector tandem (ST) and 3 dB for an inter-end-office trunk less than 200 miles.

4. Feature Group C is provided to only AT&T on an interim basis and is to be replaced by Feature Group D for equal access.

5. A 2.0-dB minimum ICL is acceptable for very short LATA tandem-connecting trunks without gain provided balance objectives are met. Any trunk whose ICL would otherwise be below 2 dB is to have a 2-dB pad included.

6. If one or both ends of an analog sector intertandem trunk terminates in either a tandem switch serving as a combined sector tandem/LATA tandem/access tandem (ST/LT/AT) or in an ST that meets through-balance requirements, the trunk loss design objective should be 0.5 dB.

functions. In this case, the combined function trunk should meet the more stringent transmission requirements.

8-6   RETURN LOSS BALANCE

Trunk or office balance limits are given in Table 8–5 for current intraLATA trunks [8] and in Chapters 9 and 18 for LATA

Short inter-end-office trunks

Figure 8–9. Examples of trunk loss design.

Tandem–connecting trunks

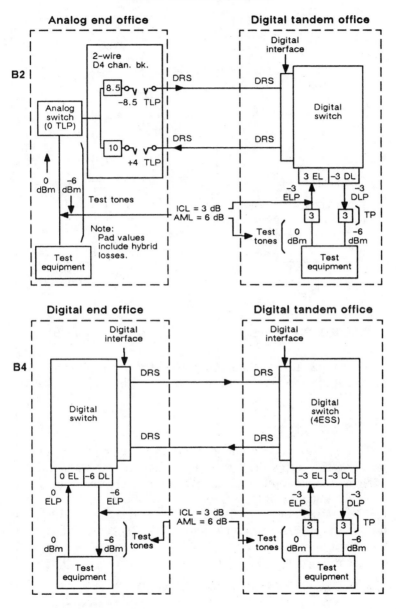

Figure 8–9. Examples of trunk loss design (continued).

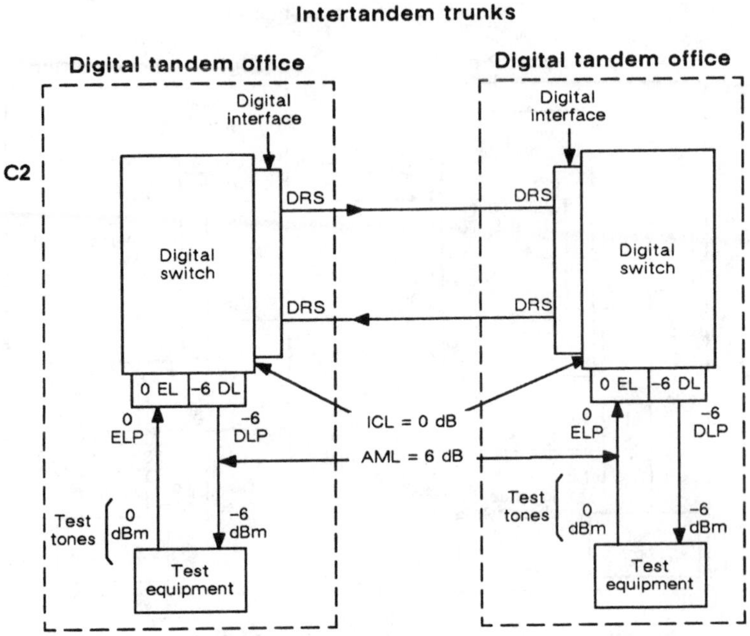

Intertandem trunks

Figure 8-9. Examples of trunk loss design (continued).

access trunks. It should be noted that there are many underlying tests that are made in order to ensure meeting balance limits.

8-7 NOISE

Trunk noise requirements depend on the type of transmission facility (cable or carrier), the type of carrier (noncompandored analog, compandored analog, or digital), and circuit length (for analog). A sample of the acceptance and immediate action limits for C-message noise and C-notched noise is shown in Table 8-6 for the various transmission facilities [8]. C-notched noise is an important parameter for voiceband data transmission.

8-8 LOSS DEVIATION (1004-Hz) LIMITS

Loss deviation is the difference between design loss (EML) and the actual loss at 1004 Hz, usually measured with a 0-dBm0

Table 8-5. IntraLATA Office Balance Limits

Measurement Type	ERL (dB)		SRL (dB)	
	Preservice Limit	Immediate Action Limit	Preservice Limit	Immediate Action Limit
TERMINAL BALANCE Analog Switch 2-wire facilities:				
Interbuilding	18	13	10	6
Intrabuilding	22	16	14	10
4-wire facilities	22	16	15	11
Digital Switch 2-wire facilities:				
Interbuilding	18	16	13	11
Intrabuilding	22	16	15	11
4-wire facilities	22	16	15	11
THROUGH BALANCE	27	21	20	14
2-wire switch	27	31	20	14
No. 1ESS 4-wire trunk ckt*	40	35	39	34
Digital switch	NA	NA	NA	NA

* With hybrid build-out resistance.

signal. Table 8-7 gives the loss deviation limits [8] for various intraLATA trunks.

8-9 ATTENUATION DISTORTION (SLOPE)

Attenuation distortion is the change of the channel loss response across the voiceband. It is usually measured at 404 Hz and 2804 Hz relative to a −16 dBm0 signal at 1004 Hz. Table 8-8 shows the acceptance and immediate action limits for trunks with various facilities.

8-10 IMPULSE NOISE LIMITS

The voiceband impulse noise limits are indicated in Table 8-9 and are based on the 6F voiceband noise-measuring set or *IEEE Std. 743-1984* equivalent. A sample of the trunks in a group is

Table 8-6. Message Trunk Noise Limits

A. For C-Message Noise (dBrnc0)					
Facility Type	Circuit Mileage				
	0-50	51-100	101-200	201-400	401-1000
VF cable	25/36*	–	–	–	–
Noncompandored analog carrier	31/40	33/40	35/40	37/42	40/44
Compandored analog carrier	26/34	28/34	30/34	–	–
Digital	28/34	28/34	28/34	28/34	28/34

B. For C-Notched Noise (dBrnc0) with a -16 dBrnc0 Holding Tone of 1000 Hz				
Facility Type	Circuit Mileage			
	0-100	101-200	201-400	401-1000
Compandored analog carrier (N-type)	38/41	39/42	40/42	–
Digital	45/47	45/47	45/47	45/47

* Key for x/y: x is the acceptance limit and y is the immediate action limit.

Table 8-7. Loss Deviation (1004-Hz) Limits

	Nonrepeatered Cable (Interbuilding)	All Other Trunks
Acceptance Limits	± 1.0 dB	±0.5 dB
Immediate Action Limits	± 3.7 dB	±3.7 dB

measured for these limits. The sampling plan is described in Reference 8.

Table 8–8. Attenuation Distortion (Slope)

		Digital	Any Analog	Nonrepeatered Cable	Repeatered Cable	Multifacility (Note 1)
404 HZ	Less loss than 1004 Hz	1.0/2.0*	1.0/2.0	1.0/2.0	1.0/2.0	2.0/2.0
	More loss than 1004 Hz	1.5/3.0	1.5/3.0	2.0/3.5	3.0/4.5	Note 2
2804 HZ	Less loss than 1004 Hz	1.0/2.0	1.0/2.0	1.0/2.0	1.0/2.0	2.0/2.0
	More loss than 1004 Hz	1.5/3.0	2.0/3.0	3.0/4.0	4.5/5.5	Note 2

* Key for x/y: x is the acceptance limit and y is the immediate action limit.

Notes:

1. Multifacility is defined as multiple carrier systems linked at voice-frequency or carrier systems with metallic extensions. Metallic facilities connected in tandem are treated as a single cable facility.

2. For multiple facility types, the "more loss" deviation limit is determined by squaring the applicable limit for each facility type and taking the square root of the sum of those numbers up to 4.5 dB for the limit and up to 5.5 dB for the immediate action limit.

8–11 REGION DIGITAL SWITCHED NETWORK

An RDSN transmission plan [7,10] that will transform the public switched network from analog to digital connectivity is discussed in Chapter 2. The plan proposes a digital network that begins at the originating analog/digital conversion (encode) point and ends at the terminating conversion (decode) point.

All of the interoffice trunking (no loss) would be tested digitally with the PCM DRS, where 0 dBm0 is represented by the DRS coding. The connection loss would be inserted at the final decoding point, whether that point is a central office, an

Table 8-9. Impulse Noise Limits

Trunk	Limits	Facility	Threshold
Voice-frequency and carrier trunks	No more than 5 counts in 5 minutes above threshold on 50 percent of trunks in group sample.	Voice-frequency	54 dBrnc0
	Immediate action limit: 20 counts in 5 minutes above threshold	Compandored	66 dBrnc0*
		Digital	62 dBrnc0*
		Noncompandored 0-125 miles 126-1000 miles	58 dBrnc0 59 dBrnc0

* -13 dBm0 holding tone.

integrated digital loop carrier terminal connected to that office, or the terminating customer's premises.

The plan identifies several alternatives that would allow increased round-trip delay in the LATA area. The alternatives involve better loop balancing to increase the loop echo return loss by segregation of loaded/nonloaded loops. Also, the return loss could be greatly increased by echo cancellers or other adaptive balancing techniques that are dedicated to each line or are included in access trunks.

References 7 and 10 cover the many aspects of a transmission plan such as grade of service, loss, noise, echo, delay, a new digital switch line classification to provide for bit integrity (no digital loss), operator services, maintenance considerations, etc.

References

1. *LSSGR, LATA Switching Systems Generic Requirements— Section 7, Transmission,* Technical Reference TR-TSY-000507, Bellcore (Iss. 3, Mar. 1989).

2. Members of Technical Staff. *Transmission Systems for Communications*, Fifth Edition (Murray Hill, NJ: AT&T Bell Laboratories, Inc., 1982), pp. 845–853.

3. "Echo Cancellers," CCITT *Red Book*, Vol. III, Fascicle III.1, Rec. G.165 (Malaga–Torremolinos: Oct. 1984).

4. Duttweiler, D. L., C. W. K. Gritton, K. D. Kolwicz, and Y. G. Tao. "A Cascadable VLSI Echo Canceller," *IEEE Journal on Selected Areas in Communications*, Vol. SAC–2, No. 2 (Mar. 1984).

5. *Central Office Balance and Echo Cancellation*, Special Report SR–TSY–000351, Bellcore (Iss. 1, Sept. 1985).

6. Mosley, E. C. "DOBS Keeps the Echo Out of the LATA Switched Access Network," *Telephony* (May 19, 1986).

7. *Region Digital Switched Network Transmission Plan*, Science and Technology Series ST–NPL–000060, Bellcore (Iss. 1, Oct. 1988).

8. *Notes on the BOC IntraLATA Networks—1986*, Technical Reference TR–NPL–000275, Bellcore (Iss. 1, Apr. 1986), Section 7.

9. *LSSGR Tandem Supplement*, Technical Reference TR–TSY–000540, Bellcore (Iss. 2, July 1987).

10. *RDSN Switch Generic Requirements*, TA–NPL–000908, Bellcore (Iss. 1, Oct. 1988).

Chapter 9

Through and Terminal Balance

The return-loss terms "through balance" and "terminal balance" are used in describing the process of matching (balancing) impedances in the control of echo and singing in a network having two-wire and four-wire circuits. Through balance is generally used for circuits going through a two-wire switching system when a high degree of balance is required. Terminal balance is used for tandem-connecting trunks terminating at two-wire end offices. Digital (four-wire) end offices have their own balance requirements for two-wire interfaces. Modern switching systems are arranged to test for return-loss balance automatically, by command, and by maintenance systems.

This chapter discusses general balancing concepts and gives simplified test connections to measure balance for circuits associated with direct access or via current two-wire tandem access to interLATA (local access and transport area) connections. Balance requirements are also given. The situations covered are: (1) a two-wire tandem, (2) direct access from a two-wire end office, (3) via two-wire tandem access, and (4) two-wire ports of a digital end office.

It is expected that most two-wire access tandems will be phased out (or changed to end offices), thus leaving only digital access tandems. Obviously, a digital access tandem that switches four-wire circuits does not have through-balance requirements.

9-1 GENERAL

Several switched access arrangements [1] are available for interLATA connections as shown in Chapter 6, Figure 6-2. The arrangements involve Feature Groups A, B, C, and D and differ in transmission and signalling capabilities. Feature Group D

(FGD) provides for equal access. The interexchange carriers (ICs) select the arrangement that will meet their needs. Chapter 18 details the access transmission requirements for the feature groups and transmission types A, B, and C. Access can be provided to end offices via an access tandem office or by direct trunks. Return loss balance is required for two–wire access tandems, two–wire end offices, and two–wire ports of four–wire end offices.

A two–wire access tandem is served by trunks using two–wire or four–wire transmission facilities. In the past, loaded two–wire trunks were used widely, but today most trunks use four–wire carrier facilities that must be converted for two–wire switching. In either case, impedance matching is important to prevent transmission impairment. It is discussed in the following paragraphs.

The four–wire–to–two–wire conversion is accomplished by a four–wire terminating set, described in Volume 1, Chapter 4, Part 5, that employs an electronic or transformer hybrid with a balancing network. Figure 9–1 illustrates a through connection with hybrids at each side of the two–wire switch. A two–wire trunk using an impedance compensator for a loaded pair is also illustrated. Included are several test circuits that are used in balancing electronic two–wire offices by way of tests to be described later. The hybrids and associated balancing networks are designed to permit the transfer of power in both transmission directions with little transfer from the four–wire input port to the four–wire output port. The unwanted transfer (which may cause near-singing or echo impairments) depends on the match (balance) of the two–wire impedance to the balancing network impedance. The return–loss ratio is a measure of the match and of the undesired power transfer. To minimize transmission impairment, high return–loss requirements are necessary.

Ideally, to meet the higher through–balance requirements, the balancing networks should closely match the circuit design impedance. They should compensate for the series resistances and shunt capacitances of the office cabling and circuits between hybrids. Adaptive hybrids that automatically provide for high balance are becoming available. These are based on using echo-canceller technology, described in Chapter 8, Part 3.

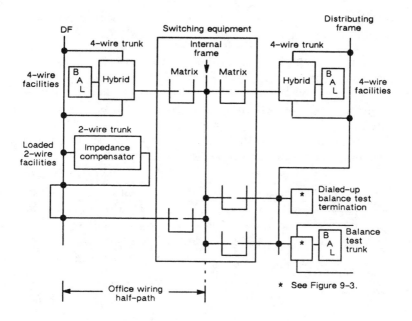

Figure 9-1. Two-wire access tandem.

9-2 ELECTROMECHANICAL OFFICES

Older hybrid balance networks used with two-wire electromechanical switching offices provided only network building-out capacitors (NBOCs) and a compromise network of 900 ohms in series with 2.16 μF. As various lengths of cabling with multiple appearances and bridged-connected circuits produced various values of shunt capacity, drop build-out capacitors (DBOCs) were required to adjust the total shunt capacity of each trunk to a common value. This value would be matched by the value of the hybrid NBOC selected for the office, which was to include an allowance for growth. The NBOC was not to exceed 0.08 μF because it would cause amplitude distortion of more than 1.2 dB at 3000 Hz relative to 1000 Hz in a 900-ohm circuit. The resistive component was controlled only by limiting the length of the two-wire 22-gauge office path in the older switching systems. Studies on these systems indicated that the two-wire loop resistance between the hybrids should not exceed 65 ohms in

177

900–ohm offices and 45 ohms in 600–ohm offices to meet balance requirements.

As an example illustrating the influence of cabling resistance and shunt capacity of the two–wire path, Figure 9–2 shows the sensitivity of both capacitive and resistive unbalance in a test arrangement. Curve A represents the return–loss performance when the circuits are well balanced. The other curves show the return–loss degradation for different values of resistance and capacitance between the four–wire terminating sets.

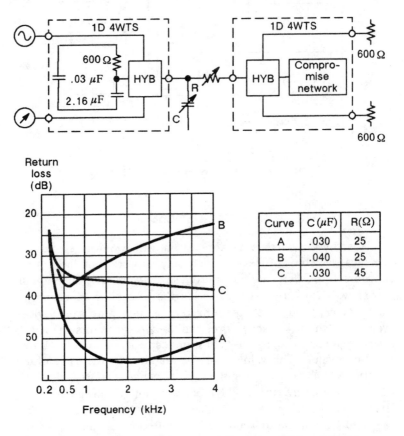

Figure 9–2. Sensitivity of capacitive and resistive unbalance.

Curve	C (μF)	R(Ω)
A	.030	25
B	.040	25
C	.030	45

Electromechanical offices cannot be used for access tandems because they lack stored program control, which is required for access–tandem features. However, they may be employed as LATA or metro tandems until replaced.

9–3 CONTROL OF ECHO AND SINGING

It has been shown by subjective tests that talker echo is a significant impairment when echo amplitude is high and delay is excessive. Another serious transmission impairment occurs when return losses are low and power is returned at a single frequency with sufficient magnitude to start self–sustained oscillation. This impairment, called singing, occurs where the round–trip gains exceed the losses around a circuit.

The voiceband frequencies in switched network connections are normally limited by the four–wire facilities to the 200–to–3200–Hz range, over which echo, singing, and near–singing impairments must be considered. The frequencies at which talkers find echo most objectionable are in the 500–to–2500–Hz range. At these frequencies, the talker usually complains of echo somewhat before singing occurs. Therefore, the balance objectives for control of return loss in this frequency range are more stringent than those for other frequencies in the voiceband. Singing generally occurs in the frequency ranges from 200 to 500 Hz and 2500 to 3200 Hz. Singing or near–singing in these ranges is usually noticed by a talker before echo becomes objectionable. Consequently, both through– and terminal–balancing procedures include separate tests to evaluate each of the impairments [i.e., echo return loss (ERL) and singing return loss (SRL)]. The results of both measurements are necessary to evaluate balance in a given circuit.

Return Loss

Return loss is a measure of the impedance match between two circuits at the point of their interconnection. It can be expressed for any frequency as

$$\text{Return loss} = 20 \, \log_{10} \frac{|Z_1 + Z_2|}{|Z_1 - Z_2|} \text{ dB}$$

where Z_1 and Z_2 are the impedances of the interconnected circuits. It can be seen that, at a given frequency, the return loss is infinite at the interconnection point when the impedances are equal (balanced) since $|Z_1 + Z_2|/|Z_1 - Z_2|$ is then infinity. Conversely, a complete mismatch (unbalance) occurs when either Z_1 or Z_2 is zero, but not both. The return loss for that frequency is then zero, since the logarithm of one is zero. The same occurs when Z_1 or Z_2 is infinite. This relationship is used to establish useful performance criteria for echo return loss and singing return loss. Return losses are usually measured using a flat noise signal shaped at the source or at the receiver by the frequency bands [2] noted below.

Echo Return Loss. ERL is a weighted average measurement of the return losses over a frequency band of 560 to 1965 Hz (3–dB points) in the echo range.

Singing Return Loss. SRL is the weighted average return loss in the singing bands of 260 to 500 Hz (SRL–low) and 2200 to 3400 Hz (SRL–high) (3–dB points).

Balance Discussion

Two–wire tandem offices were to meet balance requirements when placed into service. Balance is maintained by requirements on each trunk that is added or rearranged. Two–wire ports of four–wire digital end offices have design–oriented balance requirements of their own, as discussed later.

Through Balance. Through balance concerns the connection of trunks switched to each other on a two–wire basis. Balance requirements are to be met at all two–wire tandem switching offices and their associated operator services systems (OSSs) and, in addition, at four–wire offices that have two–wire OSSs arranged for through connections for conference calls and operator assistance. Through–balance measurements are made on test connections from the four–wire terminating set of a trunk through the switch to a termination. The termination simulates that part of the office path from the switch distributing frame to the four–wire terminating set of an outgoing trunk. The connections are established through the switching machine and may include an OSS.

Terminal Balance. Terminal balance concerns tandem–connecting trunks running to end offices. Terminal–balance measurements generally are made from a test access circuit through the switch to a termination at the end office.

9–4   TWO–WIRE APPARATUS CONSIDERATIONS

The influences of only two kinds of apparatus used in two–wire circuits are mentioned here: repeating coils and impedance compensators.

Repeating Coils

When repeating coils are present in a two–wire line to derive signalling paths, transform impedances, or provide longitudinal isolation, the degree of balance that can be obtained is limited. For instance, a 1:1 ratio coil has some leakage reactance and reduced inductance, particularly noticeable at the lower frequencies, because of saturation. The repeating coil also adds to the series resistance of a circuit. These effects modify the two–wire line impedance presented to four–wire terminating sets by different amounts over the voice–frequency range and reduce the average degree of balance obtainable.

If the coil has an impedance ratio other than 1:1, an additional limitation exists. For instance, if a 1.5:1 ratio coil is used to interconnect a circuit of 900 ohms in series with 2.16 μF and a circuit of 600 ohms in series with 2.16 μF, the capacitance components of the impedances are not in proper proportion. That is, 600 ohms and 2.16 μF transformed through an ideal 1.5:1 coil is equivalent to 900 ohms and 1.44 μF. This capacitance imbalance is in addition to that caused by leakage reactance and self–inductance effects in the repeating coil itself. Therefore, trunk terminations that use repeating coils should not be employed in through connections since these connections require a high degree of balance to satisfy objectives. The use of repeating coils in trunk relay circuits for impedance matching or signalling purposes should be limited to tandem–connecting applications.

Impedance Compensators for Loaded Cable

An impedance compensator is an electrical network used to make the sending–end impedance of a loaded cable pair more uniform over the voice–frequency range and more nearly equal to the impedance needed at the switching office, i.e., 900 ohms in series with 2.16 μF. An impedance compensator is normally used on all tandem–connecting trunks that use loaded cable.

Most loaded cables are designed so that the electrical length from the office to the first load point is equal to one–half the length of a full loading section. An analysis of the impedance characteristic of a half loading section shows that the impedance increases with frequency and results in low return losses and poor terminal balance as the cutoff frequency of the cable pair is approached.

Impedance compensators are used to build out the trunk cable–pair impedance to appear as a 900–ohm resistor in series with a 2.16–μF capacitor over the voice–frequency band when the trunk is terminated in a precision network at the end office. One such network has adjustments to provide a line build–out capacitor (BOC) from 0 to 0.101 μF in 0.001–μF steps and low–frequency impedance correction (below 1000 Hz) for 19–, 22–, or 24–gauge cable conductors. Another network includes additional features: (1) line build–out resistors (BORs) ranging from 0 to 196 ohms to correct the end–section resistance of loaded cable and (2) DBOCs with the same range as the BOC above for use in trunks that have no trunk relay equipment or where the trunk relay equipment lacks a DBOC.

9-5 TWO-WIRE ELECTRONIC OFFICES

The more recent two–wire electronic switching systems with stored program control incorporate a hybrid in their four–wire trunk circuits with a balance network having adjustable build–out resistance as well as adjustable build–out capacitance. With these adjustments plus office wiring–length restrictions and careful equipment layout, the more stringent through–balance requirements can be met without adding DBOCs to the office paths between the trunk–circuit hybrids. Trunks are also provided for two–wire operator access by means of a four–wire trunk circuit and facilities connecting to a hybrid to return to two–wire transmission at the operator location.

The balancing of the office is expedited by employing several external test circuits as shown in Figure 9–3. The method of balancing a two–wire electronic office is based on a calibrating network representing the half–path connection from the distributing frame to the center of the switch. The network is a hybrid with a balance network equivalent to 400 feet of 26–gauge cable in parallel with the usual 900 + 2.16 μF termination, which results in a balance network of 936 ohms + 2.16 μF in parallel with 0.00856 μF. The calibrating network is used to set the BOR and BOC on test equipment used in routine balance testing. This equipment includes a balance–test termination and a balance–test trunk circuit so that each represents a half–path connection.

Figure 9–3 illustrates the method used to adjust the BORs and BOCs of each of the two balance–test circuits when connected to the switching equipment to match the calibrated balance termination.

Current Balance Requirements

The balance network of each trunk circuit is adjusted when the circuit is connected through the switching network to a balance–test termination.

Simplified test diagrams indicating the connections for measuring the balance of the various trunks and trunk–circuit equipment are given in the following. In general, the purpose of the measurements is to ensure proper adjustment of the balance networks of the hybrids for three access paths. These are: (1) tandem–connecting trunks from the two–wire access tandem to end offices, (2) direct–access trunks from the interexchange carrier point of termination (IC POT) to end offices, and (3) access connections from the IC POT via the two–wire access tandem to end offices.

Current balance requirements are given in the diagrams. Also given are some values taken from the draft of an ANSI Standard

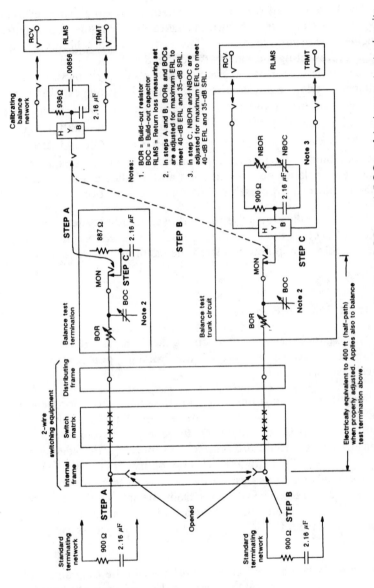

Figure 9–3. Two-wire access tandem—adjusting BORs and BOCs of balance-test circuits.

[3] on technical specifications for the switched exchange access network. The draft covers two basic exchange network arrangements, referred to as group 1 and group 2. These are modeled after the capabilities of FGD and FGC, respectively.

Tandem-Connecting Trunk

Figure 9-4 shows a tandem-connecting trunk measured at the access tandem through the switch, the tandem-connecting trunk facility, and the end-office switch to a standard termination. Figure 9-5 illustrates measurement of the trunk equipment at the access tandem.

Direct-Access Trunk

Figure 9-6 displays measurement of the end-office part of the direct-access trunk (connects the end office to the IC POT). Figure 9-7 shows measurement of the direct-access two-wire or four-wire trunk from the IC POT to a standard termination at the end office.

Connection Via Access Tandem

Figure 9-8 shows measurement of the balance of the IC trunk to access tandem. The balance-test termination at the tandem is to be adjusted to simulate 400 feet of 26-gauge office wiring plus a standard termination. Figure 9-9 displays measurement of the balance of a connection via the access tandem from the IC POT to the end office.

9-6   DIGITAL OFFICES

Digital offices [4] are inherently four-wire. When they are used with four-wire facilities there is no hybrid and hence no need for through balancing. However, a digital end office has two-wire analog line or trunk ports for connection of two-wire facilities. These ports contain a hybrid or equivalent. Each two-wire port is to meet two sets of requirements at the interfaces

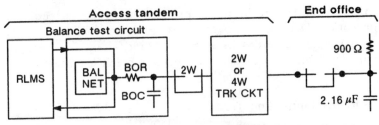

Figure 9-4. Balance of tandem-connecting trunk.

involving impedances to control singing and echo: input impedance requirements and hybrid balance requirements. The transmission interfaces are located at a distributing frame and office
wiring is included as part of the switching system, as shown in
Chapter 8, Figure 8-5.

Input Impedance Design Requirements

The input impedance of an analog line or trunk port provided
by a path through the switch is designed to meet the requirements
of Table 9-1. The connections for design testing (not routine)
for these requirements are shown in Figure 9-10(a) and (b).

Hybrid Balance Requirements at Two-Wire Ports

The impairments associated with a two-wire-port-to-two-
wire-port zero-loss connection through the digital switch, such as

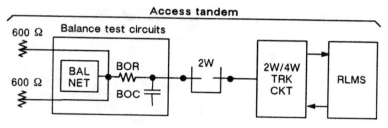

Minimum balance limits

Limit	ERL (dB)	SRL (dB)
AL	14	6
IAL	13	5

2W   = Two-wire
4W   = Four-wire
AL   = Acceptance limit
BOC = Build-out capacitor
BOR = Build-out resistor
IAL = Immediate action limit
RLMS = Return loss measuring set

Figure 9-5. Trunk-equipment balance at access tandem.

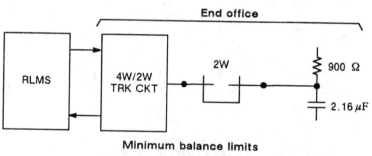

Minimum balance limits

	ERL (dB)	SRL (dB)
AL	17	12
IAL	16	11

2W   = Two-wire
4W   = Four-wire
AL   = Acceptance limit
IAL = Immediate action limit
RLMS = Return loss measuring set

Figure 9-6. Direct-access trunk balance at end office.

187

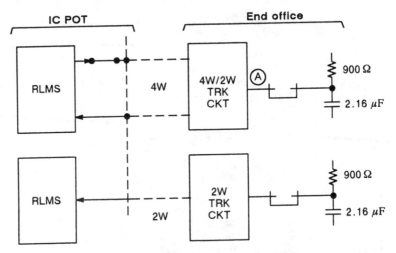

Minimum balance limits at IC POT

	AL		IAL		Notes
	ERL (dB)	SRL (dB)	ERL (dB)	SRL (dB)	
Feature Groups A & B	17	12	16	11	1
Feature Groups C & D:4W	20	13	16	11	2
2W	16	8	13	6	2
ANSI standard: 4W POT	21	14	16	11	2

2W   = Two-wire
4W   = Four-wire
AL   = Acceptance limit
IAL   = Immediate action limit
IC POT = Interexchange carrier point of termination
RLMS   = Return loss measuring set
TLP   = Transmission level point

Notes:

1. After adjustment for reading with short circuit at (A).

2. Equal level echo return loss (ELERL) and equal level singing return loss (ELSRL).

Figure 9-7.   Balance of direct-access end-office trunk (two-wire or four-wire)—IC POT to end office.

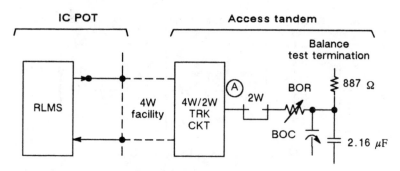

Minimum balance limits at IC POT

	AL		IAL		Notes
	ERL (dB)	SRL (dB)	ERL (dB)	SRL (dB)	
Feature Group B: 4W trunk equip.	23	15	21	14	1
2W trunk equip. (not shown)	17	12	16	11	1
Feature Groups C & D 4W trunk equip.	25	18	21	14	2
ANSI standard: Group 1	27	20	25	18	2
Group 2	27	20	21	18	2

2W   = Two–wire
4W   = Four–wire
AL   = Acceptance limit
BOC   = Build–out capacitor
BOR   = Build–out resistance
IAL   = Immediate action limit
IC POT = Interexchange carrier point of termination
RLMS   = Return loss measuring set
TLP   = Transmission level point

Notes:

1. After adjustment for reading with short circuit at (A).

2. ELERL and ELSRL.

Figure 9–8.   Balance of four–wire/two–wire trunk circuit—IC POT to access tandem.

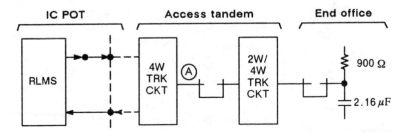

Minimum balance limits at IC POT

	AL		IAL		Notes
	ERL (dB)	SRL (dB)	ERL (dB)	SRL (dB)	
Feature Group B	NA	NA	8	4	1
Feature Group C, 4W to end office and Feature Group D	20	13	16	11	2
	13	6	9	4	1
Feature Group C 2W to end office	16	8	13	6	2
	9	1	6	−1	1
ANSI standard: 4W AT to EO trunk	21	14	16	11	2
2W AT to EO trunk	18	10	13	6	2

```
2W      =  Two-wire
4W      =  Four-wire
AL      =  Acceptance limit
IAL     =  Immediate action limit
IC POT  =  Interexchange carrier point of termination
RLMS    =  Return loss measuring set
TLP     =  Transmission level point
```

Notes:

1. After adjustment for reading with short circuit at (A).

2. ELERL and ELSRL.

Figure 9-9.   Balance of connection via access tandem—IC POT to end office.

Table 9-1. Return Loss—Digital End Office at Analog Interface

Switch Connection*		Reference Impedance	Frequency Range	Minimum Return Loss
Near-End Port	**Far-End Port †**			
2W analog loop at EO or remote terminal	any 4W	900 Ω + 2.16 μF	200-500 Hz 500-3400 Hz	20 dB 26 dB
(≥2 dB loss through switch)	any 4W	900 Ω + 2.16 μF	200-500 Hz 500-3400 Hz	14 dB 20 dB
2W analog trunk	any 4W	900 Ω + 2.16 μF	200-500 Hz 500-1000 Hz 1000-3400 Hz	20 dB 26 dB 30 dB

* See Figure 9-10 for testing.
† The far-end 4W port should be terminated.

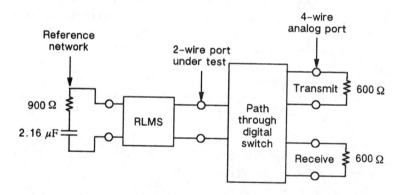

Figure 9-10. Connections for testing ERL at an analog port.

loop to loop, are singing and listener echo due to the four-wire office path for such a connection involving gain and hybrid balances at each two-wire port. A two-wire analog switch does not have gain devices or appreciable transmission delay in such a connection and hence does not have these impairments. Talker echo is associated with interoffice connections and the loop hybrid balance at the two-wire loop terminating port. Singing should never occur; to prevent this, a minimum of 4-dB listener echo path loss should be maintained at any frequency over the

200–to–3500–Hz band. Listener–echo performance will be satisfactory if the following two–wire hybrid balance requirements are
met.

Listener–echo and talker–echo performance in an all–digital
network will be influenced by the return–loss balance of the hybrid in the two–wire port of the terminating digital end–office
when connected to the general loop population. The return loss
of the loop and station set (*EIA Standard EIA–470*) referenced
to the usual balance impedance of 900 ohms + 2.16 μF can be
modeled by a normal distribution of 11 dB σ 3 dB [5]. With
such a distribution, near–singing or singing conditions may exist
for a local–to–local call through a 0–dB loss digital office; better
balance must be achieved. It was found that this could be done
by segregating the loop population into nonloaded and loaded
loops and providing a different hybrid balance network for each
[6,7]. This improves the average return loss for singing margin
and listener echo by about 10 dB and for talker echo by about 4
dB. The switching system is to provide the proper balancing network automatically. The office loop balance network, requirements, and objectives for future needs are given in Table 9–2. As
noted, the balance network and the two–wire termination network are identical for office testing. The accuracy of the termination–network components should be at least one percent. A simplified diagram of the testing circuit is shown in Figure 9–11.

A large increase in return loss can be realized automatically by
using echo cancellers (see Chapter 8, Part 3) or by using adaptive
hybrids based on canceller technology. With these applications,
echo impairment is practically eliminated and the time–
consuming routine balance tests can be simplified.

Office Wiring

Office wiring connects the ports of a switching system to a
distributing frame as shown in Chapter 8, Figure 8–5. For
engineering purposes, the wiring lengths for one–half of a connection path through the office from the distributing frame to the
center of the switching network are shown in Table 9–3 for small
offices (less than 10,000 lines) and for large offices (greater than
10,000 lines).

Table 9-2. Digital-Office Hybrid Balance Requirements

Interface Type	Balance and Termination Networks	Requirements*		
		SRL-Low	ERL	SRL-High
Loaded 2W analog loops*	1650 Ω || 0.005 μF + 100 Ω	15	20	15
Nonloaded 2W analog loops*	800 Ω || 0.05 μF + 100 Ω	15	20	15
2W analog special-services lines*	900 Ω + 2.16 μF	15	20	15
Remote terminal 2W analog loops:*				
Remote to host loss ≥ 0 dB and < 2 dB	900 Ω + 2.16 μF	15	20	15
loss ≥ 2 dB	900 Ω	10	15	10
2W analog trunks and 2W analog special-services trunks †	900 Ω + 2.16 μF	15	20	15

* Objectives are 5 dB higher.
† Objectives are 10 dB higher.

Note: SRL-low, ERL, and SRL-high are measured with a return loss measuring set conforming to *IEEE Std. 743-1984*.

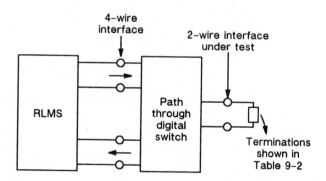

Figure 9-11. Circuit for hybrid balance tests.

Table 9-3. Office Wiring Lengths

Small Systems	(≤10,000 lines)	
Average	=	75 ft
Maximum	=	150 ft
Standard deviation	=	30 ft
Large Systems	**(>10,000 lines)**	
Average	=	275 ft
Maximum	=	525 ft
Standard deviation	=	130 ft

Note: The wiring lengths are for one-half of a connection path length from the distributing frame to the center of the switching network for a 900-ohm office.

References

1. *Voice Grade Switched Access Service—Transmission Parameter Limits and Interface Combinations*, Technical Reference TR-NPL-000334, Bellcore (Iss. 2, Dec. 1987).

2. *IEEE Std. 743-1984*, "IEEE Standard Methods and Equipment for Measuring the Transmission Characteristics of Analog Voice-Frequency Circuits" (New York: Institute of Electrical and Electronics Engineers, Inc., 1984).

3. Draft of *ANSI T1.506-1989*, "American National Standard for Telecommunications—Network Performance Standards—Switched Exchange Access Network Transmission Specifications" (New York: American National Standards Institute, 1989).

4. *LSSGR, LATA Switching Systems Generic Requirements—Section 7, Transmission*, Technical Reference TR-TSY-000507, Bellcore (Iss. 3, Mar. 1989).

5. Manhire, L. M. "Physical and Transmission Characteristics of Customer Loop Plant," *Bell System Tech. J.*, Vol. 57, No. 1 (Jan. 1978).

6. Neigh, J. L. "Transmission Planning for an Evolving Local Switched Digital Network," *IEEE Transactions on Communications Technology*, Vol. COM–27, No. 7 (July 1979).

7. Bunker, R. L., R. P. McCabe, and F. J. Scida. "Zero Loss Considerations in Digital Class 5 Office," *IEEE Transactions on Communications Technology*, Vol. COM–27, No. 7 (July 1979).

Chapter 10

Operator and Auxiliary Services

The objectives for the message loop and trunk plant have been derived on the basis of directly dialed calls; however, a substantial number of calls require dialing for operator assistance. Operator services networks are engineered so that operator–customer calls and operator–completed calls have about the same transmission quality as directly dialed calls.

10-1 GENERAL

Operator services cover the necessary activities for directory assistance (DA), call completion, billing information, intercept, conferencing, etc. These service features are provided by centralized and automated systems such as the new digital operator services systems (OSSs) for universal functions, the older automatic call distributor (ACD) systems for DA, and the Automatic Intercept System (AIS) for call intercept. The Traffic Service Position System (TSPS) remained with AT&T–C at divestiture; reversion of operator services to exchange carriers generally involved new digital OSSs, highly centralized to let the operator force work efficiently. The centralized aspects of these systems result in more tandem links and longer distances in the service connections; care must be taken to provide suitable transmission quality.

This chapter indicates some of the functions provided by the new digital OSSs, the estimated grade–of–service performance, and multiparty conferencing. It discusses the associated overall voice transmission considerations with primary attention to operator circuit parameters and to operator headset considerations. The older ACD, AIS, and TSPS systems are mentioned briefly. Finally, some auxiliary services are covered, including a brief discussion on conferencing.

Operator Services Features

Operator services systems have become highly complex in order to deal automatically with operator call–handling features, including DA, intercept, coin collection, credit recording, billing details, call completion, etc. The systems have wide application to the needs of an exchange carrier.

The generic requirements for a modern OSS consisting of a host core office (supporting operator groups) with several remote core offices (no operators) are covered in Reference 1, which includes feature requirements, a network plan, data bases, call processing, signalling, service standards, and system capacity.

Each of the core systems is to be located architecturally at the access–tandem level in the exchange carrier network. It efficiently receives, forwards, completes, or terminates calls from end offices, local access and transport area (LATA) tandems, other operator systems, and interexchange carriers (ICs). A dedicated access tandem may be established for only operator services traffic; alternatively, a combined access tandem may be used to handle operator and nonoperator traffic. Remote core offices homing on the host extend the area coverage of the system to other LATAs. The remote office performs the same call–handling functions as the host and needs access to its own data bases or those of the host. A simplified overview of the OSS is shown in Figure 10–1, with operator position groups off the host and with connections to equal– and non–equal–access end offices, to LATA tandems for intraLATA calls, and to the interexchange carrier point of termination (IC POT) for interLATA and international calls. Tie–ins to various data bases and operator systems are noted symbolically. Many of these systems are statewide, or multistate, in scope.

The system is to be capable of performing many functions. A partial list of operator services features is supplied in Table 10–1. This covers features generally related to operator call–handling, customer access, customer listings, customer call–handling, and special billing. New features are added as the need arises.

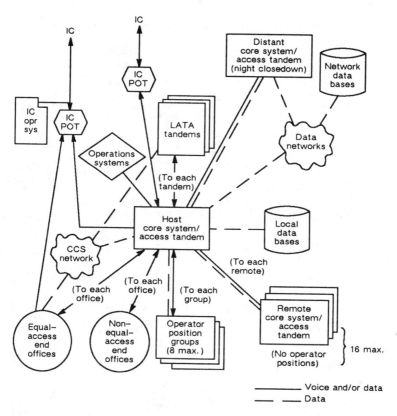

Figure 10-1. Operator services system—general.

Operator Functions

Operator call–assistance may be needed when users place calls for operator completion such as person–to–person, collect, coin, etc., and for number–service (DA) functions. Operators may perform other functions such as call intercept, verification, trouble observation, coin supervision, etc.

With older systems and equipment, some functions are performed by operators designated *outward operator* and *inward operator*. The outward operator is reached by a special code to originate calls over the network and to provide for the recording of charges. The inward operator, at the terminating end of the call, can be reached only by a distant operator. In the past, a

Table 10–1. Partial Operator Services Features

(a) Determine calling number, called number, and billed number
(b) Ascertain whether the call completion is intraLATA, interLATA, or international and connect accordingly
(c) Terminate inward calls from or originate outward calls to remote operator systems for emergency assistance, DA, call completion, busy–line verification, calling–card verification, etc.
(d) Help users to dial or establish a connection
(e) Provide for rate information, centralized automatic message accounting (CAMA) transfer (to obtain the calling number), credit recording, trouble reporting, repair service, etc.
(f) Accommodate general announcement information (weather, time, community events, etc.), including vendors
(g) Supply a ring–back test
(h) Give intercept service to indicate that a working number is no longer in service and supply a replacement number if available
(i) Handle directory assistance
(j) Provide for user and area business listing information
(k) Allow for automatic call–back when the called number becomes free
(l) Accommodate busy–line verification or conversation interruption to indicate an emergency call waiting
(m) Provide for emergency numbers and emergency call completion
(n) Supply person–to–person, conferencing, or other call completion
(o) Give calling–card, third–party, etc. billing indication.

through–operator provided assistance for manual call routing at a control switching point (CSP) and could be reached only by another operator. Except for overseas service, the through–operator function is no longer designated, although inward operators at CSP locations can function as through–operators for emergency completion of calls.

Number Services

Where *directory assistance* is required, the operator is usually reached by dialing 411 from the local area, 555–1212 from other offices in the same numbering plan area (NPA), or NPA–555–1212 from a foreign NPA for the purpose of providing telephone numbers not otherwise available.

Call intercept provides operator assistance on calls to unassigned numbers or to lines that are temporarily disconnected or in trouble. In the case of centralized intercept, an intercept operator may be reached over special trunks from the called end office. In AIS, an operator is called in only if the called number is not available or if the automatic announcement does not satisfy the customer.

10-2 TRANSMISSION CONSIDERATIONS

The general transmission objective for customer–to–customer connections completed by an operator [1] is that the loss/noise/echo grade of service (see Volume 1, Chapter 24) should not be significantly different from that of a directly dialed call. The objective for operator–customer connections is that the grade of service be at least equivalent to that provided to a customer on an intraLATA call.

The grade of service is heavily influenced by operator–room noise; consequently, room–noise limits have been established. In addition, the operator end of the circuit and operator headset are to provide suitable transmit and receive levels (acoustical and electrical), sidetone level, room noise/speech discrimination in the transmit path, and linear limiting in the receive path. These are discussed later in the chapter. The objectives assume that the operator headset meets the requirements of Reference 2, which are summarized later in the chapter, and that the customer telephone set is equivalent to the 500–type [3].

Operator Bridging

In the past, trunks that provided access to outward or inward operators on toll calls usually remained a part of the connection when the call was extended by the operator to the called customer. Whether such links were additional to the number of links in a directly dialed call depended on the class relationship between the switchboards and the offices to which they were connected.

In current analog systems, the operator connection is made at a three–way bridge located at the host or remote unit. After operator release, the bridge is disconnected and the customer–to–customer connection is cut through. By controlling return loss, delay, and noise, the transmission quality of the final customer connection is similar to that obtained on a directly dialed connection. The quality of customer–operator connections, however, may be affected. Because only the host office supports operator groups, the customer–operator connection is generally not a problem unless the operator groups are excessively remote. When a customer–operator connection is made through the remote bridge, the connection is further extended and transmission may be impaired. Such connections resulted in: (1) stringent return–loss requirements for two–wire switching offices, (2) the use of compandors to reduce subjective noise on long analog host–remote trunks, and (3) limits on added delay, as well as other restrictions.

Digital Operator Services System

The digital OSS provides operator functions by employing digital (four–wire) host and remote core switching offices, digital bridging, and digital transmission facilities. The trunks associated with the digital OSS are shown in Figure 10–2 as connecting to end offices, tandem offices, and the IC POT. To illustrate several call routings, first assume an intraLATA operator–assisted call is

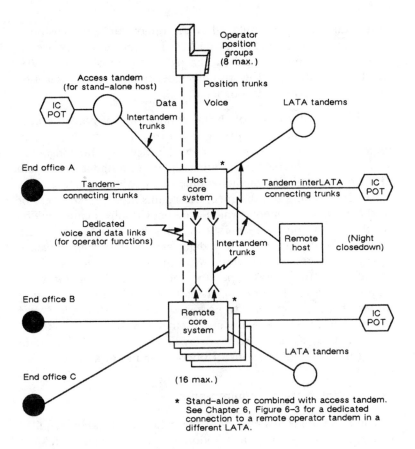

Figure 10-2. Digital operator services system trunking.

made from end office A to a party at end office B. The operator is connected through a bridge at the host office; the bridge drops off when the connection is completed. Further, the operator-dedicated trunks between host and remote offices are released and the connection is made through the paralleled message trunks. For an intraLATA assisted call from end office B to A or B to C, the bridge at the remote office is used and the bridge and dedicated trunks are released as indicated above. For inter-LATA assisted calls connected via an IC, the bridge at the originating core system associated with the originating end office is employed for the operators. The bridge and dedicated trunks, if used, are released upon call completion.

If bridges are not employed at the remote unit, the customer connections are extended to the host location until they are released by the operator and cut through at the remote office. Extra delay is introduced if customer–to–customer communication is allowed before cut–through.

The main advantages of digital connectivity of both the core systems and trunking facilities are: (1) the many return–loss limits and tests associated with two–wire offices are no longer involved and (2) circuit noise is independent of distance. The main effect on transmission quality is time delay, which increases with stages of switching and distance; this results in increased operator talker–echo impairment, which can be virtually eliminated by suitable echo–control techniques. Delay between the host and remote offices may also affect data transmission when message timing sequences are important. With proper echo control, digital connectivity, and average operator–room noise, the operator will be provided a satisfactory transmission grade of service with up to 80 ms of round–trip delay to an end office, as is shown in Part 4 of this chapter. Thus the limiting distance for operator trunking is more likely to be restricted by delay limits in the associated data links than by considerations of operator grade of service.

Major Transmission Requirements

The following is a summary of the major transmission requirements for an OSS [1] in addition to trunking and switching. Many of these requirements are based on average customer speech power. This power has been found from surveys to be equivalent to a −21 dBm 1004–Hz signal at a 0–dB transmission level point (0 TLP), the end–office outgoing switch appearance. Also, it is assumed that operator telephone headsets, briefly summarized in Part 3 of this chapter, meet the specifications of Reference 2 and that customer telephone station sets meet the specifications of Reference 3.

Acoustic Signal–to–Noise Ratio. The objective for this ratio is 29 dB at the operator's or customer's ear when receiving speech from an average customer or operator.

Operator Transmit Level. The objective is to have average operator speech and average customer speech appear at the same level at the core system. This provides for equal level participants for bridging connections. For this purpose, the operator speech at any zero encode level point (0 ELP) will be −21 dBm at any analog point, equal to −21 dBm0. It has been determined that the actual operator speech power is approximately −17 dBm at the transmit headset jacks when using a conforming headset. Hence the transmission level for the transmit jack is +4.0 TLP to achieve the −21 dBm0 objective. An adjustment of ±6 dB without changing sidetone response is desirable to accommodate other headsets.

Operator Receive Level. The objectives of the receive level requirements are to deliver as loud a voice signal as can be comfortably accommodated by operators and to achieve an average signal−to−noise ratio (including effects of room noise) of 29 dB at the ear. It has been determined that, with conforming headsets, a received speech power of −29 dBm at the jacks will satisfy the objectives; this represents a −8 TLP, given the speech power of −21 dBm0.

Studies of customer speech levels delivered to the operator receive jacks with conforming headsets have indicated that the eight−hour time−weighted average noise dosage is well below the Occupational Safety and Health Administration (OSHA) requirement of 85 dBA [4].

Sidetone. Sidetone is generally provided by a circuit external to the four−wire operator headset by feeding part of the operator's transmitted speech into the receive path. The acoustic level of the received sidetone should be 12 dB below the acoustic level of the operator's speech.

Echo Control by Voice−Switched Attenuator. Echo control is needed to provide additional loss in the operator receive path to decrease the effects of talker echo as heard by the operator. The loss is provided by a voice−switched attenuator (VSA) activated by the operator's speech level at about −17 dBm ±2 dB at the transmit jack. The attenuator increases the loss linearly to a maximum of 10 to 25 dB at a speech level of −9 dBm and retains the loss at higher levels. The attack and release times are contained in Reference 2.

Amplitude Limiting. Electrical amplitude limiting is required in the receive operator path to prevent acoustical signals of annoying level from reaching the operator's ear. It has been determined that, with conforming headsets, the maximum amplitude of a steady–state signal should be limited to −26 dBm at the receive jacks. This level takes into account the 3 dB receive electroacoustic tolerance so that a high–tolerance headset produces 94 dBSPL maximum. Attack and release times and other parameters are contained in Reference 2. High peak transient signals are limited by conforming headsets to levels equivalent to the OSHA's free–field limit of 115 dBA.

Headset Jack Impedances. The headset jack impedance is 50 ohms for the transmit path and 300 ohms for the receive path, both balanced to ground. Multiple jacks, as for an operator and a supervisor, are to be electrically independent.

Noise. Noise limits at the analog ports of the console circuits are 10 dBrnc0 and 18 dBrn0, 3–kHz flat weighted.

Radio–Frequency Interference. The noise requirement is to be met in the presence of a radio field of 10 volts per meter up to 10 GHz.

Room Noise. This is the largest contributor to degrading operator transmission quality. Room–noise requirements call for a maximum average room noise, usually assumed to be the average noise during the busy hour, and a maximum peak room noise, usually assumed to be a maximum five–minute average.

	Over–ear earpiece	In–ear earpiece
Maximum average room noise	55 dBA	52 dBA
Peak room noise	62 dBA	59 dBA

The noise is to be measured near the operator's head and may vary position–by–position depending on the layout of the operator office. Operator density, room acoustics, position display noise, and keyboard noise should all be considered when operator rooms are designed. The acoustic levels specified are all referenced to 20 μPa with A–weighting, with the noise meter adjusted for slow response.

206

Trunk Loss. Ideally, all trunks are digital and aligned to the digital reference signal (DRS) for a 0 dBm0 signal. The message connection loss is fixed at 6 dB and is established at the decode level points (DLPs) of −6 TLP at the end offices, as shown in Figure 10–2 and listed in Table 10–2. All ELPs at the end offices are 0 TLP. These levels are consistent with the region digital switched network transmission plan discussed in Chapter 2.

Table 10–2.   Operator Services System (Digital Tandem) Trunk Losses

TRUNK	ANALOG TRUNK W/GAIN	COMBIN. TRUNK	DIGITAL TRUNK	TRUNK DESIGN-ATION
EO–host or remote core system [2]	3 [3]	3	Note 4	Tandem-connecting
H core – R core [1,2]	–	–	DRS(0)*	Intertandem
H core or R core – IC POT [2]	0	0	DRS(0)	Tandem interLATA connecting
H core – position jacks [1]	Note 5	–	Note 5	Position
H core – LATA tandems [2]	VNL	1	DRS(0)	Intertandem
H core – H core (night closedown) [2]	–	–	DRS(0)	Intertandem

* DRS(0) = digital reference signal, no transmission loss throughout digital network. Decode level of −6 TLP at end office.

Notes:

1. Dedicated trunks.

2. Combined trunks.

3. 2 to 4, without gain.

4. 0 ELP and 6 DLP at EO.

5. +4 TLP, transmitting; −8 TLP, receiving.

6. EO = End office
   H = Host
   R = Remote.

The levels at the operator position headset jacks have been discussed for headsets meeting the requirements of Reference 2. These levels are −8 TLP for receive and +4 for transmit; they are shown in Figure 10–3 for a digital position trunk to a digital console and in Figure 10–4 for an analog trunk to an analog console.

Trunk Balance Limits. Trunk or office balance requirements are a given in Chapter 8, Part 6.

Trunk Noise Limits. Trunk C–message and C–notched noise limits are given in Chapter 8, Part 7.

Trunk Loss Variation Limits. The loss variation is the difference between the designed loss and the actual loss at 1004 Hz. Limits are given in Chapter 8, Part 8.

Trunk Attenuation Distortion (Slope) Limits. Attenuation distortion is the loss at 404 Hz and 2804 Hz relative to 1004 Hz. Limits are given in Chapter 8, Part 9.

Trunk Impulse Noise Limits. These voiceband limits depend on the facilities used; they are given in Chapter 8, Part 10.

10-3 OPERATOR HEADSET CONSIDERATIONS

Operator headsets have different electrical characteristics than customer station sets; they are intended to operate with consoles that provide the proper levels (adjustable) to interface the network at a tandem office instead of at the end of a subscriber loop. The hands–free electronic headset (on the ear or in the ear) is four–wire and does not have a sidetone circuit; a background discrimination circuit is used to reduce the transmitted room noise during operator speech pauses. Also, the impedances and levels are different and no loop equalization is needed. The generic requirements for telephone headsets to be used at operator consoles are covered in Reference 2. This covers human factors, environmental conditions, electrical characteristics, physical characteristics, etc. Consideration should also be given to using on–ear or in–ear headsets (monaural or binaural), breath spurt suppression on the transmitter, weight, hygienic conditions, etc. A summary of major electrical characteristics as measured under specified test conditions follows.

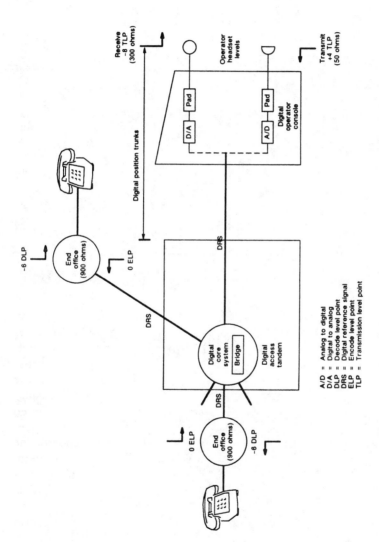

Figure 10-3. Digital operator services system transmission plan.

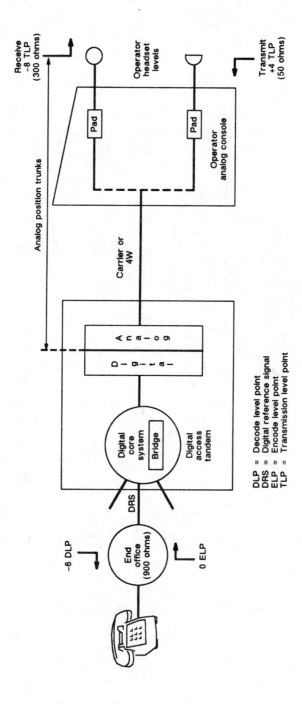

Figure 10–4. Digital operator services system—analog consoles.

DLP = Decode level point
DRS = Digital reference signal
ELP = Encode level point
TLP = Transmission level point

Major Transmit Characteristics

The major transmit characteristics of operator headsets are as follow.

(1) The electroacoustic efficiency, designated the transmit objective loudness rating [5], is to be −55.5 +4/−3 dB.

(2) The frequency response is to fall within the template shown in Figure 10–5.

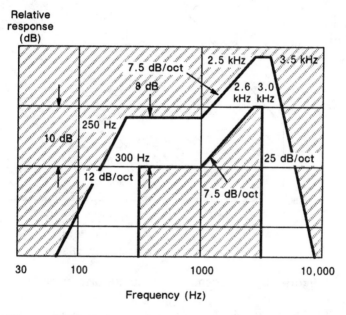

Figure 10–5. Transmit frequency response template.

(3) The background discrimination characteristic against room noise without penalizing low–volume operator talkers is shown in Figure 10–6. When the input is switched from 70 to 86 dBSPL, the time required for the output to reach 71 percent of steady value is to be within 5 to 15 ms. When switched from 86 to 70 dBSPL, the delay time to reach 29 percent of initial value is to be within 150 to 250 ms.

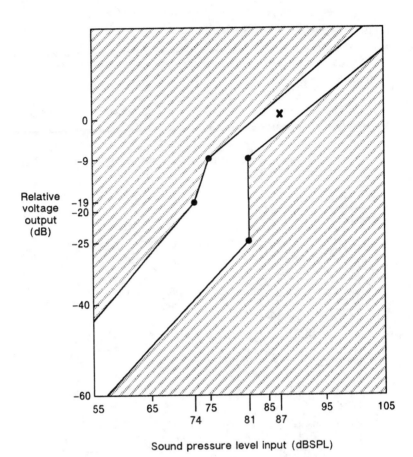

Figure 10-6. Transmit input-output discrimination.

(4) The harmonic distortion is to be less than 5 percent.

(5) The noise produced at the output is to be less than 10 dBrnc.

(6) The output impedance (300 to 3200 Hz) is to be 50 ± 20 ohms.

Major Receive Characteristics

The major receive characteristics of an operator headset are as follow.

(1) The electroacoustic efficiency, designated the receive objective loudness rating [5], is to be 36.5 ± 3 dB with some adjustable range.

(2) The frequency response is to fall within the template of Figure 10-7.

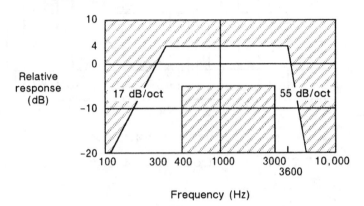

Figure 10-7. Receive response template.

(3) The acoustic limiting for steady-state is to be:

(a) For adjustable gain set to maximum, the acoustic output is not to exceed 94 dBSPL. The attack time to 120 percent of final value is to be 10 ± 5 ms and the release time to 80 percent of final value is to be 125 ± 50 ms

(b) For headsets without gain, the maximum acoustical output is shown in Figure 10-8.

(4) The input impedance (300 to 3200 Hz) is to be 300 ± 60 ohms

(5) The isolation loss between the receive terminals and transmit terminals is to be 34 dB minimum.

10-4 VOICE GRADE-OF-SERVICE PERFORMANCE

Performance studies of voice grade-of-service (discussed in Volume 1, Chapter 24) give estimates of the transmission quality

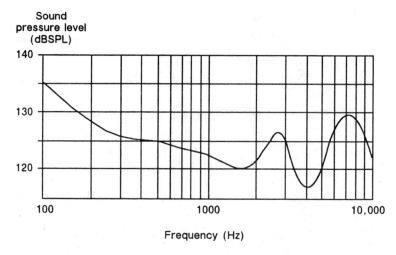

Figure 10-8. Maximum acoustic output (over-the-ear head-sets).

of telephone connections based on impairments of loss, noise, and echo. These studies depend on the electroacoustical conversion factors and distributions of the customer's telephone set and loop, the conversion distributions and levels set for the operator headset, trunk loss distributions, quantizing noise distributions, room noise, transmission delay, and return-loss distributions.

For a study of a 6-dB fixed-loss digital system, the following items have been used. For simplification, average values were used for the customer set and loop, the overall trunking loss was fixed at 6 dB, only one encode/decode step of 7-5/6 bit pulse code modulation was allowed, and only the return-loss distribution of the telephone set and loop (1973 survey) was considered as there was no other two-wire-to-four-wire conversion in the models.

For example, the study assumed a digital network model of a customer-to-operator connection and a customer-to-customer connection resulting from operator completion or from a directly dialed call. The network models for these connections are shown in Figure 10-9(a) and (b). Grade-of-service calculations were made for both connections assuming variable transmission time delay (the principal variation in digital transmission affecting

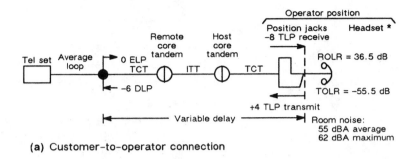

(a) Customer-to-operator connection

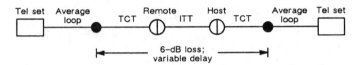

(b) Customer-to-customer connection

* Operator headset per Reference 1.

ITT   = Intertandem trunk
ROLR   = Receive objective loudness rating
TCT   = Tandem-connecting trunk
Tel set = 500-type or equivalent per Reference 3
TOLR   = Transmit objective loudness rating

Figure 10-9. Digital system grade-of-service models.

grade of service because of echo impairment). The grade-of-service results versus delay are shown in Figure 10–10 for three conditions: (A) customer to customer (customer grade of service), (B) customer to operator (operator grade of service), and (C) the same as (B), but with additional loss of 15 dB in the operator talker–echo path to simulate echo control by voice-switched attenuator, as described in Part 2 of this chapter. In practice, such loss is to be inserted automatically by the VSA. An echo canceller may be used instead, but the round–trip time delay to the most remote end office must be within the operating delay range of the canceller.

The results show rapid deterioration of the grade of service for the operator versus delay without the VSA. The operator grade of service with the VSA, however, matches the customer grade of service. This supports the incorporation of echo control.

215

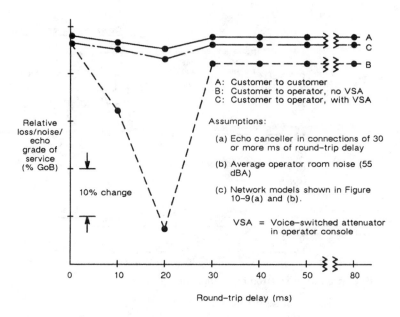

Figure 10-10. Grade of service—operator and customer.

Another grade-of-service study was made to show the effects of operator-room noise on operator grade of service. For this case, the network model in Figure 10-9(a) was assumed and two values of room noise were used: 55 dBA and 62 dBA, which are the average and maximum requirements given in Part 2 of this chapter. The results are shown in Figure 10-11; they indicate a significant deterioration in operator grade of service of 7 to 8 percent for the higher noise. This establishes the importance of maintaining low room noise, which includes the noise from operator activity. As a guide to the significance of grade of service, studies have indicated that a change of about 4 percent in the grade of service (over the range of interest) can be perceived.

10-5 THE NO. 5 AUTOMATIC CALL DISTRIBUTOR AND NO. 1 AUTOMATIC INTERCEPT SYSTEMS

Still in service, the ACD system and the AIS handle DA and intercept functions, respectively. Both have two-wire analog interfaces. The No. 5 ACD was based on No. 5 Crossbar techniques, while the No. 1 AIS was designed to use the pulse

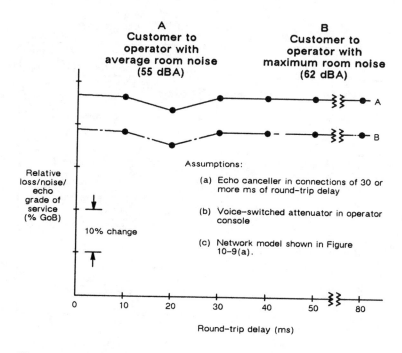

Figure 10-11. Effect of operator-room noise on grade of service.

amplitude modulation and resonant-transfer switching method of the No. 101ESS switching equipment.

The dedicated network arrangements for these systems are shown in Chapter 6, Figure 6-2. The trunk loss requirements are listed in Table 10-3. Other trunk transmission requirements are given in Chapter 8.

Digital ACDs used for directory assistance are essentially the same with regard to transmission as general-purpose digital OSSs, discussed earlier in this chapter.

10-6 AUXILIARY SERVICES

In addition to the operator assistance and number service functions, there are auxiliary services or service features that may

Table 10–3. Trunk Losses—Early Operator Systems

TRUNK	ANALOG TRUNK W/O GAIN	W/GAIN	COMBIN. TRUNK
A. Local Directory Assistance: No. 5 ACD [1]			
(a) EO–ACD	2.0–4.0	3	–
(b) EO–No. 1 TC	–	10	–
(c) No. 1 TC–ACD	–	7 (gain)	–
(d) ACD–ACD night closing	–	0	–
(e) Sect. TDM–ACD	1.5	1.5	1
(f) Dir. TDM–ACD	0.5	0.5	1
(g) ACD–IC POT	–	0	1
(h) ACD–AT	–	0.8	1
(i) ACD–Position jacks	–	{ 2 TRM { 4 RCV	–
B. Intercept: No. 1 AIS [1]			
(a) EO–AIS	2.0–4.0	3	3
(b) RAIS–HAIS	–	0	–
(c) HAIS–Position jacks	–	{ 2 TRM { 4 RCV	–

Notes:

1. Dedicated trunks.

2. HAIS = Host Automatic Intercept System
   RAIS = Remote Automatic Intercept System
   RCV = Receive
   TRM = Transmit.

be accessed by direct customer dialing or through an operator. Among these are the following:

(1) *Mobile–radio operators* for assistance in interconnecting mobile radio stations with the message network. Access arrangements for *cellular mobile carrier operators* are covered in Chapter 18, Part 11.

(2) *Coastal harbor, railroad,* and other radio–telephone services interconnected with the message network by special operators

(3) *Conferencing services* for provision of multistation message network connections

(4) *Centralized automatic message accounting—operator number identification (CAMA ONI)* for determination of calling–party telephone numbers where automatic equipment is not available or multiparty callers are involved.

Customer and operator grades of service should be similar to those for the operator services previously discussed. However, the unique facility arrangements involved in such connections often make this difficult to achieve. For each service provided, judgment must be applied to the operational feasibility and economic impact of designing the associated plant to meet network standards for any possible connection. Although space does not permit detailed discussions of design criteria for all of these services, the following transmission considerations for conferencing services and for CAMA ONI illustrate the principles involved.

Conference Bridging

Some voice/data conference bridging capabilities are provided by exchange carrier end–office or tandem–office switching systems. Conferencing services may also be provided by ICs. In general, the transmission requirements for exchange carrier digital bridging systems [6] include the following:

(1) Each port of the bridging facility should have a high return loss with matched terminations to prevent reflections.

(2) The loss between any two ports of a multiparty connection should be about 3 dB greater than that for a two–party connection without the bridge. This added loss reduces the grade of service but is necessary to reduce potential singing and echo impairments where return losses are low. If echo control (cancellation) is employed, the loss can be reduced to 0 dB.

(3) Other requirements such as loss variability, attenuation distortion, tracking error, idle–channel noise, quantizing distortion, impulse noise, overload compression, and intermodulation distortion are the same as for digital switching requirements between two–wire analog trunk

interfaces [6]. The objective is that no other transmission impairment be introduced into the connection.

The operation of the bridge should be full–duplex in nature; when two or more parties are talking at the same time, all parties should hear that more than one is talking, as occurs in a face–to–face conference.

Other desirable features to be considered are:

(1) Level contrast should be controlled by automatically adjusting the signal levels to the same level at the bridge

(2) Noise levels of a high–noise party should be controlled during idle periods when speech is not present

(3) Echo should be controlled by cancellation techniques

(4) Double–talking should be permitted

(5) Clipped speech or break–in problems should be avoided

(6) The effects of attack and release times of speech and noise detectors should not be perceptible.

CAMA Operator Number Identification

To accomplish the ONI function, a CAMA operator is temporarily bridged by the switching machine to a dialed connection. Customer–to–customer transmission is not affected since the bridge is removed prior to the call being advanced to the called customer. However, the transmission characteristics of the connection to the CAMA operator position must be controlled so that operator/calling–customer grade of service (in both directions) is satisfactory.

References

1. *OSSGR, Operator Services Systems Generic Requirements*, Technical Reference TR–TSY–000271, Bellcore (Iss. 2, Dec. 1986 with Rev. 3, Mar. 1988), Vol. 1, Section 7 and Volume 2, Section 21.

2. *Generic Requirements for Telephone Headsets at Operator Consoles*, Technical Reference TR–NPL–000314, Bellcore (Iss. 1, Dec. 1987).

3. *EIA Standard EIA–470*, "Telehone Instruments with Loop Signalling for Voiceband Applications" (Washington, DC: Electronic Industries Association, Iss. 1, Jan. 1981).

4. Occupational Safety and Health Administration. *Regulations, Title 29, Code of Federal Regulations, Part 1910* (Washington, DC: U.S. Government Printing Office, 1984).

5. *IEEE Std. 661–1979*, "Method for Determining Objective Loudness Ratings of Telephone Connections" (New York: Institute of Electrical and Electronics Engineers, Inc., 1979).

6. *LSSGR, LATA Switching Systems Generic Requirements— Section 7, Transmission*, Technical Reference TR–TSY–000507, Bellcore (Iss. 3, Mar. 1989).

Telecommunications
Transmission
Engineering

Section 4

Special Services

Much of the telecommunications business involves the provision of telephone service to residential, basic business, and coin telephone stations. These stations are connected to a serving central office through customer loops and may be further connected via the local, long–haul, and toll portions of the network to other stations. The service so rendered is referred to as "ordinary telephone service."

A second class of services, used primarily by business customers, involves the provision of service capabilities beyond ordinary exchange telephone service. This class is called special services. It includes the major category of exchange access services as used by long–distance carriers.

Chapter 11 introduces the types of special services and methods for administering them. These services are defined and classified from a usage and transmission standpoint. Switched special services are those characterized by customer dialing over either the message or private networks. Technical design considerations for switched special services are discussed in Chapter 12. Centrex service, a switched special service with wide capabilities, is discussed in Chapter 13. Switched special services may be configured into customized private networks called electronic tandem networks and switched services networks. Design criteria for these arrangements are considered in Chapter 14.

Private–line special services are those characterized by little or no switching. They include many types with a wide variety of bandwidths and transmission capabilities. Design considerations for most private–line types are discussed in Chapter 15. Transmission requirements for program and video services involve such special considerations as increased bandwidth. These are discussed in Chapter 16. The digital nature of most types of data, combined with the digital nature of modern transmission systems, offers advantages and efficiencies for data transmission. Such private-line digital data service is discussed in Chapter 17, along with high-capacity digital services at 1.5 and 45 Mb/s and the 56-kb/s public switched digital service. Finally, Chapter 18 covers the full range of exchange access services offered to long-haul carriers, along with similar arrangements provided to cellular mobile carriers.

Chapter 11

Introduction to Special Services

Special services constitute a diverse field that rivals the message network in size and complexity. These services, including channel terminals, mileage, and network usage, provide a substantial part of an exchange carrier's revenue. This large proportion of total revenues, along with pressures due to competition and deregulation, are causing rapid changes in the field of special–services telecommunications. Services and the methods of providing and maintaining them are evolving in response to existing problems and anticipated needs. Technological changes have their effect, as witnessed by the near–vanishing of certain obsolete services (e.g., telegraph) and the impressive growth of others (e.g., high–capacity digital).

Special services may be grouped in terms of transmission considerations. The classifications used in this chapter are not formally recognized but are convenient for purposes of discussion. Features and uses of the various types of service in each group are then presented. Finally, the major functions associated with implementing special services and the administrative methods of handling the high volume of orders are discussed. Fuller technical details on these services are given in Chapters 12 through 18.

11-1 CHARACTERISTICS OF SPECIAL SERVICES

Special services can be defined precisely only by citing all types of service constituting the field. Most often, specials are defined as all telecommunication services except residence, coin, and ordinary business telephone services.* This definition treats business centrex and private branch exchange (PBX) services as

---

* These nonspecial services are often referred to as plain old telephone service (POTS).

special. Special services are *special* in that they generally require engineering and operations treatment beyond that applied to ordinary services in terms of transmission, signalling, maintenance, and customer use. They are installed and removed in response to a specific customer order, as opposed to message trunks that are provided for general public use. Many of them are special in that they use interoffice facilities, or in the sense that they may not be accessible by the general public.

Special services are used primarily by business and government customers who need quick, inexpensive, and efficient communications to all parts of a geographically dispersed enterprise. Since operating efficiency depends on economical use of telecommunications, it is important to meet specific needs. Special services are considered to include certain facilities provided to interexchange carriers (ICs) to give them local access and transport area (LATA) access: nonswitched (special) access and the two end–office switched–access services known as wide area telecommunications service (WATS) access lines and Feature Group A (FGA).

While the bulk of telecommunications plant is installed for ordinary telephone service, special services use the plant in unique ways. Much of the customer loop plant has been installed under the resistance design plan with considerations focusing mainly on ordinary residential and business services. Later concepts like the carrier serving area (CSA) plan are intended to give the flexibility to accommodate specials, including digital services up to 56 kb/s, with minimal rearrangement or design effort. The prequalification of loops in bulk for digital services, and preplanned spectrum management to allow compatible use, are additional steps in this direction. Interoffice plant is largely designed for high–capacity digital and message trunk use. However, a special–services circuit may require both a loop facility and an interoffice facility that together must perform the functions of an ordinary customer loop. For some services, normal plant must be modified (e.g., loaded or deloaded) to achieve the desired transmission requirements. In other cases, the installation effort is as simple as changing plug–in channel units in carrier terminals. Another example is the construction of T1 span lines in loop plant for high–capacity digital services. For other services, it is necessary to provide supplementary plant, as with fiber installations for video services.

Transmission standards for special services are generally more stringent than for ordinary telephone service; an example is the lower allowable loss/frequency distortion of loops when tariffed and ordered for high–speed data transmission. Video and program services have unique transmission standards. Yet special services must be compatible with ordinary telephone service; neither should interfere with the other since both usually share the same cables and carrier systems.

Customer operational needs or transmission requirements that are not satisfied by an existing tariff must be covered by a special authority filed with the appropriate regulatory body. Such requirements might involve special or diverse routing for extra reliability; the provision of certain high–capacity services that, while standardized, are uncommon and are priced on an individual-case basis; or a special design to handle data at unique speeds. Sometimes special equipment assemblies may be required to provide a tariff–approved service in a nonstandard manner. While these assemblies do not require additional regulatory authorization, the unique technical requirements must be satisfied.

Ordinary telephone services are generally bulk–engineered. For example, customer loop plant in a CSA is engineered by pair–gain cable and route for all customer locations to be served in the area. Application is by the loop assignment center, usually without further engineering considerations, using an assignment data base system. Special services, however, are laid out on a per–circuit basis; individual–circuit engineering is one of the major features that distinguish special services.

The integrated services digital network (ISDN) is an underlying structure that provides 64–kb/s B channels for both ordinary and special–services use. With the ongoing proliferation of high-capacity digital special services and ISDN facilities, many of today's discrete voiceband special services will evolve into simple users of B channels and 64–kb/s slots on high–capacity facilities.

11-2 TRANSMISSION CLASSIFICATIONS AND SERVICE FEATURES

Special services may be classified in a number of ways. They may be categorized as telephone (those used only for voice communication) or nontelephone (those never used for voice). The old distinction between services (end–to–end offerings, including the station equipment) and channels–only offerings has largely disappeared in today's environment where the customer provides the premises wiring and terminal equipment. Special services have also been classified according to required bandwidth. For transmission purposes, none of these classifications is completely satisfactory.

The features of a special service are major factors in the selection of the transmission objectives and the design of the circuit or system. For instance, many voiceband services are provided simply for communication between the same two points while other services provide special access to the switched message network. The transmission objectives for these two types of service are substantially different. The transmission classifications of special services then relate to the use or nonuse of the service relative to the switched message network. Four broad classifications of service are used in this discussion.

The features of the many special services that make up the four classifications defined by transmission considerations play some part in determining the types of facility that must be used to provide these services. While the services to be described are offered by most exchange carriers, the names of the services are often unique to the particular exchange carrier. The terms defined here are those most commonly used.

Class 1 Special Services

Services that are always switched through the message network are designated class 1. In order to achieve user satisfaction, the design objectives for these services are chosen to be compatible with message network service. Thus, the loss objective for a class 1 special–services circuit is chosen to be less than the maximum loss, about 8 dB, of an ordinary loop. These services fall into two categories, depending on whether they terminate on a public or a private switch. A public switch is an ordinary end office in the network; a private switch is a centrex or PBX.

Public-Switched Services. This group of services includes FGA switched access to the LATA network for origination and termination of interLATA traffic, plus several services that may be considered as enhancements of ordinary telephone service to give the customer cost savings or special features. These services provide access to the message network for the transmission of voice, data, and digital facsimile.

Switched access provides the customer, generally an IC but sometimes a large private customer, with access to the line or trunk side of an exchange-carrier switch. Four feature groups, A through D, provide this access. As discussed further in Chapter 18, FGA provides line-side or dial-tone access and is considered a special service, while the others terminate as trunks and are not considered to be specials. Cellular mobile carriers and other radio common carriers use similar arrangements to interconnect with the local network.

911 service involves direct or tandem-switched trunking from exchange-network switches to central public safety answering points. A telephone user who dials the 911 emergency number reaches a dispatcher able to send the correct help—police, fire, medical—to the requester. With enhanced (E911) systems, the dispatcher automatically receives the name and address of the calling telephone subscriber, and other data relevant to provision of emergency aid, from a central data base [1].

Foreign exchange (FX) service treats an individual customer as a local user in an area distant from his actual location. The service is provided by an access line, called an FX line, illustrated in Figure 11-1. This service is particularly popular in cases where a metro area is served by two or more exchange carriers so that many short-distance calls are billed at interexchange rates. Some exchange carriers offer a comparable service in which the station is served by a distant central office *within* the home exchange area, for such reasons as retaining a preexisting number after a physical move between central office areas. The technical aspects of this offering are identical to those for FX service. The switching feature known as remote call forwarding provides a low–cost alternative to FX service for many businesses desiring a "presence" in a remote area, and may thus replace many FX services.

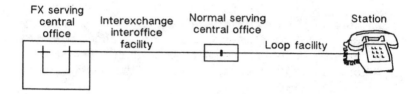

Figure 11-1. Configuration of an FX line.

Wide area telecommunications service permits a station to make calls to selected intraLATA geographical regions (called bands) for a fixed monthly charge. Access is also offered to ICs' interLATA WATS services by switching through to access trunks that send traffic to the IC. The service may be unlimited or may be restricted to a specified number of hours per month without extra billing, and may involve variable per–hour charges depending on the time of day. Separate access lines for LATA access and intraLATA service are required, terminating in a central office equipped for WATS. Most stored–program central offices (analog or digital) and many crossbar offices are capable of recording usage data for WATS, so most WATS lines today are equivalent to ordinary loops. In unusual cases where the normal serving office is not equipped for WATS, the access lines, called WATS lines, are similar to FX lines extending to the WATS office, except that they are solely for long–distance traffic.

800 service permits callers within specified geographical regions to call the 800–service customer, using an 800 area code, without incurring a toll charge. As in WATS, an access line is required between the customer station and a specified office. Only incoming service is provided by this type of line. As newer intelligent–network configurations [2] come into use, the 800–service line is more and more likely to appear as an ordinary telephone line; the 800 number is simply translated to a regular ten–digit number for switching after the originating central office consults a central data base via common–channel signalling (CCS) links.

230

Off–premises extension (OPX) service is provided by an extension telephone remote from the main station location. The main station line and the extension are usually bridged at the dial–tone central office. Bridge lifters are often required to avoid transmission losses, as discussed in Volume 2, Chapter 3.

Secretarial service provides telephone answering service. A line similar to an OPX connects the customer line to the answering–service location. The line usually terminates in a secretarial-service console for an attendant's use. These extensions, illustrated in Figure 11–2, are usually arranged for receiving calls only. As with OPX services, bridge lifters are often required. The answering service may use a concentrator system to reduce its mileage costs, combining several of its patrons' traffic onto a small group of trunks.

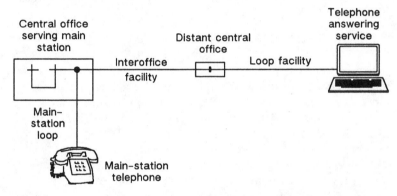

Figure 11–2. Secretarial service line.

Announcement service allows callers to connect, usually by dialing the prefix 976, to information providers for time, weather, sports scores, etc. Lines from the provider's premises to the telephone office carry the announcements. Depending on the local serving plan, the announcement lines may be local telephone numbers or may extend to a remote office.

Data operation on ordinary telephone lines provides voiceband data and digital facsimile transmission as well as voice capability over the switched message network. Local telephone loops are used. Conditioning may be required, depending on data rate and the design of the customer–premises equipment, and is often a special tariff offering.

Public switched digital service provides 56–kb/s dial–up data operation alternately with voice service. In the data mode, this service delivers most of the capacity of a 64–kb/s digital trunk for customer use. When switched to voice, the same facility is converted to ordinary pulse code modulation transmission.

A variety of other services are furnished. Access lines like those described above are used by some companies to provide services specific to the individual area. For example, a service similar to WATS provides outgoing–only service from suburban to metropolitan areas. It is possible to provide two–way WATS lines, combining WATS and 800–service features. Another example is the use of long–distance trunks to provide direct switched access from a PBX to a toll operator. This service, used mainly by hotels and motels, provides immediate toll–billing information to the hotel.

Private–Switched Services. These services are defined in terms of the placement of a centrex or PBX in the customer's network. Centrex service provides a private communications system via central–office switching or equivalent measures like remote switching modules. A PBX is a system for interconnecting telephones and/or data terminals, generally on the same premises. Connections can also be made from a PBX station to the switched message network or to other trunks terminated in the PBX. Present–generation "smart" key telephone systems concentrate a number of stations onto a smaller number of central–office trunks; thus, they are effectively small PBXs. Apart from an exchange carrier's official internal services, a PBX is normally provided by the customer under Federal registration rules, whereas centrex is an exchange–carrier offering. Either one usually has a customer–employed attendant to assist in placing a call, if necessary, or to control and administer the system.

A *main* centrex or PBX is one that has a directory number and can connect stations to the message network for both incoming and outgoing calls. Tie trunks, FX trunks, and WATS trunks may also terminate in a main centrex/PBX, but the switch does not connect tie trunks together in tandem.

A *satellite* PBX does not have a directory number; all incoming calls are routed from the main PBX via tie trunks. For outgoing service, calls may be routed directly over central–office trunks, if provided, or over tie trunks through the main PBX, connecting to its central–office trunks. The satellite PBX is usually located in the same local area as its main PBX. Centrex service to a customer's remote location, provided by a remote switching module "hosted" by a main centrex switch, is analogous to a satellite–main PBX configuration.

A *tandem* centrex or PBX is used as an intermediate switching point to connect tie trunks to two or more main centrex switches or PBXs. It usually includes the same functions as a main PBX.

Several types of trunk and line are used to provide services in private networks:

(1) A PBX is connected to the central office that normally serves it by PBX–central office (PBX–CO) trunks. These trunks may be arranged for inward, outward, or two–way operation. At the central–office equipment, they appear as station lines or, for direct–inward–dialing use, as trunks. Ground–start signalling is used on two–way trunks to prevent seizures from both ends simultaneously. PBX–CO trunks usually form "hunt groups" where traffic is spread among multiple trunks automatically. In a centrex office, connections from individual stations to the message network simply use junctors or interentity trunks within the same building.

(2) Stations collocated with a PBX are connected to it by PBX station lines, often using customer–owned cable. Those stations that are located off–premises are usually served via lines provided by the exchange carrier. The station lines can be switched through the PBX to other station lines, central–office trunks, tie trunks, FX trunks, WATS trunks, 800–service trunks, or direct–access circuits to ICs' networks. A centrex can terminate these same services.

(3) Centrex station lines connect telephone stations to the centrex switching machine. They are usually within the same wire–center area, and are often short enough to require no transmission treatment. However, they

sometimes require interoffice facilities or even custom–built digital loop carrier systems.

(4) Outgoing traffic from a switched services network (SSN) switching machine is routed to the message network via WATS trunks or off–network access lines (ONALs). These may be local, FGA access, or FX trunks.

(5) Incoming traffic from the message network is routed to an SSN through 800–service trunks and ONALs (local, FX, or FGA access). These lines usually connect through the serving switch under control of an operator console at the customer premises; occasionally the SSN switch is equipped to accept such traffic into the SSN automatically by recognizing special dialed identification digits.

(6) Centrex systems usually include trunks to the on–premises attendant; often one attendant location provides service for a multilocation customer via a dispersed group of centrex switches.

(7) Automatic call distributor (ACD) trunks connect an ACD to its serving office. An ACD is a switch that automatically concentrates large numbers of incoming lines to a smaller number of attended positions, presenting incoming calls in an orderly sequence to the available attendants. This service is in wide use by order–taking agencies such as airline reservation bureaus. ACD trunks are treated in a manner similar to PBX–CO trunks from a transmission standpoint. Where a CO switch itself provides ACD features, the system is comparable to centrex.

Class 2 Special Services

Class 2 special services are provided over lines or trunks that may be connected to the switched message network as the customer directs, but the functions of these circuits may not pertain to the message network. As with class 1 services, the loss objectives are selected so that the total loss of the several types of tandem special–services circuits required to reach the message network does not exceed the maximum loss of an ordinary loop.

The same services are provided by PBX off–premises station (OPS) lines as by on–premises station lines, except that the telephone is located remotely from the PBX as shown in Figure 11–3. Centrex and PBX OPS lines are similar to each other, except that the switching equipment for centrex is located on exchange–carrier premises, and the interoffice facility in the figure connects directly to the switch.

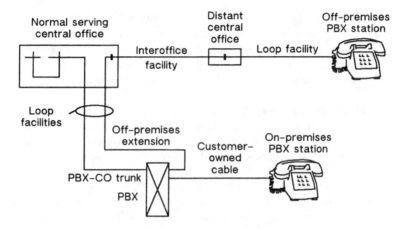

Figure 11–3. On–premises and off–premises PBX station lines.

Class 2 special services also involve the interconnection of PBXs or centrex switches by means of several types of tie trunk. *Satellite tie trunks* connect a satellite PBX to its main PBX. *Nontandem tie trunks* are used between two main PBXs; they are not switched to other tie trunks or other PBXs. These trunks are primarily intended for connection to PBX stations at both ends but may also be connected to central–office trunks, FX trunks, and WATS trunks. Simultaneous connections to central–office, FX, or WATS trunks at both ends of a nontandem tie trunk do not necessarily provide good transmission.

Class 3 Special Services

Class 3 special services are those that never connect to the switched message network. Included in this class are dedicated private lines, i.e., lines for the individual private use of a customer, both point–to–point and multipoint. Class 3 services may be voiceband or nonvoiceband. For voiceband services in this class, the overall station–to–station transmission loss objective is similar to the average station–to–station loss objective for the message network (e.g., 10 dB for two–wire operation). Following are descriptions of channels that constitute class 3 special services:

(1) Channels for remote metering, fire and burglar alarms, supervisory control, and miscellaneous signalling are, in essence, very low–speed data transmission channels. Both dc and tone transmission techniques are used, with dc slowly going out of use in favor of tones.

(2) Program channels are used by broadcasters as studio–to–transmitter links, as remote–pickup circuits, etc. They provide 3.5–, 5–, 8–, or 15–kHz bandwidths, as specified in tariffs and as ordered by the customer.

(3) Data transmission channels can be ordered for voiceband data speeds, with various optional types of conditioning, or to provide bandwidths for bit rates from teletypewriter speeds (0 to 150 b/s) through wideband data rates (50 or 230.4 kb/s). Voiceband channels are quite common; the others have been obsoleted by modern terminals and digital channels. Data channels can be used for direct access to packet switches.

(4) Channels for two–point or multipoint voice transmission are common. A wide variety of customer–premises station equipment and any of several signalling schemes may be used, depending on user needs. For example, simple telephones may be connected together, as in an "automatic–ringdown" private line from a 911 answering point to a fire station. Such voice lines may appeal to the customer because they are free of traffic blocking (a desirable feature for fire dispatching), because the distant user knows exactly which firm is calling (an important consideration for commodities traders), or because there is no call setup time. For a private mobile–radio network, the circuit may link the dispatcher's control equipment to a distant radio base station.

(5) One–way video transmission channels are provided for television broadcast service. Usually, microwave radio or fiber cable are used to provide the required bandwidth. Full–time or occasional service is provided.

(6) High–capacity digital services—usually at 1.5 Mb/s, with 45 Mb/s growing in popularity—are used by customers to build private voice–data–video networks. They are popular for large–sized trunk groups between digital PBXs, for compressed–video teleconferencing, for interconnecting local–area data networks, for submultiplexing by the customer, etc. They can be used to build customer–reconfigurable networks using digital cross–connection systems [3]. Exchange carriers offer central–office multiplexers so that ICs can combine groups of individual voice access services onto a single high–capacity digital service. With "fractional T1" service, the customer can rent use of a block of channels (8, 12, 16, or other fractions) on a DS1 interoffice facility. "Fractional T3" implies the use of fewer than the full 28 DS1s contained in a DS3 signal.

(7) Private–line digital data service is provided over the facilities of the Digital Data System (DDS) or a similar network. This system provides point–to–point digital service at bit rates of 2.4, 4.8, 9.6, and 56 kb/s; occasionally 1.2 and 19.2 kb/s also. A non–DDS variant called Basic Dedicated Digital Service (BDDS) is popular in most states, and may eventually replace DDS–based offerings.

(8) Special (nonswitched) access services are in wide use by ICs and major customers for accessing the exchange area to originate or terminate interLATA private lines.

Numerous other special applications exist to serve the specific needs of customers, but those listed cover the majority of private–line services. Most of these channels are also used internally by the exchange carrier; for example, CCS data links are based on 56–kb/s DDS/BDDS channels.

Class 4 Special Services

Class 4 special services involve the interconnection of two or more special–services circuits. Electronic tandem networks

(ETNs) and SSNs are the physical means for providing these services. Transmission objectives for this class are stringent because the services involve multiple circuit connections within the networks as well as access to the switched message network.

Electronic Tandem Network. This type of network provides end–to–end connection of tie trunks between PBX or centrex locations. Figure 11–4 illustrates a typical ETN arrangement, having nontandem and satellite tie trunks (previously discussed in regard to class 2 special services) as well as tandem and intertandem tie trunks. *Tandem tie trunks* are used between main centrex switches and tandem centrex sites. In larger ETNs, pairs of tandem centrexes and PBXs may be connected; *intertandem tie trunks* are used to make such connections. Not shown in the figure, each centrex or PBX usually has access to the local public network.

Each switch within an ETN performs the normal centrex functions but additional features result from the organization of the network. Tie trunks connected to a PBX or centrex may be switched together automatically. A caller establishes a call by dialing a fixed–length number, rather than the older variable number of digits that depended on the route selected. Such calls are advanced sequentially from switch to switch to establish the overall connection. A new dial tone is not needed for steps in the sequence. Connection to the message network is usually provided, and is used routinely, but satisfactory transmission quality is not assured.

Features included in ETN arrangements include a uniform numbering plan throughout the network, code conversion (digit addition and deletion), automatic alternate routing, message detail recording, service observing, and standard tones and announcements. In a complex network such as that of Figure 11–4, these features avoid the increasing difficulties in establishing calls due to the large number of sequential dial tones, separate dialing of routing codes, and multilink address pulsing that were formerly involved. Switching in an ETN may be on a two–wire or, especially with digital machines, a four–wire basis.

Switched Services Networks. These networks provide private–line services. They use trunks and access lines linked by

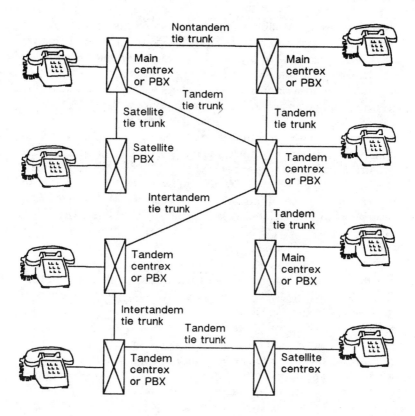

Figure 11-4. Electronic tandem network.

stored-program switching arrangements in order to connect calls between user locations. The switching equipment is located in central offices and may be shared by other SSNs, centrex lines, and the message network. The equipment at customer locations consists of attendant consoles and, often, PBXs. The network of common-control switching machines is called a common-control switching arrangement (CCSA). The central-office switching equipment is billed according to the provisions of a CCSA tariff, while trunks, access lines, and other special-services lines are billed from other tariffs.

The many features that can be provided by SSNs make these networks attractive to large businesses. Direct inward and outward dialing with an SSN provides the capability of direct station–to–station calling between network locations and reduces the need for operators. Since a fully integrated numbering plan is used, each station in the network has a unique seven–digit number. Multiswitch SSNs can provide automatic alternate routing. The administration of such matters as maintenance, traffic records, traffic engineering, and trunk–group design is the responsibility of the exchange carrier. A sample or a full copy of automatic message accounting records is provided to the customer for use in allocating communications expenses among departments. Finally, automatic completion of calls to the message network, an optional feature, allows calls from the SSN to be completed to locations off the network. ONALs provided at strategic points on the network are reached by a selective routing plan.

The usual method of organizing SSNs, a hierarchical plan based on switches in central offices, is shown in Figure 11–5. While the hierarchical plan derives from the original design of the message network, there are some differences due to service requirements. Economic restrictions may limit the number of direct (high–usage) trunks to a customer location, so that more trunks may be connected in tandem for a given connection than when direct trunks are provided as in the message network. SS–1 class switching offices appear in only the largest SSNs. Switching at SS–2 and SS–1 levels is always four–wire. Stations, including four–wire sets and data modems, may be served directly by the SS–class switch as if it were a centrex.

As ETNs have grown in features, their capabilities have become quite close to those of SSNs. The main difference is in the designation of the tandem switches: ETN functions performed in centrex switches versus dedicated switch entities as found in SSNs.

Another SSN plan, used in military networks for survivability, is called polygrid. The polygrid plan consists of overlapping grids of interconnected switching machines, each serving a particular location for inbound and outbound traffic. All switches have equal rank in the polygrid network and are interconnected in such a way that a large percentage of them would have to be overloaded or rendered inoperative before the network would be

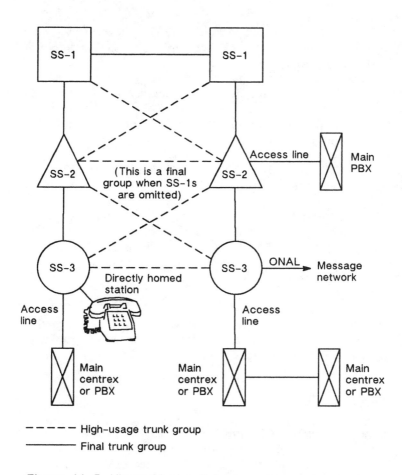

Figure 11–5. Hierarchical switched services network plan.

disrupted. Thus, the network is highly survivable and provides good assurance of call completion.

A large polygrid network is operated for the U.S. Government. This network, originally called the automatic voice network (AUTOVON) and currently evolving into the defense switched network (DSN), has a number of unique features. With only a few exceptions, transmission paths and switches are four–wire, even to the station lines and station sets. A multilevel system of priority calling is included so that certain calls are given precedence over, and may preempt, calls of lower priority.

Several types of lines and trunks are used in SSNs in addition to those defined previously. Those unique to SSN operation include the following:

(1) Access lines are circuits that connect main centrexes or PBXs to class SS–1, SS–2, or SS–3 offices in a hierarchical plan (Figure 11–5) or to switching machines in the polygrid plan.

(2) Network trunks are circuits that interconnect SSN switching offices.

(3) Conditioned access lines are circuits between stations or PBXs and switching machines. These circuits are sometimes treated for gain and envelope delay to improve their performance for high–speed voice–grade data transmission.

Major firms may build internal networks similar to SSNs, involving central–office–type tandem switches and connections to dozens of PBXs and centrex switches. Interswitch trunking is usually on DS1 high–capacity digital channels.

Universal Service

Class 1, class 2, and some class 4 special services are intended to provide satisfactory voice transmission on most universal–service connections. Universal service is defined as the interconnection between special–services facilities and the message network, at one point only, on any one call. *In this arrangement, message network connections at both ends of the special–services facility are not contemplated even though they are often used successfully.* Also, while calls originating over one private–line network may be routed via the message network to a second private–line network, this type of connection does not always provide adequate voice transmission. Lines provided for class 1 and class 2 services are generally capable of voice–grade data transmission.

Lines used for class 4 service usually provide satisfactory voice transmission on universal–service connections utilizing PBX–CO trunks or ONALs. Calls within the special network typically provide satisfactory transmission over a maximum of four analog tie trunks in tandem for voice signals and two tie trunks in tandem for data signals at rates in the 0.3–to–4.8–kb/s range. (The use of digital switching and trunking, and of automatically equalized data modems, tends to extend these figures. Specific data speeds are not formally supported in today's multivendor environment, but high–speed data and facsimile transmission are used routinely.) An SSN is engineered as an entity and should provide satisfactory voice transmission to all stations served by it.

11–3 COORDINATION AND ADMINISTRATION OF SPECIAL SERVICES

As the complexity of special services has increased, it has become necessary to improve methods and procedures for handling special–services orders. To define standard methods compatible with an integrated mechanized information system and to measure adequately the quality of the entire service–provision process, it is necessary to have standard methods specified within each operating area. The Administration of Designed Services (ADS) system, or local plans similar to it, defines these methods and procedures for intra–area processing of special–services orders.

Administration of Designed Services

While the ADS is a system of methods and procedures and may encompass the operations of several departments, it is not an organizational entity. Procedures under ADS apply to an operating area of an exchange carrier. The system may be applied to other sizes of operating units depending on the volume of special–services orders and characteristics of their geographic areas as determined by the individual company. The system is composed of five subsystems—order–writing, design, distribution, completion, and control. Figure 11–6 is a diagram of the work functions involved.

The order–writing system accepts data from the marketing negotiator, an interfunctional service coordination (ISC) team, or, for orders for access services, from the IC service center. These data encompass service inquiries, service orders for special

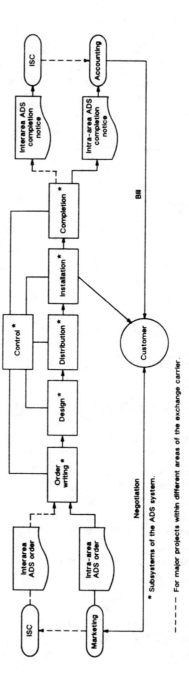

Figure 11-6. Typical ADS functions.

services, and supplements to service orders. The basic functions of the order–writing subsystem are to review the orders to make certain that they are reasonable, complete, and in the appropriate format, to forward them to the design subsystem for screening, and to enter the initial data into the control subsystem.

The design subsystem encompasses five processes—screening, circuit layout selection, station interface selection, work order record and details (WORD) or circuit layout record (CLR) preparation, and record administration. In the screening process, orders for special services are reviewed, the availability and transmission characteristics of local channels are obtained and recorded, and incoming orders are coordinated with associated documents. In circuit layout, service requests are evaluated, specific layouts are selected, and facilities and equipment are reserved or assigned. The layouts, usually prepared by machine, must be selected so that standard design objectives are met. The preparation of design documents involves automated design computations, the results of which appear on the WORD or CLR and, where an IC is involved, on the design layout record (DLR) provided to the IC. All layout data are forwarded to the distribution subsystem. Finally, the maintenance of the pending and completed circuit files for the design subsystem is accomplished by record administration.

The distribution subsystem provides for the correlation, assembly, and distribution of service orders, WORD or CLR data, and associated documents to implement the installation of the circuit and equipment. This subsystem uses internal data networks provided by the exchange carrier. The distribution of documents must be made to the required locations in sufficient time to allow for installation and tests.

The completion subsystem receives and forwards reports regarding completed installations, both internal and external, to the ADS system. Additional completion information, such as transmission measurements, may be forwarded with the completion report.

The control subsystem monitors and controls the status of service orders. The critical dates in the life of the service order are continually monitored, and control data and reports are made available to involved locations. Information is provided to assist in the administration of work activities. The acronym OSCAR refers to the functions of this subsystem—order status, control, and reporting.

Work Order Record and Details. A master circuit record, the WORD, consists of three kinds of records: (1) work authorization, (2) circuit details, and (3) test details. Other circuit–related information not included on the WORD or CLR documents (such as division–of–revenue data) is recorded separately.

The WORD document combines forms for service or circuit orders, CLRs, and circuit–order test results. There is a separate WORD for each circuit. The work authorization corresponds to and replaces the need for a carrier–system, trunk, or special-services order; the service order is not replaced for all of its present functions but the WORD document contains all the information needed to install and maintain the circuit. Circuit details contain the sequential makeup, facility assignment, and transmission level point information. Test details include the required tests, expected values and limits, and fields for recording test results. When completed, it acts as the order completion report and office record of test results. It is not necessarily a printed document; automated test systems obtain WORD records directly from the circuit–records data base.

COMMON LANGUAGE Special-Services Designations. The COMMON LANGUAGE CLCI plan provides a coded designation by which a special–services circuit may be identified. This designation is in a mnemonic form that people can remember readily; it applies to both manual and mechanized procedures. It works with related location–identifier (CLLI) and facility–identifier (CLFI) codes, all of which are used in the WORD document.

The designations for special–services circuits are written in one of three standard formats: telephone number, serial number, or message trunk. The telephone–number format is used when a circuit can be identified by a unique telephone number, possibly including an extension or trunk code. Otherwise the serial–number format is used. The message–trunk format, described in

Chapter 6, applies in special cases such as network trunks in SSNs.

The coded information presented in the telephone–number format consists of 24 alphanumeric characters entered in designated spaces for primary and secondary data, as shown in Figure 11–7. The figure shows, as an example, an FX line with the directory number 311–555–1212. By contrast, a nonswitched point–to–point line would carry a simple serial number in the same data field. The primary data are used to identify a circuit on the customer premises. The secondary data contains additional information that, in conjunction with the primary data, is used on internal company records for complete circuit identification.

The portion of the telephone– or serial–number formats relevant to this discussion is contained in positions 3 through 6, service code and modifier. Character positions 3 and 4 represent the appropriate service code, such as FX for foreign exchange, FD for voiceband private–line data, PL for voice private–line, HC for 1.544 Mb/s high–capacity, etc. Standard two–character codings have been established for all appropriate special–services circuits. The modifier, positions 5 and 6, provides added information. As an example, position 5 contains N, D, or A to signify nondata use, data use, or alternate data–nondata use, respectively, for intraLATA services. Position 6 contains T or C, as examples, to signify all–carrier–provided equipment and facilities, versus all– or part–customer–provided.

Tariffs

Exchange carriers file tariffs with regulatory agencies to describe services and to propose schedules of charges. The companies are legally bound by tariffs that have been approved by regulatory agencies. Services involving interstate jurisdiction are specified by tariffs filed with the Federal Communication Commission; intrastate services are governed by tariffs filed with state regulatory commissions. IntraLATA services may thus vary from one company and regulatory agency to another; as a result, service offerings may not be uniform for a customer who operates within the territories of several exchange carriers.

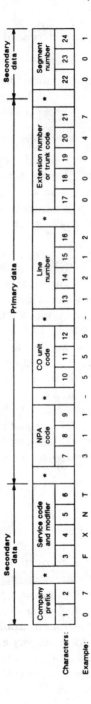

Figure 11-7. COMMON LANGUAGE telephone number format.

References

1. Dimone, V. P., D. Elsinger, and V. Perez. "911: A Good Thing Just Gets Better," Bellcore EXCHANGE, Vol. 3, Iss. 4 (July/Aug. 1987), pp. 2–7.

2. DeSantos, J. M. and B. F. Gardner, Jr. "BOC 800 Service: Offering the Customer More Choices," Bellcore EXCHANGE, Vol. 2, Iss. 1 (Jan./Feb. 1986), pp. 18–22.

3. Matthews, R. L. "Network Control in the Hands of the Customer," Bellcore EXCHANGE, Vol. 2, Iss. 2 (Mar./Apr. 1986), pp. 24–27.

Chapter 12

Switched Special Services

Most special services are switched, either through the public network or through private switching machines and networks. These services may be classified according to whether they connect directly to the switched public network. Class 1 services are always switched through the public network and are further divided into two subclasses, private branch exchange (PBX)–related and non–PBX–related. Class 1 PBX–related services use PBX–central office (CO) trunks, foreign exchange (FX) trunks, wide area telecommunications service (WATS) trunks, 800 service trunks, Feature Group A (FGA) switched access, and automatic call distributor (ACD) trunks. Class 1 non–PBX–related services include FX service lines, WATS/800 service lines, off–premises extension lines, and secretarial lines. Class 2 services are always PBX–related. They are usually, but need not be, switched through the public network, as the customer directs. These services use off–premises station lines, centrex/PBX tie trunks, and FGA switched access.

12-1 CIRCUIT DESIGN

While the details of the design process for a particular special service vary with company and organizational structure, some basic elements are common. The process is initiated by a service order that specifies the type of service desired. Tariff and technical requirements are considered in the design process. The tariff legally defines the features and rates of the service offering. The technical requirements involve such things as signalling, loss and noise objectives, telephone set current (if the service terminates in an interface to a station set), and stability. In addition, data services may require control of impulse noise, slope, and envelope delay distortion (EDD). While all are specific items, they interrelate; a change to improve a circuit from one standpoint may harm it from another.

251

Illustrative Design

Consider the design process for an FX service. The customer location sets the loop facility assignment [wire or digital loop carrier (DLC)], length, gauge, loading, and any bridged tap at the customer end. Generally, the loop layout should be determined before the interoffice facility is selected since a carrier–derived interoffice facility can compensate for loop resistance and loss. Signalling capability must be checked and station–set current must be computed to determine if a signalling extender is required.

Transmission gain requirements are considered next in the design process so that type and location of a repeater, if required, can be selected. In a marginal case, it is necessary to verify that the loop resistance of the repeater will not put the signalling out of range. Two limitations on the allowable repeater gain (stability and crosstalk) influence the location of the repeater or signalling extender and may even require a second repeater. If a signalling extender must be relocated to improve circuit stability, the design must be reviewed to verify that signalling requirements are still met. Where relocation of the signalling unit would result in signalling or supervision out of limits, or where gain and stability requirements cannot be met, a carrier channel with its signalling features may be required. When the circuit layout is complete, it should comprise facilities representing a balance among customer satisfaction, technical requirements, and economy.

A few of today's switched special–services circuits are long–haul, defined as having more than 6 ms of round–trip echo delay. Circuits having less delay are short–haul. The long–haul circuits are designed with specified minimum losses according to the via net loss (VNL) design plan in order to control echo. The short–haul circuits are designed to have a fixed loss consistent with stability, noise, and other criteria. The crossover distance on carrier facilities, in the absence of sources of additional transmission delay such as digital cross–connect systems (DCSs), is at about 200 miles. Other special services that involve switching, such as centrex, tandem tie–trunk networks, and switched services networks are discussed elsewhere in this volume.

Design and Analysis Aids

Two concepts have been developed to assist in special–services circuit design. These are the standard–design concept and the Universal Cable Circuit Analysis Program (UNICCAP).

Standard Designs. Special–services circuits of standard designs offer the advantages of thoroughly tested circuit layouts that meet requirements on net loss, transmission response, stability, and balance. The layouts are fitted to specific situations by first selecting the facility on the basis of predefined usage preferences. The number of links (loop or interoffice facilities) and the associated loss and signalling requirements for each link of an overall circuit must then be determined. Access circuits, such as FX lines and trunks, are usually composed of one loop facility and one interoffice facility; tie trunks and off–premises station lines usually use two loops and an interoffice facility. After the general facility type is selected, the detailed locations and adjustments of repeaters, signalling units, etc., can be established.

To illustrate, consider the standard design of PBX–CO trunks. Table 12–1 lists the design codes and maximum losses for various facility types. Facility losses are shown prior to the application of repeater gain. When the design code is determined, the dc

Table 12–1. Standard Designs—PBX–CO Trunks

Type of Facility			Max 1–kHz Facility Loss (dB)	Design Code
2W VF	Nonloaded	Nonrepeatered	3.5	1
		Repeatered	6.2	2
	Loaded	Nonrepeatered	3.5	3
		Repeatered	8.0	4
DLC plus nonloaded			–	5
4W VF	Loaded	Repeatered	12.0	6

resistance is calculated and signalling equipment is selected to be compatible with central–office type, interface required to the PBX, and required signalling features.

Figure 12–1 shows four possible two–wire layouts, nonloaded cable (codes 1 and 2), loaded cable (codes 3 and 4) and DLC with nonloaded cable (code 5). These layouts apply to effectively all PBX–CO trunks except those provided on a DS1 (1.544–Mb/s) basis.

Occasionally, when a longer trunk is required, a four–wire design must be employed. Figure 12–2 shows a four–wire layout (code 6) employing dc signalling in a case where carrier facilities are unavailable. The cable pairs are typically loaded in this case. Transmission paths are shown by heavy lines. Signalling leads from the repeater provide access to the signalling path for connection to signalling extenders. Not shown on the figure are access points for remote switched test access.

Universal Cable Circuit Analysis Program. This computer program, called UNICCAP, is an engineering tool in analyzing a wide variety of cable transmission problems. The program provides rapid computations of insertion loss, measured loss, bridged loss, return loss, echo return loss, singing return loss, input impedance, output impedance, EDD, peak–to–average ratio (PAR), transmission level point (TLP), and other parameters. On the basis of these data, it is possible to determine where changes can be made to optimize the circuit or meet requirements. It is incorporated into mechanized circuit design systems and thus can be invoked automatically. The result is a "flow-through" design with automatic facility assignments, machine-calculated transmission levels, a programmed layout record, and data–base retention for automatic test systems.

12-2 SIGNALLING AND SUPERVISION

Signalling is a vital part of switched special services that must be designed in to ensure proper operation. To illustrate primary signalling functions, facility features, and customer options, consider the signalling aspects of FX service, which is typical of class 1 special services. In FX service, access to the public network is

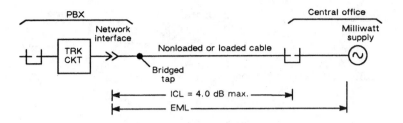

(a)   Code 1; nonloaded, nonrepeatered
          Code 3; loaded, nonrepeatered

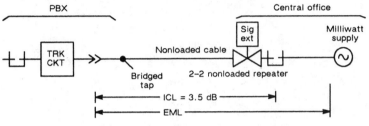

(b) Code 2; nonloaded, repeatered

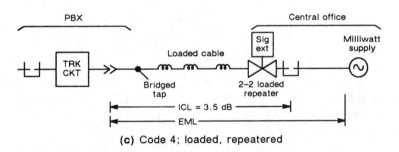

(c) Code 4; loaded, repeatered

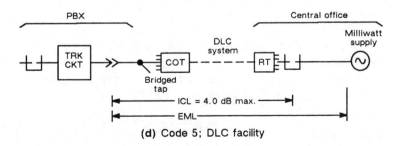

(d) Code 5; DLC facility

Figure 12-1. Two-wire PBX-CO trunk designs.

255

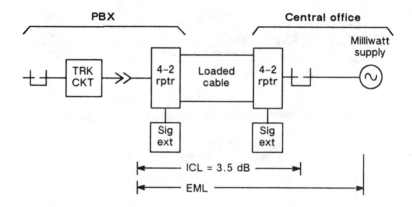

Figure 12-2. Four-wire PBX-CO trunk design, code 6.

gained through a central office other than the one that normally serves the customer's location.

Primary Functions

Signalling requirements for FX service are essentially the same as for ordinary subscriber lines. Sufficient current must be provided to operate supervisory equipment at the foreign central office when the station goes off-hook. To accomplish this, a means of repeating or reinserting the loop closure signal is required; it is provided by FX station (FXS) carrier channel units or signalling extenders. Transmission of undistorted dial pulse or dual-tone multifrequency (DTMF) signals from the station to the foreign central office is necessary and a means of repeating or regenerating dial pulses is required. The FX circuit must also be able to send ringing signals to the station set from the central office. Because FX circuits are usually carrier-derived and are relatively long, the circuit almost always reinserts ringing current. The circuit must remove the ringing current (ring-trip) when an incoming call is answered. The distant central office trips ringing when its loop-current sensor operates.

When a circuit, FX or local, terminates in a station set or forms a one-way outgoing PBX trunk, the subscriber line circuit at the dial-tone central office detects an off-hook condition by the operation of a current sensor (ferrod, resistor and

256

comparator, or relay) in series with the transmission pair. This is called *loop–start* operation. With loop–start, the only incoming–call indication received at the station is a ringing signal that has a two–seconds–on, four–seconds–off cycle. Consequently, there may be a delay of up to four seconds before the PBX receives a seizure indication. Since an outgoing seizure from the PBX end of the circuit could occur during the silent interval, loop–start operation would be unsatisfactory for two–way trunks serving a PBX.

Dual seizure (glare) can be virtually eliminated by a type of operation called *ground–start*. With this arrangement, battery is supplied to the ring side of the line through the central–office line circuit. A call is initiated from the PBX by grounding the ring lead to operate the central–office line circuit. When a dial–tone connection is established in the central office, ground is placed on the normally open tip lead. This causes the removal of the ground on the ring lead at the PBX; normal tip and ring connections are made at both ends of the circuit. Thus, with an outgoing call, the line is made busy as soon as the line circuit operates; a central–office seizure in the other direction is made impossible. On an incoming call, the central–office equipment places a ground on the normally open tip lead as soon as it seizes the line for ringing. This ground provides a busy indication at the PBX so that the circuit cannot be seized at that end. Thus, seizure is immediately indicated on both incoming and outgoing calls. In addition, ground–start operation provides central–office disconnect information to the PBX. This information is the basis for a *forward–disconnect* feature.

The forward–disconnect feature, available in many existing circuits, enables a PBX or ACD to recognize an abandoned incoming call and to release the connection. Without this option, an abandoned call would not be released and the trunk would continue to appear busy until an attendant answered. This feature is especially important for ACDs such as those used at airline reservations offices. If not provided, other callers would be prevented from using the trunk and waiting time would lengthen for the incoming call queue.

The PBX must be ready to process a new incoming call within 850 ms. This preserves traffic–handling capacity during periods of heavy calling.

One–way incoming trunks to a PBX for *direct inward dialing* (DID) involve special signalling conditions. In this operating mode, the PBX feeds battery and ground over the trunk to the serving central office. When the central office has an incoming call for the PBX, it seizes the DID trunk by closing the loop. The PBX attaches a digit register and, when ready to receive the number of the desired station, sends a signal to the central office. The indication is usually a wink–start pulse transmitted by briefly reversing the battery and ground on the trunk. (With older cross-bar central offices, the indication may be a delay–dial signal sent by reversing polarity until the PBX is ready to receive the digits.) When the attendant or called PBX station answers, the PBX passes the off–hook signal by holding the trunk polarity reversed for the duration of the call. This signalling requires signalling equipment and/or carrier channel units able to handle off–hook "winks" and steady reversals of battery. This same technique applies to one–way trunks to radio paging systems, in which the central office outpulses the directory number of a specific pager that is to be alerted.

The above discussion pertains to trunks provided on a voice–frequency (VF) basis. Where groups of trunks operate on a 24–channel DS1 basis, signalling and supervision are carried on the A and B bits of the DS1 frame.

Features and Options

Some special–services signalling functions are performed to serve the needs of transmission facilities or to provide optional features. These features may be illustrated by further discussion of FX service.

Signalling extenders may have to meet special needs. For example, it may be necessary to choose a unit capable of supplying an idle–circuit termination or disabling a repeater to prevent singing on an idle circuit.

With toll diversion, access to the toll network may be denied to certain PBX stations. When a call is placed to a destination outside the free–calling area, a battery–reversal signal is transmitted from some types of switch through the trunk back toward the PBX. The PBX detects this signal and, if the extension is denied toll access, diverts the connection to an attendant or a trunks–busy tone.

Signalling Systems

Generally, switched special–services circuits use circuit–associated signalling systems like those in the public message network. In special services, design precautions must be taken because equipment and facilities are used in a manner significantly different from public–network applications.

DC Signalling. The maximum distance over which dc loop signalling may be used is limited by the dial–pulsing range, the supervisory range, the ringing range, the ring–tripping range, or transmission considerations.

Maximum ranges have been determined for various types of signalling extender, carrier channel unit, etc. Ranges are stated in terms of circuit resistance external to the device and, where appropriate, are usually based on a minimum direct current of 23 mA supplied to the station. Figure 12–3 shows typical resistance limits for a loop–start signalling arrangement that permits the extension of the normal limit of central–office equipment on cable facilities. A current sensor in the signalling unit repeats the dial pulses toward the central office and provides a low–resistance battery feed toward the station. This circuit also reapplies 20–Hz ringing current toward the station. The maximum signalling ranges of this arrangement are shown in the figure. When VF repeater equipment is used, the resistance of the repeater slightly reduces the maximum range of the signalling circuit. No more than two signalling units may be used in tandem unless pulse correction is provided. With appropriate conversion equipment, other dc signalling arrangements [e.g., duplex (DX)] may be included.

Signalling on Digital Carrier. The available families of channel units for digital carrier, both interoffice and loop, provide a full set of signalling features for special services. They offer

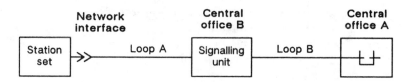

Signalling Unit Voltage	Loop A* Resistance (ohms)	Loop B Resistance (ohms)
48	0-1800	0-1600
72	1000-2900	0-1600

*The station set resistance is included in loop A.

Figure 12-3. Typical access-line resistance limits (loop-start).

loop-start and ground-start options, with forward-disconnect when needed, for the common switched services. They include DX and E and M signalling for applications like PBX tie trunks, 20-Hz ring-down signalling (manual, automatic, and code-selected) for private lines, and transmission-only models. These are applied in cases where no signalling, or single-frequency (SF) signalling, is needed. For back-to-back connections of carrier systems, DCSs or channel backs equipped with tandem channel units can be used.

AC Signalling. The dc circuit arrangements are limited to relatively short facilities because of signalling and transmission requirements, aside from the diminishing use of VF cable itself. When the circuit includes a channel on an analog carrier system, SF ac signalling arrangements are generally used. SF signalling circuits convert the loop signal to a 2600-Hz signal. This inband signal readily passes through the voice path, eliminating the need for signal converters at intermediate points when multiple carrier channels are used in tandem.

For FX circuits, the 2600-Hz signal is keyed at a 20-Hz rate to indicate ringing on a line. Since the SF tone is off in the talking condition, a band-elimination filter is not required and there is no impairment to transmission. The SF signal is used to transmit seizure and dial–pulsing information in the opposite direction in the usual tone–on–while–idle mode.

A number of SF signalling units are available for use in special–services circuits. In the current types, provision is made in one unit for battery feed, pads, four–wire terminating sets or four–wire cable extensions, and equalizers. Units with specialized functions are used at the station and central–office ends of each circuit. Different designs are used for two–wire and four–wire applications and for 600– and 900–ohm impedances.

PBX stations must be able to signal and supervise to the PBX. Normally, loop resistance from an on–premises station to the PBX is within the limits specified in the appropriate range chart and is usually on customer–provided cable in any event. Occasionally, an on–premises station is located at a distance from the PBX so that the resistance limit is exceeded; most off–premises PBX stations would exceed such limits without signalling extenders. The Federal Communications Commission (FCC) registration requirements in Part 68 of its rules [1] recognize three resistance ranges for off–premises ports on PBXs: 0–200, 0–800, and 0–1800 ohms. Thus, the range class of the port on the customer's PBX is an important input to the design process. In these cases, signalling extension may be needed to bring the station within range of the PBX.

At one time, there were numerous "cut–through" PBXs (e.g., step–by–step) where the extension drew its battery from the serving central office, and where the combined resistance of the extension loop and the PBX trunk had to be considered. The design interactions were complex and led to sizable use of signalling extenders. Both the obsolescence of this equipment and the requirements in the FCC registration rules make this consideration unnecessary.

Satellite and Nontandem PBX Tie–Trunk Signalling

A tie trunk can be arranged for manual selection at either end by attendants or for dial selection at either end from attendants or stations. Manual selection is made by operating a key on a console. With dial selection, the trunk is connected to the switching equipment of the PBX and is reached by dialing a trunk access code (e.g., 8). In all cases, one of several signalling arrangements may be provided.

Tie trunks can be described in terms of use and of method of completing incoming calls. Use is one–way only or two–way, at the choice of the customer; one–way trunks are designated incoming and outgoing at the appropriate ends. Call completion usually employs E and M signalling, which is inherently two–way, to dial the desired station or trunk group.

Other types of tie trunk (e.g., a one–way dial, one–way automatic tie trunk used between a dial PBX and a manual PBX) were common at one time. However, the registration rules focus on two–way E and M interfaces, which reflect today's usual equipment.

Both dc and SF signalling systems are used to convey information from one PBX to the other. The dc method is normally used in the trunk circuit and interfaces readily with digital carrier; where necessary, it may be converted to or from SF in a connecting circuit.

12–3 VOICE TRANSMISSION CONSIDERATIONS

In switched special services, satisfactory transmission of speech signals is maintained by observing design and operating transmission objectives. The types of circuit to which these objectives are applied include FX and WATS. The objectives are also applied to PBX–CO trunks, tie trunks, and PBX/centrex station lines. Other circuits (not discussed in detail) that are covered by similar objectives include secretarial–service lines, off–premises extension lines, 911 access lines [2], and FGA lines for local access and transport area (LATA) access.

Loss

Switched special–services circuits are designed to meet loss objectives originally based on the VNL plan. When loss objectives are met, volume, noise, stability, and echo performance are satisfactory on the majority of connections.

When short–haul circuits may become links in long built–up connections involving VNL design, they must be designed to have a loss of at least VNL + 2 dB; however, to reduce design effort and to ensure echo and stability margins, a minimum loss of 3.5 dB has been adopted as the objective for short–haul circuits. Where facility equipment is present for other reasons, it provides loss adjustment without further cost.

Loss Calculations and Measurements. The losses of switched special–services circuits are expressed in terms similar to those used in the public switched network. In special–services applications, the terminology is applied to customer lines as well as to the various types of trunk. The terms include inserted connection loss (ICL), expected measured loss (EML), and actual measured loss (AML).

Inserted Connection Loss. As with message trunks, the ICL for trunks is defined as the 1–kHz loss between originating and terminating switch appearances. For customer lines, it is the 1–kHz loss between the line side of the switch and the network interface to the station set. Included are the losses resulting from connections between different impedances, e.g., between a 600–ohm PBX and a 900–ohm local office.

Expected Measured Loss. To assure that measured loss values agree with design values, the EML is computed as the 1000–Hz loss between two readily accessible points having specified impedances. It includes the ICL plus test access loss: switching circuit losses, test–pad loss, switchable–pad losses, if any, and test-equipment connection losses. Thus, it is important to specify the originating and terminating switches properly since different specifications may result in different losses.

Actual Measured Loss. The AML is the measured 1–kHz loss between the same two access points as those for which the EML is computed. Test sets and remote–test systems should have impedances equal to the nominal impedances of the switching machines or other access points at which they are located. For special–services lines, the meter or oscillator impedance at the station end should be 600 ohms resistive. Because of the high impedance of an on–hook station set, it need not be physically disconnected from the line when making the measurement.

All routine loss measurements should fall within the maintenance limits established for the EML. When measurements fall outside the maintenance limits but do not exceed the immediate action limits, maintenance action is indicated. If the AML falls beyond the immediate action limit, corrective action must be taken to clear the trouble condition as soon as possible.

Objectives. Design objectives for class 1 and class 2 special–services circuits are given in Table 12–2. Values of EML must be derived from the ICL by adding the losses incurred in the test equipment connections.

Table 12–2. ICL Objectives for Special Services

Circuit Type	ICL Objective (dB)		
	Short–Haul		Long–Haul
	With Gain	Without Gain	
Lines			
FX	3.5	0–5.0	VNL + 4.0
WATS	3.5	0–4.0	VNL + 4.0
Off–premises extension	3.5	0–6.0	VNL + 4.0
Secretarial	3.5	0–6.0	–
On–premises PBX station*	–	0–4.0	–
Off–premises PBX station	4.0	0–4.5	VNL + 4.0
Trunks			
PBX–CO	3.5	0–4.0	–
ACD–CO	3.0	0–3.5	–
FX	3.5	0–4.0	VNL + 4.0
ACD–FX	3.0	0–3.5	–
WATS	3.5	0–4.0	VNL + 4.0
PBX tie †	VNL	VNL	VNL

* For circuits provided by an exchange carrier.
† Assumes customer's PBX includes 2–dB switchable pads or equivalent.

The loss design of PBX trunks assumes that the tie trunks can be switched to other tie trunks or can be used in universal–service connections. While the overall grade of service in a network of customer–provided PBXs is the customer's responsibility, the FCC rules [3] and EIA standards [4] limit the amount of gain that the PBX may supply between tie–trunk ports to a maximum of 1.5 dB.

Grade–of–Service Considerations. To illustrate the importance of meeting ICL objectives, the grade of service for the connection of Figure 12–4 may be compared for two assumed values of ICL on the FX line. In the first case, the FX line is taken as having an ICL of 5 dB, which just meets the maximum objective. In the second case, the FX line is assumed to have an ICL of 8 dB. Assumed noise and loss values for various metropolitan network connections in a VNL (worst–case) situation without digital switching are given in Table 12–3. The connections include up to three links comprising two tandem trunks and one intertandem trunk. The percentages for good–or–better (GoB) and poor–or–worse (PoW) grade–of–service ratings were determined for left–to–right transmission by the use of a software program. The grade of service for transmission in the opposite direction might be slightly different due to differences in station-set efficiencies and noise effects.

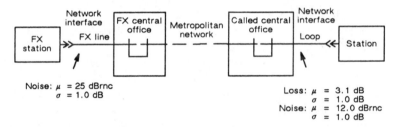

Noise: μ = 25 dBrnc
σ = 1.0 dB

Loss: μ = 3.1 dB
σ = 1.0 dB
Noise: μ = 12.0 dBrnc
σ = 1.0 dB

Figure 12–4. Special–services connection.

Table 12–3. Loss, Noise, and Grade of Service

FX Line ICL (dB)	Network Links	Loss μ (dB)	Loss σ (dB)	Noise μ (dBrnc)	Noise σ (dB)	Grade of Svc. GoB %	Grade of Svc. PoW %
μ = 5.0	Intrabldg.	8.1	1.0	22.5	3.3	84.7	3.7
σ = 0.1	One	12.4	1.4	20.5	3.9	90.3	3.5
	Two	13.7	1.4	20.0	3.5	88.3	4.3
	Three	15.1	1.8	20.6	2.6	84.8	5.8
μ = 8.0	Intrabldg.	11.1	1.0	22.3	3.2	91.4	3.1
σ = 0.1	One	15.4	1.4	20.4	3.6	83.5	6.4
	Two	16.7	1.4	20.4	3.4	80.0	8.1
	Three	18.2	1.7	20.5	2.7	74.3	10.9

Table 12-3 shows that the grade of service actually improves slightly on intrabuilding connections when the FX line loss is 8 dB. This improvement (1.7-percent increase in GoB and 0.6-percent decrease in PoW ratings) is due to fewer observations of "too loud" on these short connections. All other configurations show significant deterioration in the grade of service, due primarily to the added overall loss. The differences in noise for the two cases are only a few tenths of a decibel.

Similar effects occur when several special-services circuits are used in a built-up connection. If ICL objectives for the special-services circuits are exceeded, the grade of service for such built-up connections deteriorates rapidly with relatively small increases in ICL.

Return Loss

All special-services circuits that use gain devices must have adequate margins to avoid singing or near-singing. These circuits must be stable while in the idle state as well as in the talking condition. Idle-state stability is obtained by the transmission-cut feature of carrier channel units and SF signalling units (which opens the transmission path when the circuit is on-hook), by limiting repeater gains, by use of repeater disablers, or by applying idle-circuit terminations. In the idle condition, only enough singing margin is required to satisfy changes of cable attenuation due to seasonal temperature variations.

Some difficulty may be encountered in meeting both gain and idle-circuit singing-margin requirements in a circuit that does not have repeater control but that uses a signalling extender that repeats an open-circuit condition when idle. The extender then presents a 0-dB return loss. The allowable gain of the repeater is severely limited if the repeater is not located properly. Figure 12-1 shows desirable relative locations of signalling extenders and repeaters. However, with any design, a check of both signalling and stability must be made. Extreme cases may require a carrier interoffice facility or use of loop carrier to avoid the need for the signalling extender.

For an established connection, a computed singing margin of 10 dB is a reasonable minimum to allow for expected differences between the computed and actual results, to accommodate variations from assumed line conditions, and to avoid near–singing.

Noise

Circuit–order requirements, maintenance limits, and immediate action limits for noise are given in Table 12–4 for circuits that have one or more links of VF or carrier trunk facilities. For wire loop facilities, the circuit–order requirement and maintenance limit is 20 dBrnc and the immediate action limit is 36 dBrnc. The noise limits apply at the network interface at the station for WATS, off–premises extension, secretarial, and on–premises and off–premises PBX station lines. The limits apply at the PBX or ACD interface for PBX–CO, WATS, and ACD–CO trunks. Circuit–order requirements specify the maximum acceptable noise when the circuit is placed in service. Maintenance limits have the same values as circuit–order requirements. They specify the maximum acceptable noise when measured routinely or in response to a trouble report.

Table 12–4. Station Noise Limits for Switched Special–Services Circuits

Facility Length (Miles)	Maximum Noise (dBrnc0)				
	Immediate Action Limits	Maintenance Limits			
		VF Cable	Broadband Carrier	N–Type Carrier	Digital Carrier
0–50	40	25 dBrnc	31	26	28
51–100	40	–	33	28	28
101–200	40	–	35	30	28
201–400	42	–	37	–	28
401–1000	44	–	40	–	28
Loop	36 dBrnc	20 dBrnc	–	–	28

A circuit with measured noise less than the circuit–order or maintenance limit does not require maintenance; where noise exceeds the limit, the circuit may be placed in service or allowed to remain in service only after remedial action is taken. Under no circumstance should a circuit whose noise exceeds the immediate action limit be allowed to remain in service. These noise limits allow for the use of analog facilities, or combinations of analog and digital. Where the transmission system used is purely digital, these limits are normally met with substantial margin.

Telephone Set Current

Transmission objectives, expressed in terms of 1–kHz losses, are based on an optimum station set current of approximately 50 mA. Currents smaller than this provide less output from the transmitter while larger currents reduce the efficiency of the receiver. The typical telephone set [5] automatically adjusts the efficiency of the set according to the amount of current flowing in the loop. The output power of the tone generator in a DTMF telephone set is an inverse function of the loop current; i.e., the minimum output occurs with maximum loop current.

The battery supply may be located at any of several points in the circuit depending on the type and location of the equipment used. When transmitter battery is supplied from the normal serving central office or a DLC remote terminal, there is generally no problem in maintaining satisfactory loop current. However, the location of signalling–extender equipment must be considered from a loop–current standpoint as well as from supervision aspects. The location of the signalling unit and the battery voltage (48 V or 72 V) are chosen so that the current fed to the station set is in the range of 36 to 65 mA. In no case should it be less than about 23 mA.

Bridge Lifters

Where a special service is provided by bridging one cable pair onto another, the transmission performance may be seriously degraded by the effect of the parallel impedance. The degradation is avoided by using bridge lifters at the bridging point. The two special services that make heaviest use of bridge lifters are secretarial–service lines and off–premises extension lines.

A bridge lifter is used on the main station line to improve transmission on the special–services line when the sum of the lengths of the main station line, any bridged taps on it, and the special–services line exceeds 6000 feet of nonloaded pairs or when either line is loaded. A bridge lifter is used on the special-services line under the same conditions. However, the typical carrier channel unit presents an open–circuit condition toward the office when idle, and thus does not require a bridge lifter.

12-4 DATA TRANSMISSION CONSIDERATIONS

The VF facilities of the switched public network are used to transmit a variety of data and digital–facsimile signals. VF data services are classified according to the signalling rate of the transmitted data signals. The classifications and rates are: type I, signals transmitted at rates below 300 bits per second (b/s); type II, signals transmitted at rates of 300 to 2400 b/s; and type III, signals transmitted at rates above 2400 b/s.

Type I operation has become rare with the general move toward higher speeds. Most modern facsimile machines fit type III. The terminal equipment and FCC–registered data sets are usually provided by the customer.

Transmission Objectives

Facilities used for data transmission are generally ordinary residence or business loops, with digital performance unspecified because the exchange carrier has no control over the performance of the data set (modem). However, a few exchange carriers have an explicit tariff offering of a loop that is specially conditioned to handle high–speed data. Table 12–5(a) covers the typical requirements for these loops. These requirements include intraLATA WATS lines handled from the customer's normal central office, as most WATS lines are served today. Requirements for WATS access lines for interLATA use are given in Chapter 18.

Transmission objectives for FX and remotely served WATS data lines for type II or III use are summarized in Table 12–5(b). For type I application, only the 1000–Hz loss limits apply. The distortion objectives of a tandem–connecting trunk are allocated to WATS lines on the premise that a central office adjacent (or digitally connected) to the tandem office can be chosen as a

Table 12-5. Data Transmission Limits

(a) Local Loop Limits

Parameter	Ordinary Loop	Specially Conditioned Loop
Insertion loss (1000 Hz)	9.0 dB	
EDD (1000 to 2800 Hz)	Not specified	100 μs
Max. signal power at office MDF	-12.0 dBm	
Impulse noise threshold for 15 counts in 15 min. (dBrnc0)	Not specified	On cable: 59 On DLC: 65*
Message circuit noise	20 dBrnc (voice limit)	
Slope (1000 to 2800 Hz)	Not specified	-1 to +3.0 dB

*With -16 dBm0 holding tone.

(b) FX and Remotely Served WATS Lines-For Data Use

Parameter	WATS Line	FX Line
Insertion loss (1000 Hz)	9.0 dB	8.5 dB
Slope (1000 to 2800 Hz)	-1 to +4 dB	
Intermodulation distortion: Second order Third order	48 dB* 51 dB*	
EDD: Band (Hz) Limit (μs)	900-2500 400	800-2600 1250
Impulse noise threshold for 15 counts in 15 min. (dBrnc0): Cable, any length Digital or N-carrier Broadband carrier, 0 to 125 mi VF cable, any length	53 65† 57 53	

* On present-vintage digital carrier; 48/49 dB on broadband analog carrier. Other limits apply to less common carrier facilities.
† With -16 dBm0 holding tone.

serving office and the distortions of a connecting trunk can then be ignored because they are insignificant. The objectives would apply to the line from the data station to the serving office. However, this is true only where the relationship described between the serving office and the tandem switching center exists. If the serving office must be at a distance from the tandem switch and trunked on analog facilities, the distortions of the connecting trunk must be included in the objectives for WATS lines given in

Table 12-5(b). In such cases, the impairment of facilities encountered in tandem-connecting trunks must be added to the loop impairments to verify that the connection from the station to the tandem office is within limits.

The performance of FX lines is approximated by including the distortions of a tandem-connecting trunk and one or two intertandem trunks with those of the local loop. By subtracting the results from the long-haul contribution, the remaining distortion (which is approximately equivalent to that of a short toll circuit) is such that the overall transmission objectives for data service can be met on calls terminating within a radius of 200 miles of the FX serving office, assuming analog facilities with their dependence on distance. Performance is not specified beyond this distance.

General Applications. Attenuation/frequency distortion (slope) is the decibel difference in circuit attenuation at two specified frequencies. In data circuit design, the slope is measured between 1000 and 2800 Hz. Typical high-speed data sets are designed to compensate for slope by means of built-in equalization. The AML at 1000 and 2800 Hz should be recorded for each loop on which data conditioning is ordered, and for all FX and remotely served WATS lines regardless of the data set used. The AML should be within 1.0 dB of the EML at 1000 Hz and within 2.0 dB of the EML at 2800 Hz. The 3-dB slope objective applies regardless of the difference between the EML and the AML. Attenuation/frequency distortion objectives are given in Table 12-5(a) and (b). For type 1 data on WATS and FX lines, the slope limits are each 2 dB greater than the type II/type III limits in Table 12-5(b).

Envelope delay distortion can cause serious impairment of data signals. An EDD objective is not specified for type I data loops, but specially conditioned data loops must meet the objective given in Table 12-5(a). Objectives for FX and remotely served WATS lines are given in Table 12-5(b). Data sets are designed to tolerate some EDD. In addition, many sets have built-in compromise or adaptive equalizers that compensate for EDD. Since the loop may consist of loaded or nonloaded cable, or of carrier facilities, the amount of EDD varies. If the EDD exceeds the objectives on conditioned circuits, equalization is required.

Data signals are especially susceptible to impulse noise, particularly at the data station where received data signals are at their lowest levels. Thus, impulse noise measurements are made at the station; this procedure ensures that impulse noise from all sources is included in the measurement. Large step–by–step switching systems with rotary out–trunk switches, where they still exist, may not be acceptable because of excessive impulse noise. Fortunately, the same building usually contains a newer type of switch to which the data line can be assigned. Within an electro-mechanical central office, each path through the switching machine may exhibit a different impulse count at the specified threshold. If the contribution of the loop facilities is constant, the variation in counts registered during 15–minute measurement intervals depends on the intraoffice path of the connection. If the specified objective is barely met, it may be expected that the limit is being exceeded in a large percentage of the calls through the office. In marginal cases, it is recommended that four 5–minute measurements be made. If three of the four measurements register five counts or less at the specified threshold, the circuit can be accepted. By contrast, analog electronic and digital offices are relatively quiet; at least one digital switch is manufacturer–specified as meeting ten impulse counts in 30 minutes at the low threshold of 42 dBrnc.

Typical data modems are designed to tolerate echoes that are at least 12 dB below the minimum received signal power. Since echo requirements can usually be met without special loop treatment, no specific return loss measurements are required on data loops. If trouble is indicated by a consistently high error rate, measurements are necessary. Such troubles can usually be attributed to impedance irregularities in trunk circuits, poorly balanced terminating sets, improperly adjusted loop repeaters, or poor impedance presented by the second modem.

Design Considerations

Special–services loops and lines that are intended to transmit data signals must be designed according to criteria that are somewhat different from those applied to ordinary telephone circuits. The related parameters of loss, signal power, and transmission power, along with TLPs, must be considered.

Loops. A data loop consists of all facilities and line equipment between the connecting jack at the customer's premises and the main distributing frame at the serving central office. The loop is usually composed of cable pairs but often contains a DLC–derived feeder section. In any case, the loop loss should not exceed 9.0 dB.

The average signal power measured at the central office must not exceed −12 dBm in order to avoid overloading analog carrier facilities. The TLP at the outgoing side of an analog tandem office is −2 dB. Since the loss of an analog tandem–connecting trunk is approximately 3 dB, the serving office can be thought of as a +1 dB TLP with respect to carrier facilities at the tandem office. Therefore, the data signal power is −13 dBm0 on the carrier channel. The FCC registration rules reflect the need for this restriction. While the Commission's limits are based on what is now old technology, the good noise performance of digital facilities and modern switches implies little need for higher data levels.

Cable Facility Treatment

To meet data transmission requirements, it is sometimes necessary to improve loop characteristics. Normally, nonloaded loops up to 9 kilofeet (kft) in length meet requirements without additional treatment. However, nonloaded loops longer than 9 kft tend to have excessive slope, which must be corrected. Loaded loops with end sections longer than 9 kft also have excessive slope; those having more than three loading coils have excessive EDD for conditioned loops.

The slope characteristic of nonloaded cable pair loops can be improved by the use of a repeater equipped with the appropriate networks, as illustrated in Figure 12–5. The curves were derived from measurements of 26–gauge cable pairs terminated in 900 ohms. One curve shows the slope characteristic of the cable alone, while the other two curves show the slope improvement resulting from the use of a repeater. A repeater set for the appropriate slope equalization can be used to correct the slope of non-loaded loops of any length that meets the revised resistance loop design plan.

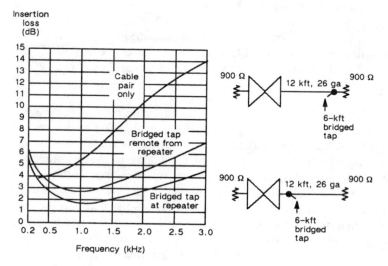

Figure 12-5. Slope   improvement   on   26-gauge nonloaded cable.

If the slope and envelope delay requirements cannot be met by the above methods, a trouble condition requiring corrective action may be indicated. Loading coils may be incorrectly spaced or there may be an excessive number of load points (more than three). The end section of a loaded line may be excessive (more than 9 kft long). A transfer to DLC facilities may be effective and easy to arrange. Finally, there may be one or more bridged taps that can be removed to improve the loop characteristics. On very long loaded loops, delay equalization applied at one end is generally necessary.

PBX Considerations

Because of the presence of an added coding–decoding step or higher impulse noise, error performance in data circuits that have dialed access to the switched message network through a PBX may be poorer than that on direct loops to the central office.

Where a choice exists, a direct loop is desirable, especially for high–speed services. By contrast, a modem or facsimile machine connected to a centrex line is likely to have the same service quality as any station, centrex or otherwise, served from the same central office.

Arrangements that permit alternate use of the switched message network with a special–services circuit (such as a PBX tie trunk) require special consideration. The data signal power requirements must be met for satisfactory performance on the switched message network but, unless compensated for, the data signal power may be too high on the special–services circuit. The difference in power occurs because the PBX switch is considered to be a +4 dB TLP for local access but is considered to be a 0 TLP for FX trunks. The 4–dB difference can be compensated for by a variety of techniques depending on economics and local design.

Data use to off–premises stations is discouraged. Satisfactory error performance often occurs but cannot be assured because data–loop design applies to conditioned trunks between the PBX and the serving central office but not to the facilities and circuits serving the off–premises station. Operation at moderate speeds is usually satisfactory.

References

1. Federal Communications Commission. *Rules and Regulations, Title 47, Code of Federal Regulations, Part 68* (Washington, DC: U.S. Government Printing Office, Oct. 1987), Figure 68.3.

2. Dimone, V. P., D. Elsinger, and V. Perez. "911: A Good Thing Just Gets Better," Bellcore EXCHANGE, Vol. 3, Iss. 4 (July/Aug. 1987), pp. 2–7.

3. Federal Communications Commission. *Rules and Regulations, Title 47, Code of Federal Regulations, Part 68* (Washington, DC: U.S. Government Printing Office, Oct. 1987), Section 68.308(b)(5)(i).

4. *EIA/TIA Standard EIA–464–A*, "Private Branch Exchange (PBX) Switching Equipment for Voiceband Applications" (Washington, DC: Electronic Industries Association, 1989).

5. *EIA Standard EIA–470*, "Telephone Instruments with Loop Signalling for Voiceband Applications" (Washington, DC: Electronic Industries Association, Iss. 1, Jan. 1981).

Chapter 13

Centrex

Centrex service offers users the advanced features of modern private branch exchange (PBX) systems, but without the need for the customer to supply secure air–conditioned building space or to finance a major equipment purchase. Centrex uses switching systems that are often shared with the message network, in the form of either regular central office (CO) equipment or remote switching units placed near the customer site. The flexible nature of centrex means that, unlike the case with PBXs, there is little danger of a specific customer outgrowing the capacity of the switch. Moves of users between buildings are often easier with centrex than with a PBX; off–premises station lines are easier to administer. Centrex offers least–cost routing, giving preprogrammed advancement of a call from a low–cost path (e.g., a tie trunk) to progressively more costly routes [e.g., foreign exchange (FX) trunks, then wide area telecommunications service (WATS), then the toll services of one carrier or another] as multicall traffic occupies the less expensive routes. The preferential selection may depend on time of day. Centrex also offers access to modem pools for data use, to voice–mail systems, and to electronic directory services. A current estimate [1] is that about six million telephones in the U.S., or about ten percent of all business phones, are served by centrex.

Originally positioned solely for major users, centrex is a successful offering for users with ten or fewer lines. It thus addresses the markets for service to dormitories, hospitals, and businesses of most sizes [2]. It even reaches into the price and feature range of larger electronic key systems.

The 2B+D basic access lines used in integrated services digital network (ISDN) technology are a natural offering in the centrex family: they afford each user location a voice and a duplex

64–kb/s data line on one pair of wires without additional or special data cabling.

13-1 CENTREX FEATURES AND ARRANGEMENTS

Each centrex installation must meet a variety of service demands. These demands are satisfied by flexible service offerings derived from software features in the serving switch [3]. Many optional features are available. The original centrex offerings provided defined packages of capabilities. However, sizable "unbundling" of features has taken place in the interests of matching the needs of specific customers. A modern switch allows features to be applied to individual lines rather than a "centrex common block."

Service Features

Each package of centrex services includes the basic features of a PBX. The attendant position for centrex is a console or group of consoles where incoming calls to the listed directory number or calls requiring assistance are answered and completed.

The following are illustrative of basic centrex features. *Direct outward dialing* (DOD) offers direct access to the network without the attendant, generally on a dial–9 basis. A *station hunting* feature directs calls to a prearranged alternate station when the called line is busy. *Station restriction* denies the ability of specific stations to place outgoing calls and certain trunk calls without assistance from the attendant. *Call transfer* enables the called party, while connected to the incoming line, to transfer the call to another station within the system. *Night service* directs calls to a station in the absence of the attendant. Centrex offers *direct inward dialing* (DID) to permit calls from the message network to reach the called station without attendant assistance. *Automatically identified outward dialing* (AIOD) identifies the calling station on outgoing toll calls for billing purposes. Enhanced *station message detail recording* (SMDR) furnishes real–time call billing details from the CO to the customer's premises over a data link for immediate allocation of communications costs to user departments.

Centrex can include numerous additional features. *Add—on* enables a station user to add another station to an existing incoming call, thereby establishing a three—party conference. *Consultation hold* allows a station user to hold an incoming call and originate, on the same line, a call to another station. After consultation, the user may add the third party to the original call or return alone to the original call. The *trunk answer any station* feature permits any station user, by dialing a special code, to answer incoming listed directory—number calls when the attendant position is on night service. These features require only simple non—key telephone sets, although key systems may be used for additional features if desired. Selected lines in a centrex may be equipped for public switched digital service at 56 kb/s.

Additional and specialized optional features are available for centrex service. FX service, WATS, direct access to interexchange carriers' (ICs') networks, tie trunks, access to a private virtual network (PVN), and switching as part of an electronic tandem network are all available.

Some optional features are offered only by the versatile electronic switching systems that are the usual servers for centrex. *Speed calling* allows the station user to originate a call by dialing an abbreviated code. *Call forwarding* enables the station user to have all calls rerouted automatically to an alternate station. *Call forwarding—busy line* permits all incoming calls to a busy station to be routed automatically to the attendant. *Call forwarding—don't answer* reroutes all calls to a station that doesn't answer within a prescribed time automatically to the attendant, a central message center, or a voice—mail system.

Other optional features provide interface facilities for customer equipment. *Paging* allows attendants and station users to connect to and page over customer—provided sound systems by dialing a special access code. *Recorded telephone dictation* permits access to and control of dictation recorders. The dictating equipment may be controlled by either voice or dial. *Code call* permits attendants and station users to activate signalling equipment by dialing a special code. The called party can then be connected to the calling party by dialing another special code. The centrex switch may provide automatic call distributor features as well as basic attendant service.

With message–desk service, a semipersonalized capability is available. Unanswered calls are forwarded to the message desk center, with the called station identified and certain other call–history information directed to the attendant. A remote message–waiting indication can be activated at the customer's premises. Conversely, the centrex switch can provide access to an electronic mailbox system.

Because the centrex switch is CO equipment, it has the usual protected battery supply, so centrex service continues during power failures. The customer need not be concerned with the expensive details of supplying backup power on ordinary lines. (ISDN station sets do require individual backup power supplies, just as with an ISDN PBX.) Where the attendant console on the premises does not have uninterruptible power, the centrex can direct incoming main–number calls to a selected group of stations.

The basic nature of centrex provides for administration of the switch and the surrounding network by the exchange carrier. The customer is freed from having to provide testing or maintenance; the switch uses duplicated processors, and is monitored from a switching control center that provides 24–hour coverage. The exchange carrier handles updating of routing tables in the software translations for the switch to match newly opened numbering plan area (NPA) and end–office (NXX) codes, requiring no attention by the customer.

At the same time, the customer may wish to administer routine rearrangements of the system. A software offering called the CCRS® system [4], and others like it, give this control to the user. The customer can directly manipulate an electronic centrex system, avoiding the need for service orders or action by the exchange carrier, from a dial–up terminal on–premises. The software updates the switch translations on either a priority or an overnight basis. The customer can change telephone numbers to accommodate internal moves, add or delete stations, activate or deactivate custom–calling features, update directory information, and make on–line queries. This feature addresses the needs of

---

CCRS is a registered service mark of Bellcore.

major customers, who typically have a large ongoing volume of station rearrangements and feature changes.

Provision of centrex service to multiple locations with a single directory number can be handled several ways. For limited numbers of remote locations, the stations can be extended in the same manner as off–premises extensions, via integrated digital loop carrier (IDLC) or other facilities. For larger networks, remote switch units hosted from one switch can be applied. Multiple locations served by different switches can share a central attendant location by use of attendant trunks and data links. For truly transparent area–wide centrex, it is becoming feasible to use several switches, tying them together via common–channel signalling (CCS). With digital switching to avoid varying trunk losses, it then becomes practical to share a single NXX code among thousands—groups of numbers in widely separated switches. Outside callers become unaware that multiple locations are involved.

ISDN–based centrex networks combine voice and data functions. Data on one of the B channels from a station can be circuit–switched to a distant terminal on a 64–kb/s basis, or can be packet–switched into a CO–based local area network (LAN), a tie into a distant X.25–based packet network, or a modem pool affording access to data via the analog public network. These approaches eliminate intrabuilding coaxial cables or other special premises wiring. By reducing the need for LAN controllers and intersite data links, they appeal to multilocation customers. The ISDN centrex can use a digital switch, a remote switching module (RSM) homed onto a distant digital machine, or an ISDN adjunct controlled by an electronic analog switch.

Equipment Arrangements

Each centrex station is served by a loop to the CO and by its own line appearance. A switching machine may provide only centrex service, for one or many customers, or may provide both centrex and ordinary telephone service. The switching machine normally is treated as an end office in the message network. Where a portion of an analog centrex machine switches tandem or intertandem tie trunks, terminal or through balancing is required to limit echoes on long connections.

Since the inception of centrex service, many improvements in technique and capability have been incorporated into switching hardware and software. As a result, there are several vintages of equipment and generic programs. Thus, it is necessary to verify that the features under consideration for a given application can be provided by the available equipment, especially for an advanced application like ISDN.

13-2 CENTREX TRANSMISSION CONSIDERATIONS

Most centrex customers are heavy toll users. The centrex is usually part of a switched services or electronic tandem network. In addition, centrex stations normally use special features such as conferencing and add–on. To provide a satisfactory grade of service, transmission losses must be maintained near objective values.

Station Lines

Centrex station lines are similar to ordinary loops. However, the maximum 1000–Hz loss of a centrex station line is limited to 5.0 dB, well below the maximum for an ordinary loop. This limit recognizes the need to maintain a good transmission grade of service between users on the same premises, who naturally have a large community of interest and who make heavy use of the previously mentioned service features. The resistance limit depends on the switch, and is typically 1300 to 1900 ohms in addition to the assumed resistance of the customer's house cable and station sets. For ISDN applications, the normal digital subscriber line loop limits for basic–rate access apply, e.g., nonloaded cable up to 18 kft or specially equipped digital loop carrier.

Most centrex installations involve the central office that normally serves the customer's location. However, some larger customers have multiple locations to be served by the centrex switch, with some of the sites located outside the wire–center area. It is often necessary to use remote–exchange techniques to bring small groups of remote–served station lines in through interoffice facilities to the switch. In these cases, IDLC techniques are potentially useful, both economically and in terms of meeting the station–loop loss requirements.

In pre–ISDN CO–LAN applications, station lines may carry data at 4.8 to 19.2 kb/s in addition to the ordinary speech signal. A small data loop carrier terminal at or near each voice/data user's location, coupled with a matching terminal in the CO, derives the data channel. A single nonloaded cable pair connects the two. For ISDN centrex, the pair of B channels on a basic–rate access line provides this voice/data function.

Where an RSM is used to provide centrex service, the RSM is generally close to the customer's main location. In these cases, the directly served station loops are quite short and present no transmission problem, but station lines from the RSM to distant or off–premises sites still need transmission–and–signalling treatment to meet limits. Groups of off–premises lines can be handled by remoting IDLC terminals from the RSM, as if it were a full–sized CO switch.

Customers use centrex lines heavily for data and digital–facsimile traffic. Such lines appear to the switch as any line served out of the same office, ordinary or centrex. There is thus no transmission advantage in using ordinary business lines outside of the centrex group. By contrast, including them in the centrex group allows giving them any of the operating features of centrex, including customer–controlled rearrangement.

Attendant Facilities

Attendant service with consoles is provided on a released–link basis. This means that calls are switched to the console for attendant assistance and automatically released from the console when assistance is no longer required. The signals and controls on the console are such that the attendant can either monitor the associated connection or split it and talk to either party privately as if the circuit were looped through the console for direct control of its continuity. A three–way bridge in the switch is added to or dropped from the connection under control of a data link (or ISDN D channel) from the console. The loss through the bridge, analog or digital, is usually three decibels, a satisfactory value to ensure stability and good echo control without undue attenuation. This is superior to the early switchboard arrangements, which required three loop facilities in tandem to gain outside access—from calling station to CO, from CO to switchboard, then from switchboard back to CO.

The console attendant completes calls requiring assistance (for example, dial–0, non–DID, or transfer) by keying through the centrex machine. When the called party answers, the attendant normally disconnects, leaving the through connection unbridged by the console circuit. However, the attendant has two other options. First, the call can be monitored to see that it is properly completed and then released. Second, the call may be held after dialing the connection, thus permitting the attendant to handle other calls and still monitor the held call at intervals. The attendant facilities are bridged onto the through connection only while monitoring. Digital switches typically use a digital conference circuit and a simple two–wire loop to the attendant for ordinary service, or a basic–rate access line for ISDN applications. Both two– and four–wire consoles are available for use with analog electronic switches. The four–wire consoles provide better transmission performance and are normally used for centrex operation in tie–trunk networks or centralized attendant operations. Two–wire consoles may otherwise be used but care must be exercised to control the impedance of the operator loop at the switch to allow use of a negative–impedance converter in the three–way conference bridge that provides operator access to customer connections.

When the console circuit is bridged to the through connection in an analog centrex, volume and return loss are affected only negligibly. The attendant loop circuit and negative–impedance converter, shown in Figure 13–1, permit bridging on a high–impedance basis and provide some gain in the transmission path to the console. The balancing network and termination are adjustable to provide adequate return loss at the hybrid, including the case where split access is used for console connections to either direction of the through path.

The proper attendant transmitting and receiving volumes are obtained by use of amplifiers whose gains are set to provide average transmitting volume at the centrex switch equivalent to that which would be received from a 500–type telephone set at the same location as the console. Average received volume at the console is maintained at preferred values regardless of loop loss.

The gain settings are also based on considerations of sidetone at the console. The sidetone depends on the transhybrid loss between four–wire legs of the hybrid, the amplifier gains, and the resulting attendant–trunk loss. The preferred value of sidetone, 12 dB below operator voice [5], is normally achieved under working conditions. When the circuit is idle, sidetone level is controlled in the console circuit by an idle–circuit termination.

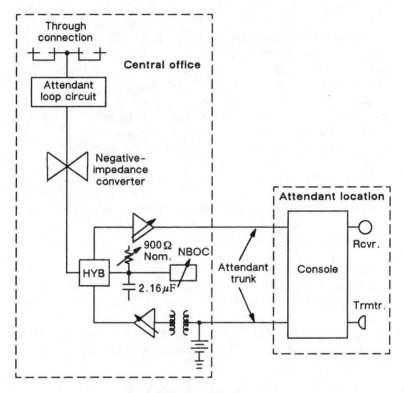

Figure 13–1. Console bridging to through connections.

A console may be equipped with an amplifier to increase the transmitted speech volume. The transmitting and receiving efficiencies of the two–wire console vary with loop resistance in a manner similar to those of a 500–type telephone set, and are comparable on short loops. For a zero–length loop, the console with collocated 400–ohm battery supply is approximately 1 dB less efficient in the transmitting direction than the 500–type set with battery supplied from the central office; the efficiencies are about the same in the receiving direction. For a loop resistance of 1000 ohms, the typical console is approximately 2.5 dB more efficient in the transmitting direction and about 2 dB less efficient in the receiving direction.

Tie Trunks

Tie trunks may be provided between centrex installations or between a centrex and a PBX. The transmission objectives for centrex tie trunks are the same as those for PBX tie trunks. The inserted connection loss (ICL) objectives apply between the termination of the tie trunk in the centrex switching machine and the interface to the distant PBX. Digital switches offer the simplicity of four–wire switching; analog units usually provide two–wire operation and require more transmission consideration.

The pertinent loss and balance objectives, derived from the via net loss (VNL) design plan, are often met for tie trunks through analog switches by the use of switched 2–dB pads. Switched–pad operation can be provided at centrex installations for calls handled on a directly dialed basis or by a console attendant.

Pad Control in Digital Switches. Digital centrex switches provide zero–loss trunking and through switching, with the end–to–end loss controlled by a variable decode level at the receiving switch. A translation table invokes the desired loss in either an analog or a digital pad, e.g., 2.0 dB from an off–premises station to a satellite tie trunk or 3.0 dB from an on–premises station to a CO trunk.

Pad Control in the Attendant Trunk Circuit. The trunk circuit used with an attendant console has two line–link appearances in addition to a trunk–link appearance in a crossbar machine. Each line–link appearance has a different class–of–service mark. On dial–0 calls into the attendant trunk circuit, the class of service of the calling party determines which of the line–link appearances is used to extend the call.

If the transmission loss of the path to the attendant trunk circuit is VNL + 2 dB, the call is extended on the attendant trunk line–link appearance that is not equipped with a 2–dB pad. If the loss of the transmission path on the originating end of the attendant trunk is VNL, the call is extended on the appearance that has a 2–dB pad, resulting in an overall loss of VNL + 4 dB.

Trunk Circuit with Pad Control. The operation of switched pads on trunk circuits is controlled by the type of circuit (centrex station line or tie trunk) connected to the originating and terminating ends of the trunk. The type of circuit is known from the class–of–service indications at the originating and terminating ends of the trunk; the switched pad is controlled from these indications. For example, if a connection is being made to a centrex station from a tie trunk that is designed to an ICL of VNL + 2 dB, the trunk circuit switches in the 2–dB pad to achieve an overall loss of VNL + 4 dB. The pad switching for various connections is shown in Table 13–1. Note that the loss between PBX interfaces and/or centrex locations is VNL + 4 dB for all connections except station–to–station intracentrex.

Table 13–1. Pad Switching on Trunk Circuits

Call	Designed Loss of Connected Circuit		2–dB Pad
	Originating End	Terminating End	
Station–station	0–5 dB	0–5 dB	Out
Station–tie trunk	0–5 dB	VNL + 2	In
Tie trunk–station	VNL + 2	0–5 dB	In
Tie trunk–tie trunk	VNL + 2	VNL + 2	Out

Tie trunks are normally two–way. They may be individual circuits, using E and M signalling, or may be on DS1 high–capacity facilities between digital switches, using the A and B signalling bits in the DS1 frame. Addressing information can use dial–pulse or multifrequency signalling or, if connecting to a PBX, may be on a dual–tone multifrequency basis. CCS is available for area-wide centrex, multiswitch ISDN centrex, or PVNs.

FX and WATS Trunks

Foreign exchange trunks for centrex customers terminate on the centrex machine. On the usual dialed basis, a centrex station may reach an FX trunk by dialing a three–digit code or, for restricted stations, by attendant dialing.

Access to WATS is provided the same way. The centrex station line then connects to an outgoing trunk between the centrex machine and the local access and transport area (LATA) tandem or an IC switch.

Conferencing

Two conferencing arrangements are available for use with centrex. One permits a station user to dial–originate conference connections up to six stations. The other provides for conferences to be established by a console attendant to a maximum of 30 stations.

Add–on service, a form of conferencing, is provided in centrex as an extension of the station transfer feature. The bridging loss associated with this arrangement is largely overcome by the use of a digital bridge or a three– or four–port analog bridge circuit with gain.

Nongain conference bridges are not used with centrex in analog offices because losses would be excessive. In order to accommodate conference connections of multiple stations, gain–type conference bridges must be used.

References

1. Abrahams, J. R. "Centrex versus PBX: The Battle for Features and Functionality," *Telecommunications* (Mar. 1989), pp. 27–32.

2. Valenta, R. A. "Centrex Makes a Comeback," *Telephony* (Dec. 26, 1988), pp. 30–31.

3. *5ESS Switch—The Premier Solution—Centrex Applications,* Customer Information Release 235–300–030, AT&T Technologies, Inc. (Iss. 1, Dec. 1986).

4. Pasternak, E. J. and S. A. Schulman. "Centrex Customer Rearrangement System: Taking the Pain Out of Change," Bellcore EXCHANGE, Vol. 2, Iss. 3 (May/June 1986), pp. 28–32.

5. "Common Capabilities," *OSSGR, Operator Services Systems Generic Requirements*, Technical Reference TR–TSY–000271, Bellcore (Iss. 2, Dec. 1986 with Rev. 3, Mar. 1988), Section 21.4.3.5.

Chapter 14

Private Switched Networks

The general features and service capabilities of private switched networks are described briefly in Chapter 11. Those descriptions are expanded in this chapter. Network arrangements and transmission designs are also presented in greater detail.

Switched special–services circuits are often configured into networks interconnecting switches at different locations. Stations served by two centrex switches may call each other via the public network. Alternatively, tie trunks between the two centrexes may provide a convenient and economical way to handle the same traffic. As customer requirements and switch capabilities have grown, private networks of tie trunks have evolved into electronic tandem networks. In these networks, switching is done at a private branch exchange (PBX) at the customer premises or by a centrex machine in a central office (CO).

Another private–network arrangement of switched special services involves switching machines located on carrier premises. This arrangement, known as a switched services network (SSN), may be organized as a two– or three–level hierarchy or, in special cases, as a polygrid in which all switching machines have equal class status. Customers may build their own networks with SSN– type switches on their premises.

As the availability of DS1 digital service has grown from a relative handful of circuits in the mid–1970s to a commonplace offering, it has become widespread practice to complement digital switches with digital trunking and packaging of groups of off– premises lines into DS1s. That trend is continuing into DS3 services with their 672–channel capacity, as voice trunking, data lines, and video transmission share a common high–capacity facility.

The expanding data–base and signalling capabilities of the public switched networks have also made it possible to build private networks on a "virtual" basis, with the actual traffic being handled via public message trunking.

14-1 ELECTRONIC TANDEM NETWORKS

Tie–trunk networks can be very small, involving only a few centrex switches or PBXs connected by direct trunks, or can be highly complex, involving a large number of customer locations. As the number of switches increases, the cost of renting private lines for trunking encourages the use of intermediate switching and tandem operation, which permit two tie trunks to be switched together into a through connection.

The tandem network is often an assembly, not a single–source service offering. The tie trunks connecting two switches are furnished by appropriate exchange or interexchange carriers (ICs). Tie–trunk terminal equipment at a centrex, like the centrex itself, is furnished under intrastate tariffs. PBXs are normally customer–premises equipment. Switches may be provided by a mix of exchange carriers and ICs. Thus, a number of suppliers are often involved in the total network.

Network Layout

Figure 14–1 is a network layout for an electronic tandem network (ETN) showing the interconnection of multiple locations. The locations are served by centrex switches, PBXs, and "smart" (PBX–like) key telephone systems of various types, interconnected by tie trunks. The serving switch at each location is classified according to the functions performed, i.e., main, tandem, and satellite. For a network as large and complex as that shown, satisfactory operation can be achieved only if the centrexes and PBXs provide routing intelligence to permit automatic interconnection of tie trunks [1].

Switch Classifications. A PBX or centrex (PBX/CTX) having its own listed number and an attendant console is commonly referred to as a main PBX/CTX. It can connect stations to the public network for both incoming and outgoing calls. A PBX/CTX served through the main switch having the same listed

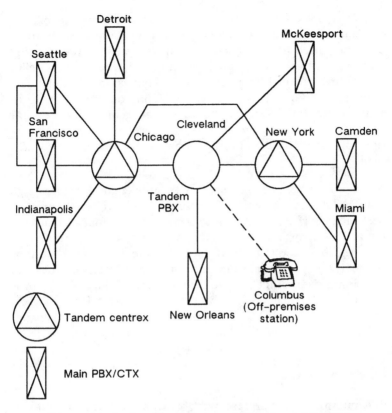

Figure 14-1. A typical electronic tandem network.

number but with no attendant is a satellite PBX. All incoming calls are routed from the main switch over satellite tie trunks. The attendant position serving the main switch cares for the satellite by means of data links. A satellite is usually located in the same exchange area as its main PBX.

A main PBX/CTX is arranged to connect stations or attendants to the ETN. Where the PBX/CTX interconnects tie trunks, as well as connecting tie trunks to station lines or attendant lines, it becomes a tandem PBX/CTX.

Trunk Classifications. Trunks that connect a main or satellite PBX/CTX to a tandem PBX/CTX are called tandem tie trunks; those that join two tandem PBXs/CTXs are intertandem tie

293

trunks. Trunks between two PBXs/CTXs without the capability of tandem operation are called nontandem tie trunks. Such trunks are often used as an economical arrangement between points having a high volume of traffic. Tie trunks that connect a satellite switch to its main PBX/CTX may function as nontandem, tandem, or intertandem tie trunks, depending on the switching arrangement. With customer network–management features, the role of a given trunk in network routing may be changed with time of day or with congestion controls in response to traffic load.

For a network provided entirely by an exchange carrier [e.g., intraLATA (local access and transport area)], voice transmission performance is assured on up to a maximum of four tie trunks in tandem, with digital switching providing additional performance margin. Connection of a tie trunk with the public network is done routinely, but transmission performance for either voice or voiceband data is not assured. Connections from a PBX to the public network may be via PBX–CO, wide area telecommunications service (WATS), customer–provided, or foreign exchange (FX) trunks.

Service Features

A number of features are available with ETNs. Station–to–station calling between locations uses tie–trunk switching on a dialed basis. Sequential advancement of a call between machines is under control of translations in the switch software. A fixed number of address digits (a uniform, closed numbering plan) is used regardless of the location called. Only one connection with the public network (universal service) is preferred; however, off–network connections at both ends can take place.

Network operating features such as code conversion, digit addition or deletion, and automatic alternate routing are available. Billing data for allotting usage costs to individual stations are included. Route advancement under heavy traffic is a normal feature: a given call may be offered to a "free" tie–trunk group, an FX or WATS line, or a public–network connection in increasing order of cost. The recording of traffic data, provision of service observing, and standard tones and announcements are other features. Supervision is not always received from the public network on connections between a tie trunk and a local or FX central office or between a tie trunk and a WATS trunk. An attendant may have to monitor and release connections of this type, or supervision–simulation devices like timers or speech–energy detectors may be used. Substantial network management features are popular to allow the customer's personnel the ability to test, busy–out, and restore defective circuits and to change traffic routings in response to network loads.

Transmission Design

The provision of good transmission performance in an ETN requires sufficiently low loss to provide satisfactorily high received volume; sufficiently high loss to ensure adequate performance with talker echo, noise, and singing; and minimum contrast in received volumes on different calls.

Echo. Echo control in a digital ETN relies on a fixed value of end–to–end loss to meet loss, echo, and stability requirements. Echo becomes a controlling design limitation with round–trip delays above 6 milliseconds (ms). Analog trunks in an ETN use the via net loss (VNL) concept to provide loss as a function of echo delay in the facilities.

For carrier facilities, the delays of the carrier terminals must be taken into account. The lines in Figure 14–2 include the round–trip echo delays of the carrier terminals and line facilities but do not reflect intermediate delay–inducing devices like digital cross–connect systems (DCSs), which produce delays roughly comparable to pairs of channel banks. An estimate of round–trip delay for two carrier links in tandem may be obtained by adding the appropriate values from the lines.

Where exceptionally long delays are encountered, as with communication satellite facilities, echo cancellers are usually necessary.

Loss. Tie trunks must be designed to have certain minimum losses to control echo. The VNL concept was traditionally used for design, but in an all–digital ETN it becomes preferable to use

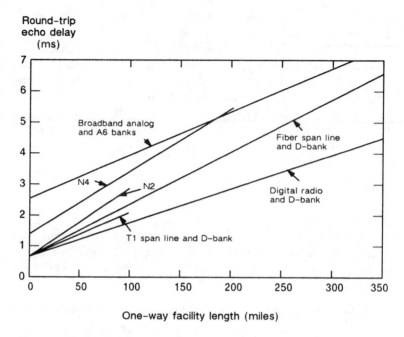

Round-trip
echo delay
(ms)

One-way facility length (miles)

Figure 14-2. Approximate round-trip echo delays of carrier
facilities.

a fixed-loss plan comparable to the 6-dB region digital switched
network (RDSN) concept discussed in Chapter 2.

If a tie trunk can be switched to other tie trunks on a two-wire
basis or to PBX-CO trunks, it is often equipped at one or both
ends with 2-dB switchable pads in an analog switch, or with selectable digital gain/loss in a digital switch. These may be switched
into the tie trunk to protect against echo for terminating connections and removed for lower loss on through connections and on
certain universal-service connections.

The primary loss objective for analog tie trunks of all types is
VNL + 2S + 2S dB, where 2S denotes the 2-dB loss of a switchable pad. For customer-premises PBXs, this equates to VNL dB
from one network interface to the other. However, the variety of
connection types in which these tie trunks may be used has led to
a number of alternative objectives that depend on the usage of
the trunk and whether it is short- or long-haul. For example,

296

tandem tie trunks may be designed to a loss of VNL + 2S + 2 dB; i.e., the switched pad is used only at the tandem PBX. The pad is controlled by the switch processor and is called in when the connection is to a station line or to a short–haul two–wire tie trunk with less than 2–dB inserted connection loss (ICL). Since an intertandem tie trunk has switched 2–dB pads at both ends, the switching control rules for tandem tie trunks are applied at each end. The application of pads to tandem and intertandem trunks is illustrated in Figure 14–3. The pads must be switched out (yielding lower trunk loss) on any through connection to another tie trunk in the network. Balance objectives are not specified when the pad is switched in. The pad may be analog or digital; however, the use of digital pads in tandem is undesirable because

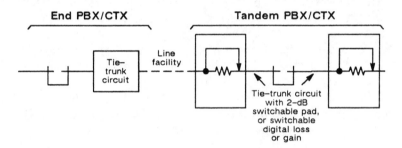

(a) Tandem tie trunk

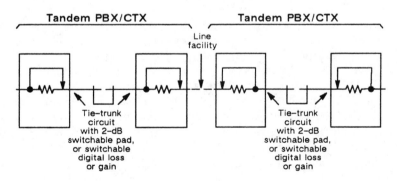

(b) Intertandem tie trunk

Figure 14–3. Switchable pad arrangement on tandem and intertandem tie trunks.

there is some cumulative buildup of distortion. Balance objectives for these connections must be met so that the margin against singing and echo is sufficient.

For PBXs, the Federal Communications Commission (FCC) registration rules [2] specify the maximum allowable net gain between switch ports [tie–trunk, off–premises station (OPS), or public–network] to prevent facility overload. Similar specifications are available in the Electronic Industries Association (EIA) PBX equipment standard [3].

For digital PBXs, an involved loss plan is specified in the EIA equipment standard. A detailed set of port–to–port losses is laid out, as shown in Figure 14–4. It gives the loss expected in each direction for connections between stations [on–premises (ONS) or OPS] and tie trunks [digital (D/TT), satellite digital (STT), and analog (A/TT)]. Entries in the figure are included for connections to a local central office [digital (D/CO) or analog (A/CO)] or to a "toll" office [digital (D/TO) or analog (A/TO)]. The D/TO configuration is useful for direct–inward–dialing (DID) trunks in 24–channel groups from the trunk side of a digital end office, with the PBX providing the decoding loss. The plan assumes that analog tie trunks, analog–terminated at the PBX, will have VNL dB of loss and that digital trunks, digitally terminated, will give zero loss; the switch then supplies loss or gain. An implementation guide [4] provides assistance in applying the loss plan. The overall focus is on evolution to a 6–dB fixed–loss concept while meeting the port–to–port gain limits in the FCC rules. Digital centrex switches are capable of the same loss values, but usually cannot (and do not need to) insert gain.

Balance. Tandem tie–trunk networks are designed for an overall ICL of VNL + 4 dB in an analog context, or similar values in a digital environment. The design objectives for tandem and intertandem tie trunks assume that the trunks involved in a tandem connection meet balance objectives. Through–balance objectives must be met at two–wire analog tandem centrexes where tie trunks operating at VNL are switched together. Therefore, balance tests are necessary as part of the lineup and acceptance tests for these trunks operated at VNL. Digital centrexes are, of course, inherently four–wire and involve no issue of through balance.

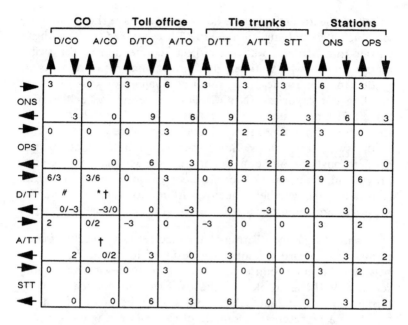

		CO		Toll office		Tie trunks			Stations	
		D/CO	A/CO	D/TO	A/TO	D/TT	A/TT	STT	ONS	OPS
ONS	▶	3	0	3	6	3	3	3	6	3
	◀	3	0	9	6	9	3	3	6	3
OPS	▶	0	0	0	3	0	2	2	3	0
	◀	0	0	6	3	6	2	2	3	0
D/TT	▶	6/3 #	3/6 *†	0	3	0	3	6	9	6
	◀	0/–3	–3/0	0	–3	0	–3	0	3	0
A/TT	▶	2	0/2 †	–3	0	–3	0	0	3	2
	◀	2	0/2	3	0	3	0	0	3	2
STT	▶	0	0	0	3	0	0	0	3	2
	◀	0	0	6	3	6	0	0	3	2

* The –3/3 dB pair should be provided for connections between an A/CO port and a D/TT port interfacing a combination tie trunk to a satellite PBX.

† The low-loss option (0/0 or –3/3 dB) is desirable when the loss of the PBX–CO trunk is 2 dB or more, the ERL is at least 18 dB median and 13 dB worst-case, and the SRL is at least 10 dB median and 6 dB worst-case, measured into a 900 + 2.16 μF termination at the CO.

The 0/6 dB pair is required; the –3/3 dB pair is a desirable option for internetwork applications where no significant configuration will encounter echo, instability, or overload because of the reduced loss. With the –3/3 dB pair, station end-to-end DTMF signals through the DCO into the private network may be mutilated because of the 3-dB gain.

Figure 14–4. EIA digital PBX loss plan.

At one time, PBXs were engineered to match a nominal impedance of 900 ohms. Later, it was found that a 500–type telephone set on a short loop presented an impedance closer to 600 ohms. Consequently, current designs use a nominal impedance of 600 ohms. Centrex offices use the normal central–office value of 900 ohms.

Through Balance. The most critical balance requirements at two–wire analog PBXs/CTXs are those for through or intertandem tie–trunk connection via electromechanical switches. Electronic switches, being smaller, have shorter and less variable office cabling, particularly 1A centrexes that use digital carrier trunk frames with their direct connection to the switch matrix. In two–wire switches, through–balance measurements are made to determine the required value of the network building–out capacitor (NBOC) and to determine echo return loss (ERL) and singing return loss (SRL). The NBOC in the four–wire terminating set is selected to balance the capacitance of the switch and wiring. Since there are numerous connection paths through the office, a compromise capacitance is selected to provide adequate balance for any connection. The ERL and SRL tests determine whether objectives are met or whether corrective measures are required.

Terminal Balance. Terminal–balance tests, applicable mainly to analog two–wire switches, are designed to check the balance between the compromise network in the four–wire terminating set of a VNL tie trunk and the two–wire impedance of a test termination. Terminal–balance tests may be made from the switch to a representative sample of ONSs covering the range of station line lengths. However, connections of other types, especially those in which the 2–dB pad may be switched out of the tie trunk, should be tested individually. These include non–VNL tie trunks, PBX–CO trunks, FX trunks, and OPSs. These tests detect irregularities that result in inadequate balance and inferior transmission on built–up connections.

Where through–balance tests are required, they should be completed before any terminal–balance tests are attempted. The NBOC values determined from the through–balance tests are to be used in the networks of all four–wire terminating sets.

In a digital centrex switch, the degree of terminal balance is largely preset by the choice of balance networks (loaded versus nonloaded) to match the station lines. The balance usually meets ETN requirements without further treatment.

Facilities

Any of the carrier or voice–frequency facilities that are used for the public network may also be used for tie trunks. Centrex–

to—centrex tie trunks usually involve either DS1 facilities between the switches themselves, or carrier terminals located in the same office. Large PBXs may also involve direct termination of DS1s, or may have carrier terminals (channel banks or digital loop carrier remote terminals) located on or near the premises. Minimum—sized trunk groups entering into smaller PBXs may use voice—frequency facilities as extensions of carrier. The design objectives covered in this chapter apply to the entire trunk including the carrier channel, any end links, and any terminal or intermediate equipment.

14-2 HIERARCHICAL SWITCHED SERVICES NETWORK

Where communication needs are large, it may be economically advantageous to use a private switched network similar in design and operation to the public switched network. These switched services networks use combinations of two— and four—wire tandem switches, PBXs, and centrexes. SSNs were historically called common—control switching arrangements (CCSAs). All—digital SSNs are commercially available; they interface subtending PBX/CTX switches directly on a DS1 basis, relieving many of the transmission issues discussed below.

Network Plan

The basic three—level hierarchy plan is shown in Figure 11–5 of Chapter 11. While it is similar to that of the public network, there are some differences due to the customer service requirements; for example, economics may limit the number of direct (high—usage) trunks between a particular pair of switches. Thus, on the average, more trunks are connected in tandem for a given connection than would be necessary if direct trunks were provided as in the public network.

The network plan permits a three—level hierarchy, with the switching offices designated classes SS–1, SS–2, and SS–3. All switching at class SS–1 offices is four—wire. At class SS–2 offices, switching may be four—wire (using digital switches, analog electronic systems with the HILO four—wire feature, or a four—wire crossbar system) or two—wire. At SS–3 offices, switching may be either two–wire or four–wire, with four–wire becoming common as digital switches become more widespread. Class SS–1 switching offices are justified only in the very largest networks. Smaller networks may have only one class SS–2 office or none at all.

Dedicated access lines connect each customer location and the serving switching center; there are dedicated network trunks between switching centers. A given switch is usually shared by a number of independent SSNs. An SSN switch can serve as a centrex and can also be a central office in the public network.

As with electronic tandem networks, automatic alternate routing is usually provided in an SSN. The originating switching machine routes all calls over available direct trunks. When the direct–route trunks are busy, additional calls are routed to first–alternate–route trunks. If both the direct and first–alternate–route trunks are busy, the originating switching center directs all additional calls to a second alternate route, if available. Overflow traffic can also be directed to FX, WATS, and public network trunks, generally in that order. In addition to network management by the carrier, the customer can be provided with a network management center that gives real–time information on traffic loads and permits application of overload controls. Automatic detection of trunks that show continuously busy or never busy is available. Such trunks can be remotely removed from service pending repair. Mechanized routine transmission testing of trunks, automatic transmission testing via centralized automatic reporting on trunks (CAROT), is included. Optional "meet–me" conferencing is offered, along with automatic identification of the calling station, identification codes, and a make–busy arrangement for switch ports.

Service Features

SSNs provide the features available in centrex services and the public switched network. In addition, a number of unique service features are offered.

The numbering plan provides a specific seven–digit address for each network station, NXX–XXXX, where N can be any digit 2 through 9 and X can be 0 through 9. The arbitrarily assigned

NXX portion of the address identifies the customer location where the station is homed but cannot be the same as the NXX digits assigned to the same switch for public network use. The XXXX digits are the numbers of the individual station at the customer location and are generally the same for both the SSN and the public network.

A network station served by a PBX/CTX, or a station directly terminated on a network switching machine, can be called by dialing its seven–digit address. When a station is served by an attendant for calls in and out of the network, the seven–digit address is assigned to the attendant access line instead of the station. As with centrex service, individual or attendant–assisted transfer of inward calls is a common feature. Any nonrestricted PBX/CTX station or any station served directly by an SSN switching machine can dial a network call.

All calls in electronic offices can be recorded by automatic message accounting equipment. The data are available to the exchange carrier to aid in the engineering and administration of the network and to the customer to use in allocating costs among user departments. (In crossbar offices, the recording is limited to a sample of up to 20 percent.)

One–way or two–way access to the public network (universal service) may be provided by off–network access lines (ONALs) to the same or other central offices. Connection via IC access, WATS trunks, etc. may also be provided. Calls to an SSN station originating off the network are screened by an attendant or automatic identity–code screening system and, if accepted, completed to the network station. Transmission performance may not be satisfactory when calls originate in the public network, connect to and traverse the private network, and reenter the public network.

Switching

PBX and centrex sites reach the network switching machine(s) via access lines. A tributary PBX is attended, homes on a main PBX/CTX for SSN access, and has PBX–CO trunks for access to the public network. A satellite PBX is unattended and may home on either a main or a tributary PBX/CTX. All incoming calls from the public network to the satellite PBX route through the PBX/CTX on which it homes. A satellite PBX homing on a tributary analog PBX/CTX tends to degrade transmission performance via an increase in loss and delay, and is not recommended: the satellite should have direct tie trunks to its main PBX/CTX for SSN service.

Transmission Performance Requirements

Transmission considerations in an SSN are similar to those of the switched public network. Figure 14–5 illustrates a fully developed SSN showing the types of office, the design and maximum losses for various types of trunk and line, and the application of echo control. Where the PBX/CTX units shown in the figure are digital, the switched pads illustrated are replaced by selectable losses per the PBX loss plan covered earlier. Loss, noise, and echo are considered in the layout of the network and the design of trunks, access lines, and station lines. The design of the transmission paths was originally similar to the VNL design used for the switched public network. As switches in the network are replaced with digital units, it becomes feasible to convert to a fixed–loss plan in which trunks operate at zero loss and loss is introduced only at the final decoding point. This is particularly true in those SSN offerings where all trunking and switching are guaranteed to be digital.

The basic transmission plan does not require or include envelope delay equalization. Attenuation/frequency requirements are similar to those in the switched public network and are of no concern with digital facilities between digital switches. However, when voiceband data requirements are more stringent than can be met by the basic plan, special conditioning in the form of equalization (delay, attenuation, or both) can be provided on analog access lines or trunks. In particular, C3 conditioning is designed specifically to handle these SSN access line and trunk requirements. To control intermodulation distortion, D–type conditioning may be applied. Phase jitter is controlled, on analog network trunks and access lines intended for data use, to a typical limit of 8 degrees peak–to–peak, 4 to 300 Hz.

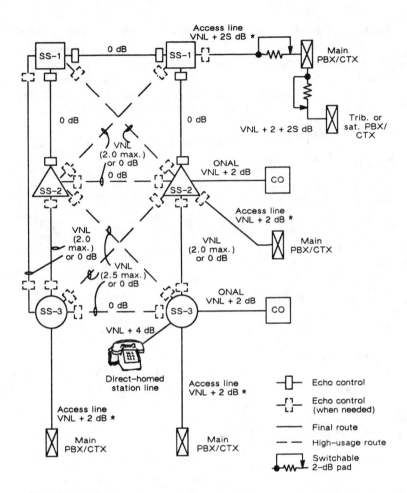

Figure 14-5. Transmission plan for an SSN: pre-fixed-loss.

Loss. The original design loss objective was developed by considering the loss between PBXs/CTXs as comparable to that between end offices in the public network, VNL + 4 dB. Fixed-loss transmission plans deliver comparable performance.

Balance. To meet ERL and SRL objectives at a two-wire (analog) SS-3 office, it is necessary to perform balance procedures similar to those required in the public network.

305

Through-balance requirements must be met on network-trunk-to-access-line and access-line-to-access-line connections. Where network trunks or access lines are connected to directly homed two-wire station lines, terminal-balance requirements apply. While the average ERL and SRL values are the same as those specified for the public network, minimum allowable values are somewhat more stringent. Terminal-balance requirements at a main PBX/CTX are similar to those for a two-wire tandem office in the public network.

When VNL-designed access lines are connected to station lines at a PBX/CTX, 2-dB switched pads or digital loss are required in the access lines at the main PBX/CTX to improve the return loss. The loss is switched out under conditions similar to those pertaining to ETNs. If terminal-balance objectives are to be met with the pad switched out on connections between network trunks or access lines and PBX tie trunks, the tie trunks must be adjusted to meet ERL and SRL requirements similar to those of tandem-connecting trunks. If the return loss were deficient, it would be necessary to switch in the 2-dB pad.

Noise. Message circuit noise requirements for trunks and access lines are stated on the basis of mileage. In general, if all trunks and access lines meet requirements, an overall connection can also be expected to meet requirements.

Objectives for impulse noise provide signal-to-impulse-noise ratios that permit data transmission with an error performance comparable to that achieved in private-line service. Because of the random nature of impulse-noise exposure on switched connections, control is provided by establishing conservative standards on individual circuits.

Frequency Response. To meet end-to-end attenuation/frequency distortion requirements, not more than seven analog trunks should be connected in tandem via analog switches for voice transmission within the network. The trunking plan should be arranged so that no more than 11 circuits may be analog-connected in tandem on any connection. If the requirements for individual links are not met, excessive distortion may be encountered on maximum-link connections. Naturally, digital trunks between digital switches are transparent with regard to attenuation and envelope delay distortion, and digitally derived trunks between analog switches have much less envelope delay distortion than trunks on broadband analog facilities.

Trunk Design

Among the factors that must be considered when designing a network trunk are the type and class of office at each end of the trunk, the type of facility, and the possible need for echo cancellers. Network trunks between analog switches are designed for 0 dB or VNL dB depending on the place of the trunk in the network involved (e.g., SS–1 to SS–1 versus SS–3 to SS–3) and whether echo cancellers are provided. All network trunks should be designed from the outgoing switch of the originating office to the outgoing switch of the terminating office. The transmission level point (TLP) at an outgoing analog switch is –2 dB as in the public network under VNL design, or –3 dB in a digital switch not operating in a fixed–loss plan. If the trunk extends between class SS–1 and SS–2 offices, the loss at 1004 Hz should be 0 dB. Trunks between other offices are designed for VNL dB. To keep the overall loss between stations within limits, the loss of the network trunks should not exceed certain values. The 1004–Hz losses for trunks in the final route between class SS–3 and SS–2 offices may not exceed 2.0 dB. High–usage trunk losses between class SS–3 and SS–2 or between class SS–3 offices may not exceed 2.5 dB. If VNL design exceeds the maximum, echo cancellers should be provided and the losses reduced to 0 dB.

Facilities. Carrier facilities are used universally for network trunks. To avoid any need for equalization, trunks should use only one link of carrier, should employ carrier tied back–to–back through a DCS, or preferably should terminate directly on a DS1 basis on digital switches. The present (μ –law) designs of digital carrier systems are widely used for network trunks.

Echo Cancellers. Network trunks require echo cancellers in accordance with rules that apply to all networks in which a two–wire station can be connected at either or both ends of a connection. Cancellers are required on all trunks between SS–1 and SS–2 offices. They should have enough time–delay capability to handle echoes from the remotest part of the network on their drop sides, or from the remotest public switched network point that is likely to occur. Echo suppressors are also found in the older SSNs; the associated switch activates disabler leads to prevent double echo suppression when two suppressor–equipped trunks are connected in tandem.

Access Lines

The access lines that interconnect a main PBX/CTX with the serving switching center are analogous to tandem–connecting trunks when interconnecting PBX station lines with SSN trunks, and comparable to tandem trunks when interconnecting PBX/CTX tie trunks and SSN trunks. Access lines are of four–wire design and operate at VNL with switched 2–dB pads (or equivalent) located at the PBX/CTX. The pads are bypassed when access lines are connected to tie trunks. Where there are no tie trunks terminated at a main PBX/CTX and where pad switching is not needed for universal–service calls, an access line may be operated at VNL with the 2–dB pad inserted at all times or may be operated at VNL + 2 dB without the pad.

The maximum loss (exclusive of switched pad) should not exceed 2.5 dB. This loss corresponds to approximately 1650 miles of analog broadband carrier facilities, or 1050 miles of fiber. Where the type of facility and the length of the circuit would result in a VNL of more than 2.5 dB, efforts should be made to rehome the access line so that the loss requirement can be met.

All SSN trunks, access lines, and non–centrex station lines terminate in a nominal impedance of 900 ohms at a two–wire class SS–3 office. At four–wire analog switching centers, the nominal impedances are 600 ohms; at CTXs, 900 ohms.

A directly homed station line served by a class SS–3 office is a form of access line since connections can be made to any part of the SSN. This type of line, terminating on a four–wire switch, is designed to a loss of 6 dB. A line terminating on a two–wire switch is nondesigned (local–loop loss) if it is within the normal wire–center area of the switching office; otherwise, it is designed to VNL + 4 dB.

PBX/CTX Tie Trunks

In a switched services network, tributary and satellite PBX tie trunks and OPS lines must be considered as integral links in the network. It is often more difficult to provide good transmission on these tie trunks and station lines than on the access lines and network trunks.

Ideally, all PBXs and CTXs would have access lines. Each PBX/CTX could then be treated as equivalent to an end office in the public network with access lines designed for VNL + 2 dB. However, in practice, access lines may terminate at a main PBX/CTX; other PBXs/CTXs become tributaries or satellites and home on the main PBX/CTX as shown in Figure 14–5.

Off–premises station lines from tributary and satellite PBXs/CTXs should home onto the main PBX/CTX wherever practical.

The designed loss of tie trunks is VNL + 2 dB where traffic terminates solely at the centrex or PBX; otherwise, it is VNL + 2 dB with a 2–dB switched pad or equivalent. This establishes losses of VNL + 2 dB for connections to stations at the main PBX and VNL dB for through connections to the network.

Station Lines

Most station lines are two–wire connections to an SSN through a PBX or centrex. However, four–wire lines, called subscriber lines, are provided to connect four–wire station equipment to a four–wire switch. A subscriber line is considered to extend from the outgoing switch of the serving four–wire switching office (−2 dB TLP) to the network interface at the customer premises. It provides improved performance for voice or data services by reducing the number of trunks in tandem and by taking advantage of the duplex transmission inherent in four–wire operation. The station may be a telephone or a data set; they may be used together for alternate voice–data use. The customer–premises equipment may also transfer the subscriber line between a station set and a PBX. This feature, called dual use, is provided by operating a transfer key at the customer premises.

Universal Service

The SSN and switched public network may be interconnected at a main PBX/CTX or at an SSN office. Universal–service calls may originate in either network.

Satisfactory performance on universal–service calls depends on certain limitations. The transmission objectives established for SSNs minimize the transmission penalty on universal–service calls but cannot guarantee satisfactory transmission on all calls. The degree of satisfaction on universal–service calls is difficult to predict because the transmission quality depends on the relative locations of the stations in the two networks and the routing, often multivendor, used to make the interconnection. It is preferable that universal–service calls be limited to the serving area of the central office associated with the main PBX or ONAL. Although the losses encountered on toll connections may be no higher than those on local connections, the effect of echo may be an added consideration. A reasonable quality of voice transmission can be expected only if the interconnection is restricted to one point on any given call. High–speed data services should preferably terminate in the same network as that in which they originate.

Where a tributary PBX/CTX is in a toll–rate area different from that of the main PBX/CTX, it may be economically desirable to complete universal–service calls through the tributary PBX/CTX. However, the concentration of universal–service traffic at the main PBX/CTX may justify an ONAL directly to the distant exchange. This eliminates the loss of the tie trunk between the main PBX/CTX and the tributary PBX/CTX.

Machine–switched universal–service connections should be made at SSN offices, rather than via PBX–CO trunks, because this type of connection results in as much as 7 dB more loss than a connection over an ONAL from the SSN office. SSN trunks are designed for VNL; therefore, ONALs designed for VNL + 2 dB should be provided to a local central office and the pads in the access lines should be switched out. These facilities must meet terminal–balance requirements at the SSN office. Calls to network stations from the public network are routed from the local central office to a main or tributary PBX over ONALs. The

310

PBX operator then completes the calls to the network stations unless an automated system is used for checking of account codes, etc.

14-3 POLYGRID SWITCHED SERVICES NETWORK

Certain SSNs have unusual requirements for reliability and survivability. An example is the automatic voice network (AUTOVON), a worldwide communication system used by the United States Department of Defense, which is evolving into a defense switched network (DSN). This network uses a unique configuration termed polygrid. It is described here to show a unique departure from conventional network design.

A polygrid network employs a continuum of grids of interconnected switching centers, all of equal rank. The switching centers are interconnected in such a way that a large percentage of them would have to be rendered inoperative before network service would be disrupted. Thus, the network is highly survivable.

Network Plan

The polygrid network actually consists of two superimposed structures. The basic network shown in Figure 14-6 furnishes short- and medium-haul capability. The basic unit of this structure is the home grid, which is a set of switching centers surrounding and directly connected to a destination center as shown in Figure 14-7. Most switching centers have direct trunk groups available from several adjacent centers. Home grids are functionally discrete even though they may overlap and share a number of switching centers to allow traffic routing over many transmission paths. Home-grid arrangements are similar for most switching centers, except that those located at the periphery of the network (overseas sites) have truncated patterns.

To minimize the number of tandem links on long connections, a long-haul network is superimposed on the basic network structure as shown in Figure 14-8. To reach the destination center, the call is advanced via an exterior routing plan, so-called because it is exterior to the home grids. Ten possible exterior trunk

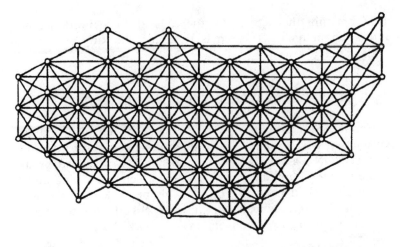

Figure 14-6. Configuration of a basic polygrid network.

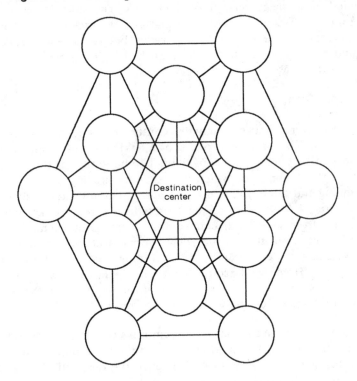

Figure 14-7. Home grid.

routes can be programmed to a destination switching center from an originating center. One is a single trunk group connected directly to the destination center. In addition to the direct route, there are nine alternate routes in sets of three, called triples. The first is the most–direct triple; it normally represents a set of three forward routes. The second group is the best–alternate triple, while the third is the second–best–alternate triple. Both the best– and second–best–alternate triples normally represent lateral routes.

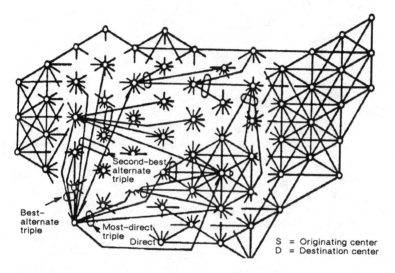

S = Originating center
D = Destination center

Figure 14–8. Long–haul routing.

An example of a set of ten exterior routes is shown in Figure 14–8 for a call originating at center S and destined for center D. The direct trunk group is the first choice as the best possible route for all calls.

Of the three triples shown, the most–direct triple and best–alternate triple are preferred because they represent forward routes that advance the call to switching centers that are considerably nearer the destination center than the originating center. The second–best–alternate triple employs lateral routes that do not advance the call but that aid survivability because they are automatically available to circumvent damaged or overloaded sections of the network. The routing program determines the route a call takes. The route is selected on the basis of the precedence of the call, its destination, and the congestion of the programmed routes. The sequence of selection of an outgoing trunk group is not fixed for precedence calls. On each new call, the hunting sequence for each triple is rotated to avoid possible repeated call blocking at a tandem office.

Service Features

All basic service features of SSNs previously mentioned, such as class–of–service and duplex operation, are provided. In addition, some unique features are available.

This network is designed to transmit data as well as speech signals. Both two– and four–wire station sets are used. Generally, stations served through centrexes or PBXs are two–wire whereas stations connected directly to a switching center are four–wire. All network switching offices are four–wire and all trunks are of unconditioned four–wire voice-grade design; conditioning is available for access lines and subscriber lines. Special features can be provided on an automatic selective basis under class–of–service or user control. Some of these are controlled by auxiliary pushbuttons or by the keying or dialing of a prefix code at the station set. Other optional features are multilevel precedence preemption, off–hook ("hot–line") service, and automatic and selective conferencing. The caller receives a busy tone, for either a station–busy or an all–trunks–busy condition, from his serving switch rather than holding up trunks to give tone from the point of congestion.

Dual–tone multifrequency (DTMF) calling provides pushbutton signalling with 16 distinct two–tone signals, 10 of which are used for regular telephone services. The * and # buttons are for special services. For example, if during the process of setting up an automatic conference call, a conferee does not answer, the conference announcement tone continues. Should it be desirable to conduct the conference even though not all conferees are connected, the originator may stop the tone by depressing a button (usually #). The four buttons on the right side of the DTMF pad are marked FO (flash override), F (flash), I (immediate), and P (priority) to designate precedence levels. Users who are authorized a certain precedence level may preempt calls assigned a lower level.

Facilities

The network is designed for satisfactory transmission over connections up to 12,000 miles in length. Four—wire facilities are used for trunks, access lines, and subscriber lines. Carrier systems are used exclusively for trunks and widely for access lines and subscriber lines to minimize delay and echo.

Four—wire station equipment is used for subscriber lines to minimize echo problems, to improve survivability in the event of failure of a local centrex or PBX, and to provide for alternate duplex data transmission. If two—wire station sets terminate subscriber lines, normally four—wire facilities with a two—wire conversion at the station location are used. Such lines must be treated as access lines at the serving switching center in order to obtain proper echo control.

Access lines are normally four—wire with two—wire conversion (if needed) at the PBX or centrex. If these lines are also used for data or other four—wire applications, they are called dual—use lines.

14-4 HIGH-CAPACITY DIGITAL NETWORKS

Use of DS1 Services

With the wide availability of DS1 services at attractive prices, the orientation of private networks has shifted from a circuit orientation to a structure based on 24—channel digroups. As shown in Figure 14—9, these may be used directly between centrex and PBX switches for tie trunks, may be applied from the trunk sides of digital central—office switches to PBXs for DID trunks, and may involve the use of customer—premises T1 network managers to provide a spectrum of capabilities. These latter devices can be optioned to serve as programmable DCSs, channel banks, drop—and—add multiplexers, low—bit—rate voice

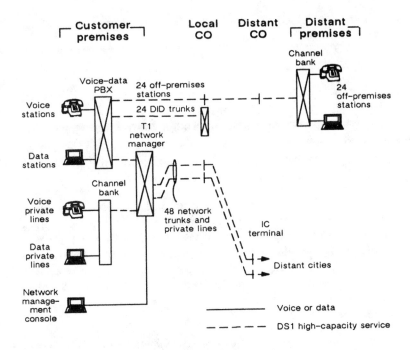

Figure 14–9. Application of high–capacity digital services.

coders, statistical data multiplexers, and remote network management systems. Via a flexible framing process, they can mix voice with data at the usual speeds (4.8 kb/s, 56 kb/s, etc.) or special speeds like 512 kb/s. In the larger networks, compressed video can be handled along with voice and data on a DS1 channel. These resource managers thus allow the integration of formerly separate PBX and data networks. Further, they can provide automatic restoration switching if one of the 1.5–Mb/s channels in a multicircuit network fails. Even without advanced hardware, groups of OPSs can be handled by extending a DS1 service from a PBX or centrex to a channel bank at or near the off–premises location. For long–haul networks where it is desirable to spread the cost of a DS1 service as widely as possible, DS1 codec devices are available to derive 44 trunks by using 32–kb/s adaptive differential pulse code modulation (ADPCM), or 96 trunks by applying ADPCM plus digital silence removal. The major ICs and a growing number of exchange carriers also offer "fractional T1" services with fewer than 24 channels' capacity for small trunk groups, an offering of wide popularity based on DS1 facilities and DCSs.

Exchange carriers offer customers multiple DS1 services that terminate in DCSs located in central offices, with customers remotely controlling the cross-connection of channels on their DS1s and thus able to do network reconfiguration from hour to hour [5]. The capabilities offered include the ability to add and delete legs from multipoint voice and data circuits, to do reservations of circuit facilities, and to perform switching on either a preprogrammed or an ad-hoc basis. The only important limitation on these networks is the avoidance of certain types of circuit where switching is incompatible with the use of the A and B signalling bits.

Use of DS3 Services

The next common step beyond DS1 high-capacity services is DS3. A DS3 service (T3, in popular usage) with its capacity of 672 voice circuits represents a substantial unit of capacity, but the larger private units can make use of it for bulk access to IC terminals, for backbone facilities to exchange carrier offices for demultiplexing to groups of 28 DS1s, for multicircuit video, for bridging of separated local area data networks, for interlocation backbones with drop-and-add DS1-DS3 multiplexers, etc. Figure 14-10 shows such a multiplexed hubbing application. Resource managers comparable to similar T1 devices are becoming available, but with DS3 interfaces.

Performance

Transmission issues in these digital-based networks are relatively simple. The transmission plan is controlled by the customer, within the limits of FCC constraints on port-to-port loss of PBXs. Switching is four-wire, eliminating through-balancing. The relatively low propagation velocity of fiber facilities (about 8.4 μs/mi one way, versus 5.3 μs/mi for radio) increases echo delay, as aggravated by delays in resource managers, other DCSs, ADPCM transcoders, and multiplexers. However, many such

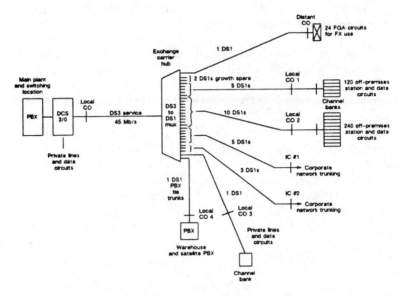

Figure 14-10. Application of DS3 high-capacity service with hub multiplexer.

delays would also figure in the performance of traditional analog networks. The good error performance of DS1 channels, enhanced by the error correction available with Dataport channels, aids the integration of voice and data networks. These networks serve as an evolutionary step toward primary-rate access to integrated services digital networks (ISDNs), and to DS1 channels provided by broadband packet switching, as in the prospective Switched Multi-Megabit Data Service (SMDS).

14-5 PRIVATE VIRTUAL NETWORKS

In the presence of common-channel signalling (CCS) between central offices and centralized data base access at a service control point, it is feasible to replace a dedicated inter-centrex/ inter-PBX trunk network with a private virtual network (PVN) based on regular central-office switches and use of message-network trunking [6]. The customer uses the equivalent of an ETN numbering plan for dialing; the CCS network and central data base specify a translation from customer–network numbering to message–network numbering, much as is done with 800 service numbers. An interswitch call is then trunked through the public network. Intelligence in the data base specifies the size of simulated private trunk groups and defines the route–advancement treatment needed if the simulated trunk group overflows. Additions and deletions of trunks, lines, switches, etc. in the virtual network are a matter of data updates; traffic measurement, network management, and billing come from the data base system. Switched access to the customer's attendant is included. Control by the customers themselves is feasible. The concept has the potential of making private networks attractive to smaller users. The PVN concept is related to and supportive of the centrex enhancements termed area–wide centrex, which uses similar "intelligent network" capabilities to give the user identical service, as if all locations were served by the same switch.

Transmission aspects of PVNs (also known as software–defined networks) are minimal; the performance of the private network is essentially that of the public system. Service reliability is likely to be improved because the alternate routing built into the message network is potentially available to PVN users as well. Some customers are likely to use both a PVN and a dedicated network. Interconnection between the two is analogous to ordinary connections between a dedicated network and the message network.

References

1. Shay, N. "A Better Way of Designing Tandem Tie Trunk Networks," *Bell Laboratories Record*, Vol. 55 (Sept. 1977), pp. 220–226.

2. Federal Communications Commission. *Rules and Regulations, Title 47, Code of Federal Regulations, Part 68* (Washington, DC: U.S. Government Printing Office, Oct. 1987), Section 68.308(b)(5), pp. 186–190.

3. *EIA/TIA Standard EIA–464–A*, "Private Branch Exchange (PBX) Switching Equipment for Voiceband Applications" (Washington, DC: Electronic Industries Association, 1989).

4. "Digital   PBX   Loss   Plan—Application   Guide,"
   TR-41.1.1/89-07-017, Draft 5, EIA TR-41 Engineering
   Committee on Telephone Terminals (Washington, DC: Elec-
   tronic Industries Association, July 1989).

5. Matthews, R. L. "Network Control in the Hands of the Customer," Bellcore EXCHANGE, Vol. 2, Iss. 2 (Mar./Apr.
   1986), pp. 24–27.

6. Pierce, L. and R. M. Weiss. "Meeting Private Needs with the
   Public Network," Bellcore EXCHANGE, Vol. 4, Iss. 1
   (Jan./Feb. 1988), pp. 8–13.

Chapter 15

Private-Line Channels

A private-line telephone circuit designed to connect two stations is similar to a message-network connection. In addition to the usual talking and signalling uses of private lines, other applications are common. Some of these additional uses are: telemetering, remotely controlling operation of radio transmitters or receivers, automatically regulating pipeline pumping stations, extending alarm or power-control circuits from unattended to attended locations, providing facsimile transmission, transporting digital secure voice, connecting computers or high-speed file servers with other computers or input/output devices, interconnecting local area networks (LANs) for data, transmitting wired music, serving remote radio broadcast pickups and studio-to-transmitter links, and supporting television transmission. Channels are commonly used for voice, voiceband data, and high-capacity digital networks; occasionally for wideband data, telegraph, and telephotograph transmission. Transmission aspects of program and video services are covered in Chapter 16. The Digital Data System (DDS), Basic Dedicated Digital Service, and digital high-capacity services are described in Chapter 17.

The engineering considerations applicable to toll circuits and local plant are generally relevant for designing private-line circuits. However, private-line service is often requested connecting three or more stations so that each station may signal and communicate singly or collectively with the others. The layout of such multistation private lines involves unique engineering to reflect the type of station equipment that the customer plans to install, signalling arrangements, connection with other private lines, access to remote locations, and the alternate use of the voice channel for data or facsimile transmission. The resolution of these basic considerations into a final circuit layout ranges from a routine automated process to a highly challenging problem.

There is no easy way to categorize the possibilities offered by private-line services. Perhaps the most exact method of classification is that offered by local tariffs which group private–line services on the basis of bandwidth and use.

15-1 PRIVATE-LINE SERVICE CATEGORIES

During the 1970s, considerable effort was expended to modernize private–line tariffs. Included were the consolidation of a number of tariffs and the restructuring of service descriptions. The process of restructuring intraLATA (local access and transport area) service tariffs to meet present business demands continues today, with further consolidation likely.

Service Elements

As part of the tariff simplification, private–line services were categorized by type of channel and assigned a series of four– or five–digit numbers. Channel charges, for the most part, are based on channel mileage between exchanges within the LATA. This mileage is the airline distance between appropriate rate centers; pricing is on a per–mile–per–month basis with the per–mile rate generally decreasing with distance. (For intraexchange services, of course, there is no interexchange mileage.)

Each interexchange channel requires a terminating arrangement (a service terminal or channel terminal) that usually consists of the central–office equipment, the local loop, and the network interface at the station.

Station equipment is normally customer–provided; however, analog channels normally include network channel terminating equipment (NCTE) as part of the channel terminal. This can include channel conditioning equalizers, bridging arrangements, remote loop–back controls or tone generators, signalling arrangements, etc.

Service Descriptions

As mentioned, the service offerings in a typical state tariff are divided into a series of numbered descriptions, each of which applies to one type of service. Excerpts from this tariff are shown in Table 15-1. Within each series is a number of more specific services fitting the general description.

Table 15-1. Private-Line Services in Typical State Tariff

Series 1000 (Sub-Voice-Grade)
1005   – Transmission up to 75/150 baud. Furnished for remote operation of radiotelegraph, teletypewriter, teletypesetter, data, supervisory control, and miscellaneous signalling purposes; for half (or full) duplex operation, two-point or multipoint service, intra- or interexchange.
1011   – Transmission up to 30 baud. Furnished for remote metering, supervisory control, and miscellaneous signalling purposes; for two-point or multipoint service, intra- or interexchange.
Series 2000 (Voice Grade)
2001   – For private-line voice, supervisory control, remote operation of mobile radiotelephone systems (two-wire); for two-point or multipoint service, intra- or interexchange.
2002   – For remote operation of mobile radiotelephone systems (four-wire); normally for two-point service, intra- or interexchange.
2005   – For connection to exchange access service from a foreign central-office district (FCOD) office.
2006   – For connection to exchange access service from a foreign exchange (FX) central office, except for Type 2006A services.
2006A – For connection to exchange access service from an adjacent FX central office; for residence service only.
2011   – For connection to exchange access service from an off-premises [non-PBX (private branch exchange)] extension line; intra- or interexchange.

Table 15-1. Private-Line Services in Typical State Tariff (Continued)

Series 2000 (Voice Grade) (Continued)
2012   – For connection to exchange access service from a company-provided centrex off-premises extension station line; interexchange only [nonadjacent central-office district (COD)].
2014   – For PBX off-premises extension station line use, also furnished for...nonadjacent COD applications, off-premises extensions; intra- or interexchange.
2021   – For tie-line use between PBXs...or customer-provided communications systems; for two-point service, intra- or interexchange with [signalling].
2025   – For tie-line use between PBXs...or customer-provided communications systems and a centrex; for two-point service, intra- or interexchange with [signalling].
2026   – For tie-line use between company-provided centrex systems; for two-point service, intra- or interexchange.
2040   – For use as a direct secretarial line terminating in either cord-intercept equipment or customer-provided concentrator/identifier equipment; intra- or interexchange.
2043   – For use as a concentrator channel. This channel (associated with telephone answering service) is used to connect a concentrator to an identifier.
Series 3000 (Data Transmission)
3002   – For half- or full-duplex data transmission and audio tone protective relaying (only when equipped with Type C-2 conditioning); for two-point or multipoint service, intra- or interexchange.

Table 15-1.   Private-Line Services in Typical State Tariff
(Continued)

Series 3000 (Data Transmission) (Continued)
3040   —   For Dataphone Select-A-Station (DSAS) use, for access lines (two-wire) between the data station selector (DSS) and a remote premises (protected patron).
3041   —   For DSAS use, for access lines (four-wire) between the master station and the DSS, between the DSS and a remote premises (protected patron), or between DSSs.
Series 6000 (Program Transmission)
6005   —   A two-wire interface with effective two-wire facilities engineered for a 1000-Hz maximum preequalization loss of 32 dB and equalized to ±3 dB of the 1000-Hz loss from 100 to 5000 Hz, arranged for transmission in one direction only; for two-point service only, on an intra- or interexchange basis; normally suitable for audio transmission.
6007   —   A two-wire interface with effective two-wire facilities engineered for a 1000-Hz maximum preequalization loss of 32 dB and equalized to ±3 dB of the 1000-Hz loss from approximately 50 to 8000 Hz, arranged for transmission in one direction only; for two-point service only, on an intra- or interexchange basis; normally suitable for audio transmission.
6009   —   A two-wire interface with effective two-wire facilities engineered for a 1000-Hz maximum preequalization loss of 32 dB and equalized to ±3 dB of the 1000-Hz loss from approximately 50 to 15,000 Hz, arranged for transmission in one direction only; for two-point intraexchange service only; normally suitable for audio transmission.

Table 15-1.   Private-Line Services in Typical State Tariff (Continued)

Series 6000 (Program Transmission) (Continued)
6011 — A two-wire interface with effective two-wire facilities engineered for a 1000-Hz maximum loss objective of 12 dB without equalization, arranged for transmission in one direction only; for two-point service only, on either an intra- or interexchange basis; normally suitable for audio transmission.
6020 — A two-wire interface with effective two-wire facilities engineered for a 1000-Hz maximum loss of 14 dB without equalization, arranged for transmission in one direction only; for multipoint service only within the same COD area; normally suitable for wired-music distribution.
6021 — A two-wire interface with effective two-wire facilities engineered for a 1000-Hz maximum preequalized loss of 34 dB and equalized to ±4 dB of the 1000-Hz loss from approximately 50 to 5000 Hz, arranged for transmission in one direction only; for multipoint service only within the COD area; normally suitable for wired-music distribution.
6022 — A two-wire interface with effective two-wire facilities engineered for a 1000-Hz maximum preequalized loss of 34 dB and equalized to ±4 dB of the 1000-Hz loss from approximately 50 to 8000 Hz; arranged for transmission in one direction only; for multipoint service only within the same COD area; normally suitable for wired-music distribution.
Series 9000 (Local Area Data Channels)
9001 — A two-wire interface with effective two-wire facilities with transmission characteristics specified [following].
9002 — A four-wire interface with effective four-wire facilities with transmission characteristics specified [following].

Table 15–1.   Private–Line Services in Typical State Tariff
(Continued)

Series 9000 (Local Area Data Channels) (Continued)	
End–to–End Facility Length in Route–Miles	Maximum Insertion Loss at 1000 Hz, in dB, with 135–ohm Resistive Terminations
1	9.0
2	13.5
3	17.0
4	20.0
5	23.0
6	25.5

Series 10,000 (Entrance Facilities)

10,000 – For the purpose of extending customer–provided communications systems to a customer's premises. They are furnished for half– or full–duplex operation on a two–point basis only. They are furnished for service 24 hours per day, 7 days per week for a minimum period of one month. A type–10,000 channel has an approximate bandwidth of 300 to 3000 Hz. These particular channels are [similar to] Series 1000/2000/3000 channels. The customer's premises must be located within 25 airline miles from the point at which the customer–provided communication channel is connected to the company's entrance facility.

The 1000 series includes channels capable of transmitting binary signals at rates up to 150 baud. These channels allow half–duplex or duplex operation over two–point or multipoint facilities. (In half–duplex operation, the full channel capacity is available in only one direction at a time; full–duplex operation gives full capacity both ways.) The 30–baud channel in this series is widely used for dc fire and burglar alarm purposes.

The 2000 series provides voiceband private–line service for voice transmission. Two–point or multipoint service is furnished on a half–duplex basis. Two–point service may also be duplex, by

327

use of a four–wire layout throughout. The channel bandwidth is generally the 300–to–3000–Hz band.

Channels in the 3000 series have an approximate bandwidth of 300 to 3000 Hz, which is used for voiceband data or digital–facsimile transmission, telemetering, supervisory control, and miscellaneous signalling. Two–point or multipoint service can be furnished for half–duplex or duplex operation. Special conditioning is available for these channels to improve their transmission characteristics: to flatten and widen their frequency response (C–conditioning), control their envelope delay distortion (EDD) (C–conditioning), or reduce their intermodulation distortion (D–conditioning).

The 6000 series channels are program and wired–music services for use where state jurisdiction applies, i.e., outside the broadcast industry. These circuits are covered in Chapter 16.

Offerings in the 9000–series are for intra–wire–center transmission of data using limited–distance data sets, typically at speeds between 2.4 and 19.2 kb/s.

The specialized channels in the 10,000 series are used when a "last mile" entrance link is needed to extend a local customer–provided microwave system or other facility into his main site. These channels can be as simple as cable from the top of a building to a lower floor; they can also be as complex as interexchange private lines with signalling. (Where the customer's facility crosses LATA boundaries, the relevant state or federal "access" tariff applies instead.)

Other portions of the state tariff offer digital–data channels, 1.5–Mb/s high–capacity services, etc. This arrangement of services, while simplified in recent times, is still cumbersome. Some exchange carriers have moved in the direction of "generic" channel offerings similar to those used for exchange access. The example in Table 15–2 compresses the full spectrum of normal services into only 13 offerings.

15–2 VOICE TRANSMISSION

Circuits that are common (but not unique) to private–line voice services may be arranged as either two–wire or four–wire,

Table 15-2. "Generic" Private-Line Services

UY **LOW-SPEED SIGNALLING** – A channel for the transmission of signalling rates of 0 to 150 baud. Suitable for use as a control/status, protective alarm (dc) or telegraph-grade channel.

UN **LOW-SPEED SIGNALLING CUSTOM** - A low-speed signalling channel allowing the customer to request parameters and interfaces not available on the standard low-speed signalling channel.

UC **VOICE-LINE** - A voice-grade transmission channel (300 to 3000 Hz) suitable for line-type circuits. Signalling options and data conditioning may be added to the standard channel.

UD **VOICE-TRUNK** - A voice-grade transmission channel (300 to 3000 Hz) suitable for trunk-type circuits. Signalling options and data conditioning may be added to the standard channel.

UG **ANALOG DATA** - A voice-grade transmission channel (300 to 3000 Hz) suitable for use as a basic data circuit. Optional data conditioning may be added to the basic channel.

UZ **BASIC VOICE** - A voice-grade transmission channel (300 to 3000 Hz) offered without signalling or performance enhancements. The channel is provided to an intraLATA customer on a two-point basis using metallic or carrier facilities at the [exchange carrier's] option.

UQ **VOICE-GRADE CUSTOMIZED** - A voice-grade channel that allows the customer to request parameters or interfaces not available in the standard voice grades.

UE **AUDIO PROGRAM (NONBROADCAST)** - A channel for the transmission of nonbroadcast program signals between a point of program origination and one or more other locations.

UP **AUDIO PROGRAM CUSTOM (NONBROADCAST)** - A channel for the transmission of audio programs that will allow the customer to request parameters or interfaces not available in the standard audio program channel.

US **DIGITAL DATA** - A channel for the transmission of digital data at rates of 2.4, 4.8, 9.6, and 56 kb/s.

UX **DIGITAL DATA CUSTOM** - A channel for the transmission of digital data allowing the customer to request parameters or interfaces not available in the standard digital data channel.

UH **DIGITAL HIGH-CAPACITY** - A channel for the transmission of digital data at rates of 1.544 Mb/s (DS1) or higher. Optional central-office multiplexing is available.

UM **HIGH-CAPACITY CUSTOM** - A high-capacity digital channel allowing the customer to request configurations unavailable in the standard channel offering.

with two–point or multipoint layouts and with optional selective signalling. Consideration must be given to the selection of facilities, bridging of multipoint circuits, and signalling.

The design of a private–line circuit begins from a preliminary layout of the circuit without detailed consideration of transmission and signalling. Generally, predesignated routes and bridging offices are available, reflecting overall economic studies and the locations of digital cross–connect systems (DCSs) with bridging features. The initial layout is modified to take into account available facilities, equipment, and maintenance capability.

Selection of Facilities

The facilities for two–point private–line circuits may be two-wire where loss and stability requirements can be met; however, four–wire layouts are preferred for multipoint circuits. The extent to which two–wire facilities can be used depends on the number of stations served. The multiple singing paths produced by several two–wire branches limit their use and, therefore, where a two-wire layout is desired for economic or other reasons, return–loss analysis is necessary. Some two–wire branches may be used on multipoint circuits provided the layout is predominantly four-wire. Except for special cases like Telemetry and Alarm Bridging Service (TABS), multipoint configurations involving large numbers of stations usually require completely four–wire design, even including four–wire stations.

Types of Facilities. The facilities used in a private–line voice-grade circuit are the normal carrier channels and voice–frequency (VF) repeatered lines. A complex circuit involving numerous branches may require extensive facility mileage when bridging points are remote from the stations.

Loop Repeaters. VF repeaters are often needed, either to reduce the loss of a loop to meet objectives or to match the impedances of a nonloaded loop and a loaded trunk cable. Repeater spacings on four–wire loops are defined by the maximum sending transmission level point (TLP) and minimum receiving TLP. Maximum spacings are 21 dB for nonloaded cable (limited by equalization) and 15 dB for loaded (limited by crosstalk and noise).

Transmission Plan

The transmission design of multipoint circuits is generally based on a 0–dB backbone route (center of bridge to center of bridge) with the allowable net loss assigned to the loops. The net loss is 0 to 10 dB for circuits using two–wire stations; it is fixed at 16 dB for those using four–wire stations. The 6–dB difference allows for the assumed hybrid losses of two–wire station sets. The transmission design for circuit–layout purposes is typically based on 0 TLP at the station and a +7 dB TLP into an analog bridge. Loss is inserted after the transmitting port of the bridge to meet the −16 dB TLP at the transmitting side of a four–wire carrier channel unit. The receiving TLPs at the station interface are 0 to −10 for two–wire voice circuits and −16 for four–wire. (Bridges contained in DCSs operate digitally, without TLPs; however, they typically allow for digitally inserted loss to make up for misalignment of levels at the distant coding point [1].) Interface TLPs for data circuits are covered below.

TLPs at the outputs of repeaters feeding cable facilities are controlled to avoid interference to and from other circuits that would result from near–end crosstalk. Minimum TLPs at cable facilities connected to repeater inputs are also considered so that the value of signal power is well above noise and crosstalk. TLP considerations for use of cable facilities are covered in more detail below, with regard to voiceband data transmission.

Bridging Arrangements

Multipoint private–line arrangements require the interconnection of legs at a common bridging point. Each line or branch connection constitutes a leg of the bridge. The more elaborate multipoint layouts use more than one bridging point. The complexity of a bridging design depends on whether the legs to be bridged are two–wire or four–wire, on the impedances of the interconnected circuits, and on any need for switching of the multipoint arrangement. To preserve transmission stability and constant level, bridge inputs and outputs must be properly terminated when not otherwise connected.

Two-Wire Bridges. The bridging arrangements most commonly used for two−wire layouts are the straight bridge and the resistance type.

The Straight Bridge. This arrangement, shown in Figure 15−1, is simple and inexpensive. As shown in Table 15−3, it has less loss than the other types. It is sometimes used where the higher loss of another type of bridge would require a repeater. However, the straight bridge has severe limitations that must be considered. Lines or loops of different types (e.g., a mix of loaded and non-loaded), connected together directly through a straight bridge, limit the degree of balance obtainable across any repeater or hybrid connected to other bridge legs. Also, the calculated losses of the straight bridge arrangement assume 600−ohm terminations on all appearances. In actual use, these terminations usually vary appreciably from 600 ohms and, consequently, the real loss of

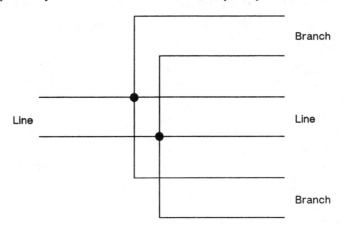

Figure 15−1. Straight bridge.

Table 15−3. Losses in Straight Bridge

Total Number of Bridge Legs (n)	Insertion Loss (dB)
3	3.5
4	6.0
5	8.0
6	9.5

the bridge arrangement may differ from the computed value. This suggests another limitation. If one of the bridge legs is exposed to a low impedance, short circuit, or open circuit, all other legs are affected and the balance of an adjacent repeater may deteriorate so as to cause singing. Where the backbone circuit is repeatered but the branch circuit is not, a 600–ohm pad of 5 dB or more should be inserted in the branch circuit adjacent to the bridge, provided the loss of the branch is low enough to permit it.

The insertion loss of the straight bridge, if all legs are terminated in the same value of impedance, can be computed from

$$\text{Insertion loss} = 20 \ \log\frac{n}{2} \ \text{dB}$$

where n is the total number of legs. The losses for several numbers of legs are shown in Table 15–3.

The Resistance Bridge. In this bridge, illustrated in Figure 15–2, the loss is several decibels greater than that in a straight bridge with the same number of legs; thus, the resistance type is generally used where gain is available on the circuit branches. The resistance bridge is usually preferred to the straight bridge. While use of this bridge does not always improve the singing margin, trouble on one of the circuit branches is less likely to affect the other branches or the singing margin of the overall circuit.

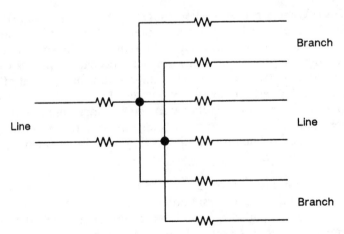

Figure 15–2. Resistance bridge.

The insertion loss of the resistance bridge of n legs, with all legs properly terminated, is

$$\text{Insertion loss} = 20 \log (n-1) \text{ dB.}$$

The value of R, the series resistors in Figure 15–2, can be determined from

$$R = \frac{R_T(n-2)}{2n} \text{ ohms}$$

where R_T is the nominal impedance of the bridge and n is the number of legs. The values of R and the corresponding insertion losses for several commonly used 600–ohm bridges are shown in Table 15–4.

Table 15–4. Resistors and Losses in Resistance Bridge

Total Number of Bridge Legs (n)	R (ohms)	Insertion Loss (dB)
3 (BR23)	100	6.0
4 (BR24)	150	9.5
5 (BR25)	180	12.0
6 (BR26)	200	14.0

Four–Wire Bridges. The 44–type (four–wire, four–way) bridge is a resistance network that interconnects four–wire lines. As shown in Figure 15–3, each of the four sides has an input and an output. Within the bridge, a transmission path connects each input with the outputs of the other three sides; thus, a total of 12 paths link the desired inputs and outputs. These paths, however, also provide transmission between each bridge input and the other inputs, and between each bridge input and the output on the same side of the bridge. Generally, the paths between inputs are not important.

The return currents are controlled by the six individual paths from any input to its corresponding output. Each of the six consists of a direct path from the input to the other three outputs; from each of the three outputs there are two paths in series back to the output associated with the input. Turnovers in the bridge

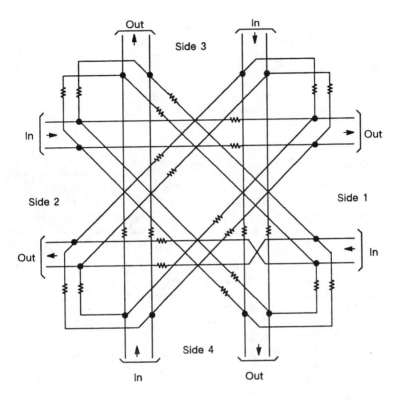

Figure 15-3. Schematic of 44-type bridge.

network, as indicated in Figure 15-3, cause three of the six paths to be out of phase with the other three. Since these six transmission paths theoretically have the same loss, they cancel each other in pairs. Thus, this circuit uses the balance-to-null principle of the Wheatstone bridge. The resulting loss between an input and its corresponding output is theoretically infinite. In practice, these paths do not have identical losses because of manufacturing tolerances in the resistors and because of minor differences in the impedances of the lines or repeaters connected to the other three sides of the bridge.

With all bridge ports terminated in 600 ohms, the transmission loss between any input and the outputs of each of the other three sides is approximately 15 dB; the loss between any input terminal and the output terminal of the same side is in excess of 75 dB.

For some combinations of gross imbalance, such as short–circuiting or opening two or more terminals of the bridge, this loss may fall to as little as 38 dB, which would cause objectionable echo in the presence of enough transmission delay.

The impedance of the bridge is about 650 ohms, with a nominal termination of 600 ohms. Since the bridge has a relatively high loss, the impedance of a port does not vary greatly with varying terminations on the other ports.

Although the 44–type bridge is, for practical purposes, a symmetrical circuit when properly terminated, it is asymmetric with respect to reflected or return currents. Extraneous currents entering the bridge may be as much as 9.6 dB higher at output 4 than at any of the other outputs. For this reason, side 4 is either assigned to a branch circuit or left spare. On a service where a spare side is likely to be needed for connection to an additional bridge for circuit growth, it is preferable to assign side 4 to a branch and leave a different side spare.

Bridges of similar design are employed for six–way (46–type) bridging. These have 20 dB loss between input and output. A 43–type bridge is also possible but is not commonly used, as is a 48–type device with 23 dB of loss.

Talk–Back Features

Since a four–wire station has no connection between the transmitting and the receiving sides of the circuit, no sidetone path is provided. In addition, communication is not possible between stations bridged together at the same location. It is necessary, therefore, to provide a "talk–back" path, arranged so that it does not cause objectionable echo on the main circuit. Where talk–back is inserted at a central office, it must not be separated from the station by facilities having more than a few milliseconds of time delay, since this would cause the talk–back to sound like echo rather than sidetone. Talk–back is provided at a bridging point or as part of the four–wire station arrangement. This function may be accomplished by the use of a talk–back amplifier or a resistance–type talk–back bridge.

Talk–Back Amplifier. One method, shown in Figure 15–4, uses an amplifier to connect the transmitting and receiving sides of the branch. With this arrangement, there is no transmission from C to B since the amplifier is a one—way device. In each of the two main transmission paths, A to B and C to D, there is a loss of about 3.5 dB due to double terminations. Thus, the gain of the talk—back path from A to D is the gain of the amplifier less 7.0 dB. In general, the amplifier gain is set so that the transmission from the station talker to its own receiver (and other receivers at the same location) is equivalent to transmission from distant talkers.

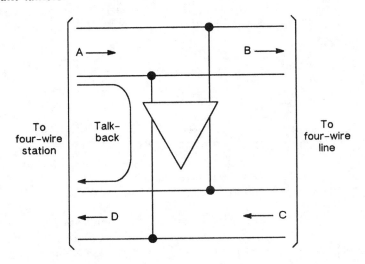

Figure 15—4. Talk—back amplifier.

Resistance Talk—Back Bridge. A second method of obtaining talk—back is shown in Figure 15—5. This arrangement consists of a resistance network inserted in the four—wire branch. Talk—back bridges cannot be used unless a repeater (or the gain of a carrier channel) is associated with the branch, since the loss would be too great for satisfactory transmission, even with zero—loss loops. Since the bridge is a two—way device, the main circuit as well as the branch is subjected to feedback in the form of echo. A talk—back bridge having 21 dB of loss is sometimes provided for circuits using 44—type bridges. The talk—back signals transmitted into the bridge are then 36 dB below the normal signals due to the 15—dB loss in the 44—type bridge. This amount of feedback or echo on the main circuit is not objectionable for most voice applications; however, on large multistation networks, it is

337

desirable to keep the number of resistance talk–back bridges to a minimum and use talk–back amplifiers instead.

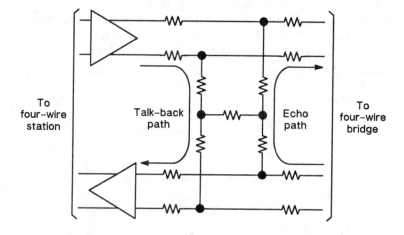

Figure 15–5. Resistance talk–back bridge.

Switching Arrangements

Arrangements are available for switching multistation private lines at a bridging point. Switching is accomplished by four–wire relays operating between four–wire bridges. Relay operation is controlled by a private–line station using a control channel between the central office and a station, by a selective signalling system, or by a dial–in data link and a microprocessor controller. From a transmission standpoint, a switching arrangement is similar to a plain bridging arrangement since, in the switched condition, two or more circuits are interconnected and appear as one circuit to the bridge. In the nonswitched condition, the switchable legs are terminated in resistors.

With a digital cross–connect system, the control of individual channels on DS1 systems passing through the DCS may be given to the customer. In this case, switching commands enter the DCS controller from the customer's site, allowing substantial ongoing reconfiguration of the customer's network. Security features limit the customer's rearrangement capability to only his own circuits [2].

Station Features

Telephone private lines usually terminate in an interface to a station set, a PBX, or key equipment. The customer location contains network channel terminating equipment supplied by the exchange carrier. It is the private–line termination common to all stations at the customer location. The station equipment is the customer's telephone set, etc. NCTE packages are available for use with both two–wire and four–wire loops, with numerous options such as idle–circuit termination or loudspeaker, bridged station extensions, alternate–use transfer key, loop–back testing, equalization, level control, talk–back, and various types of signalling.

One typical four–wire NCTE is shown in Figure 15–6. It consists of loop transformers, amplifiers, a loop–back tone detector, station transformers, a loop–back pad, and a talk–back amplifier. The transformers are selected to match the loop and the nominal 600–ohm impedance of the station equipment. Connections to the loop sides of the transformers provide for 20–Hz ringing, a dc control channel, or a sealing–current path as required. The transformers on the station side are used so that the four–wire loop may serve more than one station. One arrangement, commonly used, presents a low–impedance termination to the station sets. With this arrangement, the loss due to

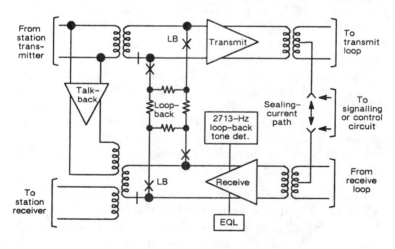

Figure 15–6. Four-wire station termination.

impedance mismatch decreases as the bridging loss increases, thus providing a nearly uniform station loss for the varying numbers of stations that might be bridged.

Loop–back is an arrangement used on a four–wire facility to interconnect the transmitting pair with the receiving pair for the purpose of remotely testing the facility. Either loop–back is provided at a point where the TLPs of the transmitting and receiving pairs are equal, or a pad is provided with loss equal to the difference in TLPs. Loop–back control is by tone on the loop or, in older arrangements, by dc current. The tone is usually 2713 Hz, chosen as falling in the band between 2450 and 2750 Hz, which is protected as to customer signal power yet able to exist with 2600–Hz signalling on the same circuit.

Signalling Systems and Associated Arrangements

Many types of signalling can be applied to a multistation private line; the choice depends largely on customer needs. These vary from no signalling (or signalling in one direction to only a few stations) to the extreme case in which a large number of stations may signal any of the other stations individually, collectively, or by preselected groups.

Loudspeaker Signalling. On many circuits, one station controls the operation of all of the other stations. Circuits operated in this manner often use a simple loudspeaker signalling system. All remote stations are equipped with a loudspeaker so that they can be paged. The control station can have a loudspeaker or can be signalled in any other way. A station equipped with a loudspeaker can monitor the circuit without restricting personnel activities. However, the audible range of loudspeaker signalling is usually limited to the confines of a small area unless auxiliary speakers are used. The efficiency of loudspeakers varies inversely with room noise. Furthermore, loudspeakers can be reduced in volume or turned off and forgotten.

Manual Code Ringing. Manual code ringing employs simple ringers or bells that operate when a signal is being received; 20–Hz signalling is usually employed. The codes are normally assigned by the customer and consist of a combination of long and short signals.

This system is inexpensive but has several disadvantages. For example, all stations receive all signals and attention is necessary in order to identify received codes. When loud horns, gongs, etc., are used, the receipt of unnecessary signals is annoying. Also, improper manual signalling may result in the wrong station, or none, answering. Thus, manual code ringing is generally unsatisfactory on circuits having more than about ten stations.

Ringdown Signalling. Ringdown signalling from a pushbutton is applied at the customer premises and transmitted over the loop. It has its widest application on two–point circuits and on multistation circuits where one station is equipped to receive ringdown signalling and all others employ loudspeaker signalling. The station equipment is usually arranged to lock in on the first signal received. With locked–in signalling, the signal at the station continues until the call is answered or a time–out feature operates.

Automatic Ringdown. With this system, the stations (generally plain two–wire telephones) receive dc battery feed. When one station goes off–hook, closing a dc loop, an automatic-ringdown converter in the central office sends 20–Hz (or equivalent) ringing toward the other station(s). An off–hook signal from a called station stops the ringing. The signalling converter may be a repeater plug–in or part of a carrier channel unit.

Code–Selective Ringdown. The station circuits for ringdown signalling are also used for code–select signalling; the required microprocessor selector equipment is located at the central offices in the form of a special repeater plug–in or a carrier channel unit. On circuits with code–select signalling, each station is equipped with a ringing key and sends a code for the desired station. The selectors at the central office, one for each loop, count the rings; if the number of rings received matches the selector setting, a single 20–Hz ringing signal is applied to the customer loop. A signal is received at a particular location only when that location is called, but there is no provision for selecting one station out of a group bridged on the same loop. While code–selective signalling is considerably slower than dual–tone multifrequency (DTMF) signalling, it is less costly.

Selective Signalling Systems. Selective signalling systems are in wide use. An early version (SS–1 or equivalent) was designed for rotary dials. The system is capable of selectively signalling 81 individual stations. The two–digit station codes are generated by dial pulsing. The digit 1 is used only to cancel an erroneously dialed first digit. The dial pulses are converted to frequency–shifted tone pulses for transmission over four–wire facilities. Single–frequency receivers convert the tones back into dc pulses. A decoder then counts, registers, and sends a momentary signal on the corresponding two–digit code lead to ring only the called station.

The system is ready for dialing when the handset is removed from the switch hook or a key is operated. However, before dialing, the station user must monitor the line to prevent interference with other users. If the system is in use, the added party hears either speech or, if the system has privacy features, a tone. Conference calls are accomplished by dialing as many two–digit codes as desired.

A later selective signalling system (SS–3 or equivalent) uses three–digit DTMF codes transmitted over a four–wire multipoint private line. It may be arranged for up to 698 codes. A typical system involves several different locations. Equipment at each location consists of telephones, a DTMF signal receiver, signalling equipment, and four–wire line terminating circuits. Three–digit code signals are received at each location, decoded, and sent to the control logic. Every code dialed is received by the decoders at all locations. The equipment at the location where the code has been assigned responds by placing a momentary ground on the code lead of the station circuit, which rings the desired telephone. In addition to station ringing, the code response from the SS–3 system may be used for remote–control applications by providing auxiliary circuits. The system has options for privacy (cutting off stations not involved in a given conversation), group calling, talk–back, etc. This type of system is widely used for the exchange carrier's internal order–wire circuits (radio order wires, etc.), as well as general applications by customers.

15–3 VOICEBAND DATA TRANSMISSION

Private–line data circuits may be designed for either two–point or multipoint operation. Line design may be either two–wire or four-wire but four-wire design is preferred for multipoint service because of maintenance and return-loss considerations. The actual terminations at the customer's premises may be two- or four-wire, at the customer's option. Return loss normally limits the complexity and length of two-wire multipoint layouts since it decreases with the number of points served and with deviations from the desired resistive terminating impedance of 600 ohms ± 10 percent across the 300-to-3000-Hz band. If more than six points are to be served, four-wire designs are normally used. Noise, a mileage-dependent parameter, was once a limiting factor in the design of multipoint services on analog carrier. However, with the decline of analog-carrier usage, this limit has been removed.

A duplex channel may be used for simultaneous bidirectional transmission of two full-voiceband signals. A half-duplex channel may be used for transmission of one voiceband signal in either direction, but not simultaneously, or for transmission of two signals simultaneously, one in each direction, each using noninterfering portions of the voice bandwidth. With present-generation data sets, simultaneous transmission at 1200 or even 2400 b/s in both directions on a half-duplex channel is possible by band-splitter or echo-canceller techniques.

Either two-wire or four-wire terminations may be specified, subject to multipoint stability limitations; however, a request for four-wire terminations does not automatically mean four-wire design should be used. Half-duplex service may be provided with either two-wire or four-wire facilities, but duplex service requires four-wire facilities.

Bridging Arrangements for Multipoint Data Circuits

A multipoint split bridge is widely used for voiceband data services. Split (electrically independent) distribution and collection networks, shown in Figure 15-7, are normally used since the primary application is for four-wire data services between the main and multiple remote stations typically found in computer polling networks. The bridge assembly integrates the equipment and functions required at a bridge location (bridging, gain and equalization, test access, etc.) into a self-contained unit that

343

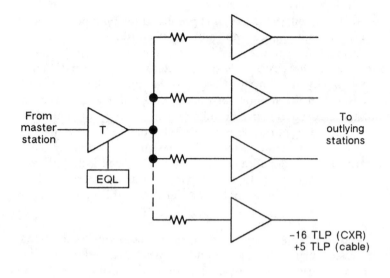

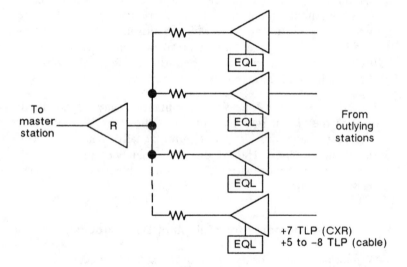

Figure 15-7. Typical split-bridge assembly.

minimizes cross-connections. The assembly includes VF amplifiers, equalizers, and test jacks. The bridge provides an interface with cable and carrier facilities and has a constant loss, regardless of the number of ports in service.

As shown in Figure 15-7, common TLPs are provided within the bridge. Both in-service and out-of-service testing access is provided. While a typical bridge shelf accommodates equipment for 12 functional bridging ports, flexibility is usually provided for both intrashelf and intershelf growth. Cascading ports provide flexibility for the intershelf connection of more than one bridge without requiring circuit rearrangements. Plug-in boards are required only for those ports of the bridge actually in service.

Bridging need not involve a packaged assembly. Pairs of two-wire bridges (Figure 15-2) can be used, with either an external four-wire repeater or the gain of a carrier channel unit to make up bridging losses.

Data Termination Arrangements

A typical NCTE for connecting a four-wire data termination to a four-wire loop is shown in Figure 15-8. The terminating equipment provides equal-level loop-back and switching arrangements for a voice-coordinating telephone when required for alternate-voice use. The terminating unit also provides equalization and gain to compensate for loss and to establish TLPs as shown. When the local loop is connected to a carrier channel at the central office, the +5 dB TLP at the output of the terminating unit and the fixed TLP internal to the carrier channel (typically −8.5 for digital systems) require a combined loss of the loop and transmitting pad of 13.5 dB. Adjustment of the receiving amplifier in the terminating unit establishes the −3 dB TLP at the loop-back point. At the data set, a transmitted signal power of 0 dBm is standard, with an overall loss of 16 dB.

NCTE can also provide two-wire terminations for four-wire facilities. Except for a two-wire termination (hybrid) and the use of different TLPs, the typical arrangement is similar to the one described.

Transmission Levels

Private data lines share facilities with message-network circuits; therefore, the design of the data channels must keep them

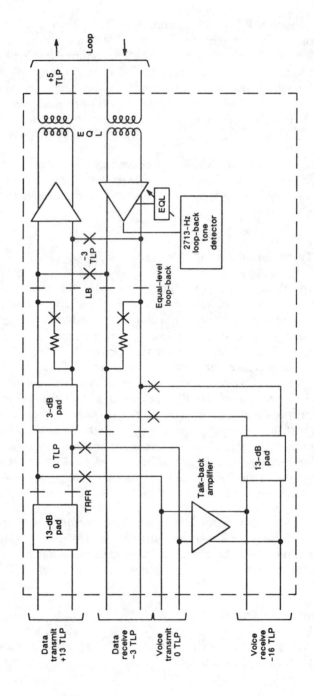

Figure 15–8. Four–wire loop to a four–wire data set and a four–wire telephone set.

compatible with the shared facilities. Many of the design criteria covered elsewhere are summarized here and some additional criteria are given to illustrate how private-line design aspects supplement message-network considerations.

The most critical design parameters involved in providing private-line service without adversely affecting the message network are the specification of TLPs and the application of signal power limits in terms of these level points. These two design parameters protect the message network from excessive crosstalk and carrier-system overload, and protects the private line from clipping in digital coders.

To be consistent with message operation, a data private line is designed to have the normal TLPs at carrier channel banks (e.g., −16 dB for analog systems, −8.5 dB internal to a typical digital system). The standard carrier system design then results in a normal TLP at the output of the channel bank (e.g., +7 dB for analog, +4 dB internal to a digital system). Several other TLPs are also defined. For example, on channels that are used alternately for voice or data, the connection of the telephone station set to the line is generally defined (for the transmitting direction) as a 0 TLP. Some variation is permissible to accommodate NCTE design restrictions, provided signal-power limits are not exceeded.

The data signal power is specified as not exceeding an average of −13 dBm0 in a voiceband channel over any three-second interval in order to meet analog carrier overload limits. The peak voltage in the signal should not exceed the equivalent of a +3.16 dBm0 sinusoid, no matter how short a time the peak exists, to meet overload limits for digital carrier. The power of the data signal (three-second average) is typically 0 dBm at the output of the data set. This point is the interface between the terminal equipment and the channel; pads, amplifiers, equalizers, etc., are considered to be part of the channel. Thus, with the TLPs shown in Figure 15-8 and with the specified signal power, a −13 dBm0 signal is realized.

Four-wire data private lines are designed for a net loss of 16 dB. The loss must be allocated in a way that satisfies signal power, TLP, and noise objectives. For example, in order that the signal-to-noise ratio not be degraded, data channels that use VF cable are usually limited to a maximum of 15 dB loss in a non-repeatered section of loaded cable.

A +5 dB TLP is typical at the input to a cable facility for voiceband private-line data circuits. The maximum allowable value is a +7 dB TLP; the minimum is -3 dB. These values limit signal power differences in order to control crosstalk between circuits using the same facility. The +5 dB TLP produces a private-line signal power that corresponds closely to the signal power applied to the average-length telephone loop. Thus, crosstalk between private-line and message-network data signals tends to be equal.

The maximum value was selected as a +7 dB TLP to be compatible with the maximum +7 dB TLP at a carrier channel output, avoiding crosstalk that might result from higher values. The minimum value of -3 dB, combined with the maximum recommended repeater section loss of 12 dB, is consistent with the -15 dB TLP recommended as the minimum input to a VF repeater. The range of -3 dB to +7 dB allows flexibility in the design of these circuits.

Some departure from the established guidelines regarding TLPs and signal power are permissible provided network protection criteria are not violated. Care must be exercised in these exceptional cases to maintain equal-level points for loop-back.

Impairments

Several types of impairment are controlled [3] in order to meet transmission objectives for an intraLATA private-line channel. Most data-signal impairments are the same as or similar to those affecting voice signals; however, there are some (such as impulse noise, delay distortion, and phase jitter) that have a more critical effect on the data signals. These impairments and the applicable objectives are reviewed here with regard to private-line data transmission.

Impulse Noise. Data error performance is seriously affected by impulse noise of sufficient magnitude and frequency of occurrence. As with other impairments, the susceptibility of data signals to impulse noise varies with the transmission rate and with the type of modulation. Impulse-noise objectives and their allocations to facilities and links on multipoint private-line data channels are given in Table 15-5.

Table 15-5. Impulse-Noise Objectives

Facility	Noise Threshold (dBrnc0)
Overall (with a −13 dBm0 holding tone)	71
Local loop	59
VF trunk cable	54
Analog broadband carrier:	
0 to 125 miles	58
126 to 1000 miles	59
N carrier	67
Two N-carrier facilities in tandem	69
Digital carrier	66
End link (multipoint circuits)	67
Middle link (multipoint circuits)	See facility limits above

The impulse-noise objective is specified in terms of the rate of occurrence of the impulse power above a specified magnitude. The objective is expressed as the threshold in dBrnc0 at which no more than 15 impulses in 15 minutes are measured by an impulse counter with a maximum counting rate of 8 counts per second. The overall objective of 71 dBrnc0 implies a 6-dB signal-to-impulse-noise ratio in the presence of a −13 dBm0 signal.

Message Circuit Noise. There are two general message circuit noise limits for voiceband private-line data channels. As shown in Table 15-6, message circuit noise is related to the length of any analog facilities that are used and is a measure of idle-circuit random noise in dBrnc0. The C-notched noise (a term derived from the method of measurement) is the measure of noise on a channel when a signal is present. A single-frequency holding tone is applied to the line as a signal; this tone operates digital coders and other signal-dependent devices. At the receiving end, the tone is removed by a narrow band-elimination filter (notch filter) and the noise is measured through a C-message filter [4].

The maximum C–notched noise limit of 53 dBrnc0 is based on a 24–dB signal–to–C–notched–noise ratio in the presence of a −13 dBm0 signal. The mileage–dependent limits for message circuit noise are facility maintenance limits; noise in excess of these limits indicates a trouble condition. The noise limits derive from the performance of analog facilities; digital facilities have constant, and lower, message noise independently of length.

Table 15–6. Message Noise Limits

Facility Length (miles)	C–Message Noise (dBrnc0)
0 – 50	31
51 – 100	33
101 – 200	35
201 – 400	37
401 – 1000	40

Single–Frequency Interference. This type of interference may appear on channels in the form of unwanted steady single–frequency tones. Most of these involve cross–modulation and other leakages in analog carrier systems. Occasional bursts of low–amplitude signals that may occur from crosstalk of multi-frequency signalling, for example, are not included in this category. The requirement for this type of interference is that, when measured through a C–message filter, it must be at least 3 dB below C–message noise limits.

Attenuation and Envelope Delay Distortion. It may be necessary to control attenuation/frequency and envelope–delay/frequency characteristics of a channel to permit satisfactory data signal transmission. Attenuation and delay distortion may be corrected by the use of equalizers when requirements cannot be met by available facilities or when the customer orders conditioning. The wide use of adaptively equalized data modems has substantially reduced the need for these controls.

The attenuation/frequency requirement specifies the allowable deviations of the attenuation characteristic over a given frequency range. There is no provision for the transmission of dc components. The allowable deviations and frequency ranges vary with the grade of conditioning; the deviation limits become narrower and the frequency range wider as the higher numbered grades of conditioning are provided. The allowable deviation is specified as the difference in loss between that measured at a specified frequency and that measured at 1000 Hz.

As with the attenuation/frequency characteristic, the allowable EDD becomes smaller and the frequency range wider for progressively higher grades of channel conditioning. The overall EDD limits are specified in terms of the difference between the maximum and minimum envelope delay within a given frequency band. As a specific example, for C1 conditioning, the EDD is calculated for the band 1000 to 2400 Hz, and must be less than 1000 μs. For the band 800 to 2600 Hz, an entirely separate calculation is made, possibly involving a different minimum–delay reference frequency, and must give 1750 μs or less.

The requirements for voiceband data circuits, with and without C–conditioning, are given in Table 15-7. The C1 and C2 grades of conditioning accommodate up to four midlinks on multipoint circuits that use analog bridging; bridging via DCSs is "transparent" and can permit larger numbers of midlinks. C3 conditioning was developed for access lines and interswitch trunks in switched services networks, as described in Chapter 14. C4 conditioning, with its upper frequency at 3200 Hz, is stringent enough to be limited to two–, three–, or four–point circuits, and C5 conditioning is restricted to two–point circuits. The C3 and C5 grades are not offered in all tariffs.

Attenuation distortion, which arises mainly in VF cable, is usually corrected at the ends of the cable section itself. Mismatch equalization via use of 150–ohm terminations is convenient and effective for nonloaded cable. NCTE units at the station, equalized carrier channel units, and VF repeater plug–ins provide adjustable low– and high–frequency equalization.

Envelope delay distortion, a product mainly of the high–frequency cutoff of loaded cable and of the filters in carrier systems, can be corrected at any point in the circuit without transmission problems; however, it is usually most convenient to equalize it at the locations of multipoint bridges. A gain-and-delay equalizer that fits a VF repeater mounting is

Table 15-7. C-Conditioning Requirements

Condi- tioning	Attenuation Distortion		EDD	
	Band (Hz)	Variation (dB)†	Band (Hz)	Variation (μs)
None	*500-2500 *300-3000	-2 to +8 -3 to +12	*800-2600	1750
C1	1000-2400 300-2700 *300-3000	-1 to +3 -2 to +6 -3 to +12	1000-2400 *800-2600	1000 1750
C2	500-2800 300-3000	-1 to +3 -2 to +6	1000-2600 600-2600 500-2800	500 1500 3000
C3 (access line)	500-2800 300-3000	-0.5 to +1.5 -0.8 to +3	1000-2600 600-2600 500-2800	110 300 650
C3 (trunk)	500-2800 300-3000	-0.5 to +1 -0.8 to +2	1000-2600 600-2600 500-2800	80 260 500
C4	500-3000 300-3200	-2 to +3 -2 to +6	1000-2600 800-2800 600-3000 500-3000	300 500 1500 3000
C5	500-2800 300-3000	-0.5 to +1.5 -1 to +3	1000-2600 600-2600 500-2800	100 300 600

* Not a tariff requirement.
† (+) means loss with respect to 1004 Hz; (-) means gain.

most often used, but specialized equalizer shelves and even hardwired fixed equalizers can be used.

The attenuation distortion of a circuit is measured at the customer station (-3 dB TLP) and adjusted after the 1000-Hz loss adjustment. Delay distortion measurements are made after the attenuation distortion has been brought within limits.

Absolute Delay. A requirement for this parameter is not specified; however, absolute delay may retard systems that use a retransmission scheme for error control from passing data at the maximum data transfer rates (throughput) desired. When satellite channels are used for data transmission, the absolute delay of several tenths of a second requires a suitable data protocol to avoid problems of this nature.

Net Loss Variations. At installation, the channel should be lined up to within ± 1 dB of the designed net loss at 1000 Hz. In operation, the net loss may vary up to ± 4 dB (maintenance limits) from the design value. These variations are caused mainly by daily and seasonal temperature changes that affect aerial cable; digital carrier channels have long-term stability of ± 0.2 dB or better.

Frequency Shift. Frequency shift in carrier channels is seldom a problem for data applications. It does not occur on digital carrier or on analog carrier systems that have a transmitted-carrier modulation scheme, such as N2; it may be found on channels employing suppressed carrier transmission, such as N4 and broadband carrier, in the event of failure of the carrier synchronization system. The 3-Hz end-to-end limit specified for voiceband data channels derives from the performance of early nonsynchronized analog systems.

Phase Jitter. Total phase jitter between customer stations within a LATA should not exceed ten degrees, measured in a 20-to-300-Hz or 4-to-20-Hz band; or 15 degrees, measured in a 4-to-300-Hz band. The objective for phase-jitter distortion on tandem broadband analog facilities is eight degrees maximum, peak-to-peak. Phase-jitter requirements for digital carrier systems have been established and are much tighter; however, apparent phase jitter as measured on these systems is generally the result of quantizing noise and is not true jitter.

Nonlinear Distortion. Nonlinear distortion is the portion of a channel output that is a nonlinear function of the channel input. Digital channel banks are one source of controlled amounts of distortion. The impairment results from circuit nonlinearity and from the quantization process used in pulse code modulation (PCM). Intermodulation distortion can be measured by two pairs of tones centered at two frequencies, A and B. Distortion products are measured using narrow bandpass filters centered at $2B - A$, $B - A$, and $A + B$ [5]. This technique provides consistent measurements for PCM systems and correlates well with higher–speed data set performance. Limits for four–tone intermodulation distortion on intraLATA data channels are 27 dB second–order (B – A and A + B) and 35 dB third–order (2B – A).

A special grade of conditioning, D1, deals with this parameter. Intended for use with the higher data speeds (9600 b/s and above), it offers a ratio of signal to C–notched noise of 28 dB or more. It controls nonlinear distortion, giving a ratio of signal–to–second–order distortion of 35 dB or more, and a ratio of signal–to–third–order distortion of 40 dB. This type of conditioning is provided mainly by exclusive use of digital carrier facilities of modern types employing 7–5/6 bit coding, and by limiting the number of coding/decoding steps.

Transients. Rapid gain and phase changes on transmission media cause data errors. Such changes occur infrequently but may be produced by switching a carrier system to a protection facility. The changes may be either transient, with the gain or phase returning to its original value after a short time, or long–term, with the gain or phase remaining at the new value indefinitely. The seriousness of a given gain or phase change depends on the type of signal transmitted and the method of signal detection. Generally, simple two–level signals are less affected than multilevel or multiphase signals. For a given type of signal, the number of errors introduced by a rapid gain or phase change depends on duration, rate of occurrence, and magnitude of the change.

A sudden change in received signal amplitude (greater than ±3 dB) lasting 4 to 32 ms is defined as a gain hit. Amplitude hits less than 4 ms are considered impulse noise. A reduction of 12 dB or more in received signal power for at least 10 ms is defined as a dropout. Maintenance limits on these impairments have been established. A sudden phase change of 20 degrees or more and in excess of 4 ms is defined as a phase hit. The limit for phase hits is a maximum of two in a 15–minute period. Phase hits of predictable size occur when slips take place on digital systems; a one–frame slip (125 μs) results in a phase change at 1000 Hz of 125/1000 of a cycle, or 45 degrees. Correcting the slip problem removes this source of impairment.

A sudden pulse of nonlinear distortion is called a distortion hit. These may be caused by some types of signalling passed through a channel bank or by power supply transients. The compressor circuits provided with first-generation digital channel banks, unless refitted, are particularly susceptible to these transients. Since such hits may adversely affect the error rate for voice-channel data services at speeds greater than 2400 b/s, a special compressor circuit was developed for these early banks.

15-4 WIDEBAND DATA TRANSMISSION

The evolution of wideband services and the transmission plan to meet the demand for high-speed data channels [6] were based on the use of existing transmission facilities or those readily available for installation. While digital facilities are generally most suitable for transmitting digital data, the transmission plan had to include the use of the then-available frequency-division multiplexed analog systems. These channels are no longer widely used, but they remain in some tariffs and illustrate some basic data-transmission principles.

Services Accommodated on Analog Facilities

Two bandwidths were made available for wideband services. These are 48 kHz and 240 kHz, which correspond to the 12- and 60-channel group and supergroup bands of the broadband multiplex. The maximum synchronous serial data rates accommodated were 50.0 kb/s for the group band and 230.4 kb/s for the supergroup band. Service terminals were also made available to accommodate 19.2 kb/s data (half-group band). Each of these signals may also be transmitted over T1 systems with special terminals.

Facility Types. A basic analog supergroup may be terminated in a wideband data modem, may be connected through to a similar supergroup, or may be terminated in group banks for subdivision into five groups. A basic group may be terminated in a wideband data modem, may be connected through to a similar group, or may be terminated in a channel bank for further subdivision into 12 voiceband channels.

Wideband modems were also used on short-haul carrier routes. The N-carrier wideband terminal translated a 50-kb/s signal in the 0.1–kHz–to–38–kHz band into the band of the N–repeatered line along with two voice channels. In digital carrier systems, several wideband banks (T1WB–1, –2, and –3) were used for translating wideband signals from up to eight group or two supergroup services into a DS1 line signal.

A baseband (analog) repeatered system was also applied to permit extension of wideband data services over ordinary cable pairs. These could span distances of up to about 10 miles from customer premises to the nearest central office having access to carrier facilities. These wideband loop repeaters included adjustable equalizers to match repeater gain to cable attenuation.

The wideband data test bay functioned as an access point for error–ratio tests and other maintenance of wideband data circuits wherever signals appeared at baseband. It provided baseband interconnection for carrier systems and a means for extending signals over repeatered cable pairs. Centralized patching and testing facilities provided access during system alignment and maintenance. A simplified version called the wideband service bay served as a common level point for both directions of transmission, which facilitated link–by–link testing on a looped–circuit basis.

A VF channel accompanied the wideband channel to permit coordination of data operations by voice communication. Alternate–voice arrangements were offered as an option so that the customer could use the full voiceband capability of the wideband channel for regular private–line service when not transmitting data, e.g., 12 PBX tie trunks during the day, 50–kb/s data at night.

Transmission Requirements. Transmission requirements for two–point service were generally specified for station lines and interoffice facilities, as opposed to overall end–to–end requirements. Where no interoffice facility was employed, the station lines on both sides of the serving test center were designed to meet station–line requirements.

Wideband channels were generally lined up for 0 dB net loss end–to–end, and between wideband service bays. The signal power at the output of the data set was typically 0 dBm.

Transmitting and receiving points at the wideband service bay were 0–dB system level (SL) points for supergroup and group–band services and −10 dB SL for half–group services.

A typical group–band circuit is shown in Figure 15–9. The transmission requirements for group–band two–point private lines are summarized in Table 15–8. The requirements for half–group service are less critical; supergroup requirements are more stringent. Attenuation/frequency and EDD requirements are given in terms of relative slope (difference in gain between extremes of the passband), sag (divergence between a straight line representing the slope and a smoothed curve representing the frequency response), and peak (deviation of the actual frequency response from the smoothed curve) over the baseband frequency range. While lining up a circuit by means of adjustable gain equalizers, a frequency–response plot was obtained to compare channel performance with requirements.

Circuit Testing. Test equipment was included in the wideband data test bay. This equipment measured digital error performance, signal or test power, interference and distortion, noise, impulse noise, and the transmission characteristics of local loops and carrier channels.

Trouble investigation was generally made by examining the eye pattern of a dotting sequence or a stream of random data at baseband frequency. The eye–pattern method of circuit observation is not an absolute means of testing for circuit malfunction; rather, it is an additional aid in trouble analysis. Transmission impairments such as attenuation/frequency or delay distortion cause regular closing of the eye while noise or phase hits cause wild traces through the center of the eye.

15–5 LOCAL AREA DATA CHANNELS

There is substantial demand for limited–distance data channels, generally used for short–range transmission at 300 baud through 19.2 kb/s via simple and inexpensive data sets. The application is usually intrabuilding or cross–campus, sometimes off–premises. Unlike the more expensive data sets used on voiceband channels, limited–distance or "line driver" data sets

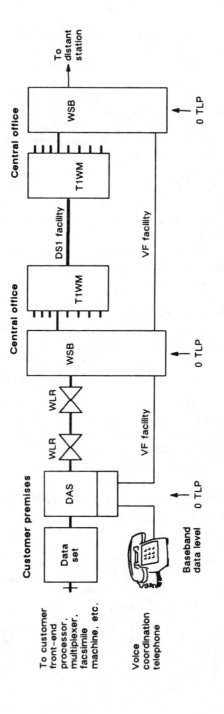

Figure 15-9. Typical group–band data circuit.

Table 15-8. Requirements: Two-Point Group-Band Data Circuit

Parameter		Inter-exchange Facility	Each Station Line
Envelope delay (μs)	– Slope	9.0	0.5
	– Sag	12.0	3.5
	– Peak	10.0	6.5
Gain deviation (dB)	– Slope	3.5	0.5
	– Sag	1.0	0.5
	– Peak	2.0	2.0
Noise at 0 TLP	Gaussian	64 dBrn	54 dBrn
	Impulse counts*	60	110
	SF interference	–30 dBm	–30 dBm
Digital errors in 5 minutes, max.		6	3
Gain at 25 kHz		0 ±0.5 dB	0 ±1.0 dB

*In 30 minutes, at 85-dBrn threshold.

use a semi-wideband signal that requires the use of nonloaded cable for the channel. A typical state tariff like the one in Table 15-1 offers the service on intra-wire-center circuits with up to six route-miles between stations at 2.4 or 4.8 kb/s, four miles at 7.2 kb/s, three miles at 9.6 kb/s, and two miles at 19.2 kb/s. There may be separate offerings of two-wire and four-wire channels. Loss limits for these channels are given in Table 15-9.

Table 15-9. Insertion Loss Limits—Local Area Data Channels

Tariffed Maximum Data Speed (kb/s)	Maximum Channel Length (Route-Miles)	Maximum Insertion Loss (dB) Frequency (kHz)					
		1.0	2.4	4.8	9.6	19.2	38.4
19.2	1	10.5	12.0	15.0	19.0	22.5	26.5
19.2	2	14.5	16.5	20.5	27.5	35.5	46.0
19.2	3	18.0	21.0	27.0	36.5	50.0	67.0
7.2	4	21.0	26.0	34.0	46.0	62.5	–
4.8	5	23.5	30.5	39.5	53.5	–	–
4.8	6	26.5	33.0	43.5	58.5	–	–

These loss figures are about 1 dB higher than the tariffed values in Table 15–1 because they are based on measuring through a channel service unit, now customer–provided, at each end. Power–spectrum limits for the data sets to be used on these channels are given in the Federal Communications Commission registration rules [7].

References

1. *Digital Cross–Connect System—Requirements and Objectives for the Digital Multipoint Bridging Feature*, Technical Advisory TA–TSY–000281, Bellcore (Iss. 2, June 1986).

2. Matthews, R. L. "Network Control in the Hands of the Customer," Bellcore EXCHANGE, Vol. 2, Iss. 2 (Mar./Apr. 1986), pp. 24–27.

3. Bell System Technical Reference PUB 41004, *Transmission Specifications for Voice Grade Private Line Data Channels*, American Telephone and Telegraph Company (Oct. 1973).

4. *IEEE Std. 743–1984*, "IEEE Standard Methods and Equipment for Measuring the Transmission Characteristics of Analog Voice–Frequency Circuits" (New York: Institute of Electrical and Electronics Engineers, Inc., 1984), Paragraph 4.3.2.

5. As [4], Paragraph 4.6.3.

6. Mahoney, J. J., Jr. "Transmission Plan for General Purpose Wideband Services," *IEEE Transactions on Communications Technology*, Vol. COM–14, No. 5 (Oct. 1966), pp. 641–658.

7. Federal Communications Commission. *Rules and Regulations, Title 47, Code of Federal Regulations, Part 68* (Washington, DC: U.S. Government Printing Office, Oct. 1987), Section 68.308(f).

Chapter 16

Program and Video Channels

Audio program and video are two of the specialized private–line channels provided by exchange carriers. While small in number, these offerings are important in the operation of the broadcast industry. Because of the high transmission quality that is needed, they involve transmission facilities and circuit designs that are not found elsewhere.

16-1 THE MARKET

Broadcasters have a number of choices in obtaining program and video channels, involving facilities that they can build themselves, that they can rent from specialized carriers, and that they can obtain from an exchange carrier. It is helpful to understand the market for these channels as well as their transmission characteristics.

Radio broadcasters do a sizable amount of "remote" programming via dialed telephone connections. Any connection involving a digital switch, a loaded cable, or a carrier facility is limited in bandwidth to the 200–to–3200–Hz range. This quality is satisfactory for most mainly–voice applications like sports pickups. Where the broadcaster wants to increase the naturalness of the transmitted program, a low–frequency extender is often used. At the originating end of the connection, the extender equipment uses analog heterodyne techniques to shift the entire audio band upward by 250 Hz. At the receiving end, the equipment converts the incoming signal down to its original frequency spectrum. The resulting overall connection then has a frequency response of roughly 50–3000 Hz or a bit more. In terms of perceived quality, the loss of frequencies at the top of the audio band is more than offset by the improved quality (two or more octaves added) at the bottom of the band. However, quality better than this is unobtainable at present via single dialed connections. Dual dialed connections are occasionally used, with suitable band–splitter and converter equipment, to give a response of 50 to 5000 Hz [1].

Looking ahead, coding standards are available to allow encoding of audio in the 40–Hz–to–7–kHz frequency range into a single 64–kb/s integrated services digital network (ISDN) B channel or a 56–kb/s Digital Data System (DDS) circuit [2,3,4]. This technique has considerable potential for high–quality broadcast use, on either a dialed–up or a private–line basis, in an ISDN environment.

For channels over distances of up to typically 30 miles, broadcasters may use radio facilities of their own [5]. Radio remote pickups are commonly done via FM radio in the 450–MHz band. Permanent studio–to–transmitter links (STLs) for radio use are often placed in the 950–MHz region. Television pickups and STLs are commonly found in the 7– and 13–GHz areas. Where a radio path approximating line–of–sight is unobtainable, broadcasters may use intermediate radio repeaters. Limitations on the broadcaster's use of radio include the large installation effort when complex setups are needed, the potential unavailability of a radio path, and the scarcity of unused frequencies in metropolitan areas.

For permanent radio facilities (e.g., 15–kHz stereo STLs), there is a growing interest in using exchange–carrier DS1 services ("T1 lines"), channelized by commercially available terminals supplied by the broadcaster. A single DS1 service can provide a 15–kHz stereo channel, an 8–kHz secondary program channel, and a group of two–way telephone and data channels.

Most network radio and television broadcasting currently uses satellite channels for long–haul transmission and distribution. The terrestrial networks that began in the 1920s [6] are long gone. Earth–station antennas of modest size (e.g., three to five meters) provide satisfactory downlinks for network use, the smaller sizes being suited to audio applications. For telecasting of major events like political conventions, the usability of temporary satellite uplinks may be limited by simple unavailability of space in which to locate the terminal vans without path blockage from buildings.

The dominance of satellite television facilities for network use is not necessarily assured. As the availability of 45–Mb/s (DS3) digital channels continues to grow, it becomes feasible to encode broadcast–quality analog television into a bit stream of this speed [7]. The resulting signal can then be sent, essentially unimpaired, over transcontinental distances via digital–radio and especially fiber facilities. These 45–Mb/s channels, connected into a remote–controlled switched network, promise the same flexibility for hour–to–hour network reconfiguration that satellite networks offer. They also offer excellent privacy compared to satellite transmission.

Compared to other alternatives available to the broadcaster, exchange–carrier program and video channels offer installation and maintenance without work on the part of the broadcaster, a significant feature for small radio stations without permanent engineering staffs. They require no capital investment by the broadcaster, involve specific assured levels of transmission performance, include maintenance at no additional charge, and avoid the need to search for or coordinate the use of radio frequencies that may not be available. They remove the broadcaster from considerations of whether a radio path can be obtained, and are available for major events where the station or network staff would otherwise be overtaxed. They are also available over distances beyond the broadcaster's ability to cover. Thus, there is a substantial market for both short–period and permanent channels.

Broadcasting is considered to be an interstate business, so program and video channels for this purpose are under the jurisdiction of the Federal Communications Commission (FCC). Rates and conditions for these services are included in the interstate access–service and special–construction tariffs filed by each regional company, by each major independent carrier, and by the National Exchange Carrier Association. Technical parameters for these services are given in the appropriate special access Technical References [8,9] and are listed in Chapter 18. By contrast, there is a small demand for audio and even video channels that are not extensions to interstate services for such uses as wired–music distribution or nonlicensed broadcasting on cable television (CATV) systems. Local state tariffs govern these applications, which do not fall under Federal jurisdiction.

16-2 AUDIO PROGRAM CHANNELS

Offerings

Four grades of service are commonly offered to broadcasters, differing mainly in terms of bandwidth. Table 16–1 lists the primary transmission requirements for these services. The channels are all offered on either a short–term or a permanent basis, as either two–point or multipoint services, for transmission in only one direction. They connect broadcasters' sites, interexchange carrier (IC) terminals, or both. For reference purposes, the table includes some designations that applied to earlier versions of these channels, and that are still encountered in state tariffs.

Channels of 3.5–kHz quality are general–purpose services for "occasional" network feeds, wired–music distribution, sports remotes, etc. The requirement for transmission up to 3500 Hz and the stringent signal–to–noise limit set them apart from ordinary voice–grade channels; it is difficult to meet both frequency–response and noise limits when using most analog carrier facilities, carrier channels connected back–to–back, or H88–loaded cable. However, digital carrier with modern digital banks, digital loop carrier (DLC), digital cross–connect systems (DCSs) for multipoint bridging, and nonloaded loops are suited to this offering.

Circuits of 5–kHz bandwidth are usable for medium–quality remote broadcasting. They are the minimum facility suitable for STLs for AM radio, where a traditional frequency–response limit of 100–5000 Hz $\pm$ 2 dB has applied to the entire radio station from local microphone to antenna. They are also used for such applications as feeding secondary programming [Subsidiary Communications Authorization (SCA)] to FM transmitters.

Services of 8–kHz quality are adapted to high–grade AM STLs, good–quality remote broadcasts of music events,

Table 16-1. Transmission Limits—Program Special Access Channels

Network Channel Code	Tariff Package	Frequency Response (Hz)	(dB)*	Max. Loss (dB)†	Signal-to-Noise Ratio (dB)‡	Total Harm. Distortion Max. (%)	Older Designations Tariff Series	Tariff Schedule
PB	AP0	300–2500	–2, +12	12	64	5.5	6001§	E§
PE	AP1	200–3500	–3, +10	32	65	3.5	6002§ 6003	D§ C
PF	AP2	100–5000	Fig. 16–1	32	64	2.5	6004§ 6005	B§ A
PJ	AP3	50–8000	Fig. 16–1	32	62	2.0	6006§ 6007	BB§ AA
PK	AP4	50–15,000	Fig. 16–1	32	71	1.0	6008§ 6009	BBB§ AAA

* Positive numbers correspond to increased loss.

† Gain conditioning for zero loss is also available.

‡ Referred to +18 dBm peak sending level; C–message weighted for PB and PE and 15-kHz flat weighted for the other channels. "Tone–on" time is limited to avoid crosstalk.

§ For part-time, recurring, or occasional use.

interstudio ties, etc. Competitive AM broadcasters in metro areas often use this quality, or even 15–kHz channels.

Channels of 15–kHz quality are the "premium" offering. They are the basic channel for FM STLs to meet a traditional overall performance of 50–15,000 Hz ± 2 dB and an FM signal–to–noise ratio from microphone to antenna of 60 dB. They are commonly ordered in pairs for stereophonic use, with care taken in layout to minimize phase differences between the left and right channels: the core phase–difference limits are 7.5°, 200 Hz to 4 kHz, rising at the band edges to 15° at 40 Hz and 20° at 15 kHz.

Figure 16–1 shows the frequency–response limits for 5–, 8–, and 15–kHz channels. These limits have been relaxed in recent years to allow an additional rolloff at the band edges; the requirements were originally ±1 dB across the entire band. (Channels offered under state tariffs for nonbroadcast use generally carry looser requirements, e.g., ±3 dB.)

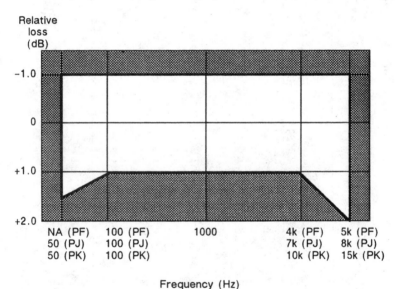

Figure 16–1. Frequency–response limits—5–, 8– and 15–kHz channels.

Table 16–1 includes a fifth offering. The PB channel is "non-equalized." It guarantees a frequency response to only 2500 Hz, a tradition dating from much earlier times when some now–vanished transmission facilities (e.g., H172–loaded cable, early open–wire carrier systems) provided frequency response barely extending above this figure. It is offered in only one regional-company tariff, partly because a broadcaster wanting a quality slightly less than the 3500–Hz PE channel may order and use an ordinary voice–grade access service (e.g., network channel code LB) for radio purposes. In addition to offerings in the table, there is a "PQ" channel with customized transmission parameters (e.g., envelope delay distortion) for specific needs. These channels are also quite rare.

Figure 16–2 shows a typical FM radio station and the network of program channels that it might use: an entrance link from a remotely located satellite receiver, an access channel from an IC terminal, various remote–pickup loops, a stereo STL for main-channel programming, and a monophonic STL of lesser quality for SCA programs.

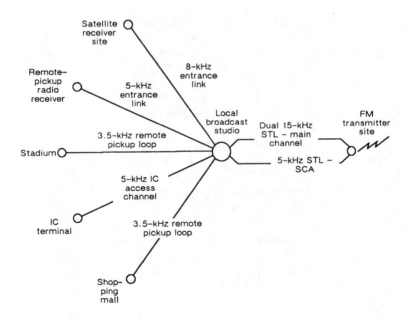

Figure 16–2. A typical layout of local program circuits.

16-3 OVERALL TRANSMISSION DESIGN—PROGRAM

The design of a program circuit starts with requirements: the points of origin and termination, the bandwidth required, the service date, and any special arrangements like multipoint operation. The type of channel ordered defines the noise, loss, and distortion limits. A review should be made of available facilities to determine any need for special construction, feasibility of using digital carrier, requirements for deloading of cable pairs, etc.

In laying out a 3.5-kHz service, the main concern is meeting the requirement of 10-dB maximum rolloff at the top of the passband. Short sections of loaded cable can be equalized with existing voiceband equalizers, but the feasible distance is severely limited. Nonloaded cable and single links of present-vintage digital carrier are preferred. On multipoint services, the bridging of circuit legs on carrier can be done best by a DCS equipped with multipoint bridge features, avoiding a second decoding step and thus preserving good frequency response.

In the use of cable facilities for 5-, 8-, or 15-kHz services, the cable sections should be short enough so that they can be readily equalized and so that good signal-to-noise ratios can be obtained. The losses in the available cable control the locations of amplifiers. The use of digital carrier facilities, either end-to-end or connected through a DCS, avoids most of the complications inherent in the use of multiple sections of cable.

Nonloaded cable pairs are used for program circuits provided they can be equalized for the bandwidth ordered. The 1000-Hz attenuation (the loss that would occur if the cable were terminated in its characteristic impedance) of a nonloaded cable pair to be used for a program circuit or for a section of such a circuit should not greatly exceed 12 dB if a satisfactory signal-to-noise ratio is to be maintained. Currently available equalizers can readily provide somewhat more than 20 dB of attenuation equalization (slope). Therefore, the loss of the cable pair should not sizably exceed 30 dB at the highest frequency to be used.

Maximum cable-section lengths, in terms of the 12-dB 1000-Hz loss and the 30-dB top-frequency loss, are shown in Table 16-2 for some commonly used types of cable. Where there

Table 16-2. Maximum Lengths—Nonloaded Cable

Cable		Maximum Section Length (mi)		
Gauge	μF/mi.	5 kHz	8 kHz	15 kHz
26	0.079	4.2*	4.1 †	3.1 †
25	0.064	5.1	5.0 †	3.6 †
24	0.864	5.1*	5.1 †	4.0 †
22	0.082	6.6*	6.6*	5.5 †
19	0.084	9.5*	9.5*	8.9 †

* Limited by 12-dB loss at 1 kHz.
† Limited by 30-dB loss at the top frequency.

is a choice, cables of coarse gauge and low capacitance are preferred. It will be noted that the length for 5–kHz use is always controlled by the loss at 1000 Hz, while the length for 15–kHz services is always limited by the loss at 15 kHz. For use of mixed gauges, the losses for the individual sections may simply be added together. A junction of two dissimilar cables produces a small reflection loss, but the reflection loss is greater at 1 kHz and below than at higher frequencies. Thus, mixed–gauge cable actually produces a small amount of self–equalization. Bridged taps of normal lengths are not a major problem; the equalizer has enough range to compensate for their effects.

Digital carrier, in either loop or interoffice form, is an attractive medium for program services. Some typical carrier channel units are listed in Table 16–3. Carrier avoids the need to deload cable pairs or do section–by–section equalization. The channel

Table 16-3. Carrier Program Channel Units

Service	Type	DS0 Slots Used	Coding Bits Used
3.5 kHz	ETO*	1	8
5 kHz	5 kHz	2	8
5 kHz STL	7.5 kHz	3	14/11
8 kHz	8 kHz	3	8
15 kHz STL	15 kHz	6	14/11

* Equalized transmission-only.

units that use 11–bit coding provide good signal–to–distortion performance on critical STLs and have very low noise with or without the presence of a signal.

Where stereophonic operation is involved, the designer must limit differences in length between the two channels. Use of a single carrier system for both ensures identical path length. Two carrier systems between the same two offices often involve considerably different span distances or dissimilar time delays in high–capacity multiplexers. Use of the same cable sheath for both channels ensures negligible difference in length; however, when this is not possible, differences in length of up to 1200 feet of 26–gauge cable (more on larger gauges) can still meet the limits for differences in phase.

Equalizers

Several types of equalizer are used with program circuits. One passive type, designed to mount in the same housing with a flat–gain amplifier if needed, can equalize nonloaded cable for 5–, 8–, or 15–kHz circuits. These equalizers are arranged for bridging across the line at the receiving end of a cable section, as shown in Figure 16–3. They are used in conjunction with a repeating coil connected for 150 ohms on the line (cable) side and 600 ohms on the drop (broadcaster) side. The equalizers are bridged onto the line side of the coil when equalizing to 5 or 8 kHz. They are usually connected on the drop side of the coil when equalizing to 15 kHz to obtain greater effect. Active equalizers, typically involving both low– and high–frequency adjustments, are incorporated into plug–in amplifiers and

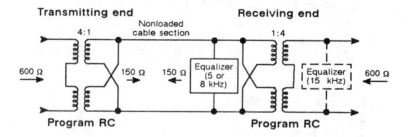

Figure 16–3. Program equalizer for cable.

digital–carrier channel units. As plug–ins, they are convenient to use and reuse; however, the amplifiers nominally require mounting shelves having shielded office cabling to ensure low noise and crosstalk. They require power when used at the customer's premises, but gain at the receiving end is essential when the customer orders a circuit with the zero–loss option. Special enclosures with radio–frequency shielding are available for use at transmitter sites.

The equalization process relies on two mechanisms. Basic equalization at low frequencies is obtained by deliberately terminating the cable in 150 ohms at both ends. At low frequencies, the characteristic impedance of the cable is quite high (e.g., 2910 – j2900 ohms for 26–gauge cable at 50 Hz). Compared with a 150–ohm load, there is a sizable mismatch. The result is a reflection loss of about 5 dB at each end. At high frequencies, the characteristic impedance is reasonably close to 150 ohms (e.g., 190 – j150 ohms for 26–gauge cable at 15 kHz), so the reflection losses become insignificant. Thus, basic equalization is obtained without adjustment. Final equalization is obtained by an adjustable passive or active equalizer at the receiving end of the cable section. The equalizer offers shaped loss to low frequencies. Thus, by careful design and alignment, an equalization tolerance of ±1.0 dB can be met routinely. When the circuit is divided into sections with amplifiers, equalization adjustment begins at the first equalizer and proceeds to each successive equalizer–amplifier so that the attenuation/frequency characteristic at the end of each section is as flat as possible. This procedure minimizes cumulative deviations.

Figure 16–4 shows the losses involved in a 14–kft section of 26–gauge cable. It shows the bare–pair attenuation, the actual insertion loss when the pair is terminated in 150 ohms but without an equalizer, and the typical results on a 15–kHz service after equalization. In this case, the equalized cable meets requirements with ample margin.

Program Amplifiers

When amplification is required, any of a number of available program amplifiers may be used. Such an amplifier must have a

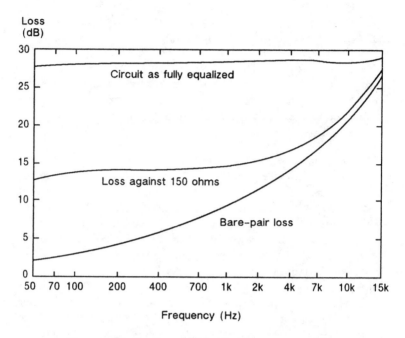

Figure 16–4. Loss—14–kft 26–gauge cable.

uniform frequency response over the circuit's required range. The output noise typically does not exceed 26 dBrn, program-weighted, when measured at full gain. The amplifier must be able to handle an output volume of at least +8 vu (+18 dBm peak) without overloading or distortion in order to pass instantaneous peak signals that are not measurable by use of vu meters. Amplifiers that meet these requirements are available in both ac- and dc-powered types. The ac-powered amplifiers are usual for locations that do not have 48-volt dc supplies (customer premises and installations on temporary or seasonal pickup loops). Where DLC facilities cannot be used, amplifiers can also be mounted in weathertight housings on poles or in manholes, using either commercial power or a central-office battery furnished over a separate cable pair. However, the addition of outside-mounted equipment tends to result in nonreusable investment after the circuit is ultimately discontinued. Amplifiers with multiple outputs are used in central locations when multipoint service is ordered.

Terminal Arrangements

A repeating coil is used at the sending end of an equalized program circuit. These coils are used so that imbalance in the termination cannot convert longitudinal noise to metallic noise, and to obtain a 600:150–ohm stepdown. They usually include electrostatic and electromagnetic shields and are of such quality that they do not cause equalization problems or low–frequency distortion.

Stereophonic Channels

Where the broadcaster, AM or FM, transmits stereophonic programs, two matched 15–kHz channels are provided. The two channels, designated left and right, are separate until they are combined at the transmitter.

If one channel is electrically longer than the other, the two portions of the signal are not in phase at the transmitter and the transmission/frequency characteristic of the combined signal is degraded by an amount dependent on the degree of phase shift. Because the difference in phase is greatest at the highest frequencies, a rolloff results at the upper end of the transmitted band of the combined signal. If the rolloff is not more than 1.0 dB at 15 kHz, the overall transmission requirements are met.

If the two channels are of equal length and if each has an attenuation/frequency characteristic that is uniform within ±1 dB over the range of 50 to 15,000 Hz, and if they are within 0.5 dB of each other throughout the band, the combination produces a response within established limits and monophonic reception of the combined signal is satisfactory. Matching for stereo use is not usually offered on 5– or 8–kHz channels.

Signal Amplitudes

Audio signals are normally delivered to the program circuit at +8 vu (or 0 vu in the case of 2.5–kHz services). This signal amplitude is acceptable for transmission in cable plant; satisfactory crosstalk coupling losses usually exist between the program circuit and other facilities. If lower values of signal were delivered to the program circuit, the signal–to–noise ratio would suffer unless the loss ahead of the first amplifier or carrier channel were correspondingly less than the maximum allowable for noise control and customized levels were used.

Gains of amplifiers and of carrier channel units are normally adjusted for a program output of +8 vu for services other than 2.5 kHz. Since program signals are not normally available or suitable for lineup purposes, it is customary to use a 1004–Hz test signal of 0 dBm at the sending end and to adjust each amplifier or carrier channel for 0–dBm output. Circuits normally have equalized losses not exceeding about 32 dB in any amplifier section. Therefore, program signal amplitudes are usually above −24 vu at all points.

16-4 PROGRAM CHANNEL NOISE

An understanding of noise objectives is essential to good transmission design of program circuits. The choice of facilities and equipment is as important in terms of noise control as from the standpoint of high–quality transmission.

It is common practice in the broadcast industry to express performance and requirements in terms of signal–to–noise ratio, whereas in the telecommunications industry noise is usually measured and expressed in terms of dBrn. In order to relate these practices, program circuit noise is expressed in dBrn referred to a +8 vu point in the circuit. This point is usually at the terminating end of a zero–loss circuit or at the output of an intermediate amplifier. It is analogous to a transmission level point in message circuit design. A noise measurement at any point on the line may be corrected to the reference point by adding the 1004–Hz loss of the facility from the reference point to the point of measurement; it may then be converted easily to a signal–to–noise ratio.

The dynamic characteristics of vu meters are such that instantaneous program signal peaks are substantially higher than the observed readings. A peak factor of 10 dB is assumed and a nominal test tone of 404 Hz at +18 dBm and having a duration of less than four seconds is used to adjust broadcast transmitters for 100–percent modulation. Similar limited durations apply in the measurement of distortion. Since signal–to–noise requirements are based on 100–percent modulation, signal–to–noise ratios are based on peak signal amplitudes, i.e., +18 dBm. With the required signal–to–noise ratio and signal amplitude known, noise limits can be expressed in dBrn. Noise requirements must be met in order to comply with FCC rules for broadcast services. These limits, referred to a +8 vu point, range from +37 dBrn to +46 dBrn depending on the type of channel and the weighting network to be used in the measurement (C–message for 2.5 and 3.5 kHz; 15–kHz flat for the others).

Program circuits usually meet noise requirements if they are short enough to be equalized. In those cases where nonloaded cable does not appear short enough to equalize, intermediate amplifiers may be added or carrier may be used instead. Where Digital Data System circuits appear in the same cable as 15–kHz program circuits, the loop–coordination rules pertinent to DDS avoid interference to the program circuit.

16–5 TELEVISION CHANNELS

Compared to audio program channels, the variety of television channels is quite small. The basic purpose of the channel is to transport broadcast–quality color signals conforming to the National Television System Committee (NTSC) format [8] within the local access and transport area (LATA). The broadcaster's signal involves nominally 60 (59.94) scan fields per second, 30 complete frames per second, 525 scan lines per frame, a horizontal sweep rate of 15.75 (15.73426573) kHz, and a color subcarrier at 3.58 (3.579545) MHz. The resulting composite video signal requires a passband essentially flat from 10 Hz to 4.5 MHz, with controlled envelope delay distortion. The television signal is described in greater detail in Chapter 15 of Volume 1.

The video signal usually carries a single or dual (stereo or bilingual) audio channel diplexed with the video by use of FM–modulated carriers at 5.8 and possibly 6.4 MHz, applied at a level 20 dB below peak video. Separate (nondiplexed) audio channels are offered but are rarely used.

Table 16-4 gives the basic characteristics of the television channels listed in the access service tariffs. Main performance limits for these offerings are listed in Chapter 26 of Volume 1. Television channels provided under state tariffs, as for nonbroadcast educational or industrial use, are essentially the same.

Table 16-4. Television Channel Offerings

Channel Code	Tariff Package	Audio Response	Audio Medium
TV	TV1	15 kHz	Usually diplexed
TW	TV2	5 kHz	Usually separate

With the "TV" channel (15-kHz audio), phase and gain differences between the sound channels are controlled in a manner similar to that for 15-kHz stereo program channels. These channels are offered as one-way, two-point only, full-time or short-period, with one or two audio channels. The end points are specified by the customer, and may include an exchange carrier "hub" office. One regional company does not offer a TW channel.

Where "special construction" is required, generally involving unusual expenses (as for temporary pickups) or nonreusable investment (as in some permanent circuits), the particular extra charges or termination liabilities are filed in the applicable special-construction tariff. In addition to the standard TV and TW offerings, there is a "custom" TQ channel comparable to the PQ audio channel with negotiable technical specifications. The required frequency response of the channel depends on route mileage, as shown in Figure 16-5. It is consistent with *ANSI Standard T1.502-1988* [11], which is intended to replace the older *EIA Standard EIA-250-B* [12], and *NTC Report No. 7* [13]. The video interface with the customer may be either 75 ohms unbalanced or 124 ohms balanced, with a signal level of 1.0 volt peak-to-peak video and 0.1 volt peak-to-peak subcarrier. The FM subcarriers use the same modulating preemphasis as is used in FM broadcasting—with a 75-μs time constant, corresponding to 17 dB of boost at 15 kHz relative to 400 Hz. Different interfaces are often used at the two ends of the channel, e.g., 124-ohm balanced at one end with diplexed audio at one end, and

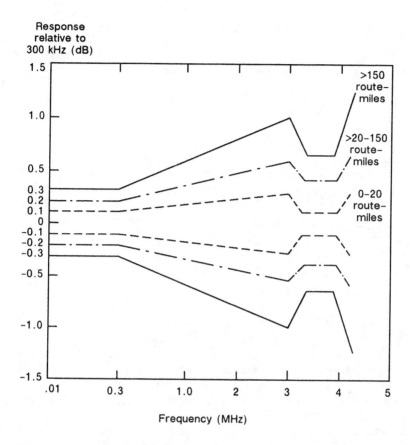

Figure 16-5. Frequency response tolerance for video circuits.

75-ohm unbalanced with two additional two-wire audio interfaces at the other end.

The audio channels included with the TV channel carry frequency response requirements given in References 9 and 11: referred to 404 Hz, the loss may not be less by more than 0.5 dB anywhere in the 50-Hz-to-15-kHz band; it may be more by up to 0.5 dB, 100 Hz to 7.5 kHz. The tolerance of added loss at the band edges is comparable in shape to that in Figure 16-1: it rises from 0.5 dB at 100 Hz to 1.0 dB at 50 Hz; likewise, it increases from 0.5 dB at 7.5 kHz to 1.5 dB at 15 kHz.

377

16-6 OVERALL TRANSMISSION DESIGN—VIDEO

The layout of a video channel enjoys only limited flexibility in the choice of facilities because of the large bandwidth. However, a number of options exist.

Coaxial Cable

Coaxial cable of 75–ohm impedance (RG–11/U, and various similar types) is usable for short links, as in acquiring signals on the grounds of a remote pickup site. With lengths beyond perhaps 200 feet, however, it is exposed to the pickup of low–frequency crosstalk and 60–Hz hum via ground loops. Special hum–reduction transformers are helpful but cannot provide complete relief.

Video Pairs

Some older line facilities for local links, designed especially for video, are 16–gauge polyethylene–insulated (16 PEVL) pairs with longitudinal and spiral copper shields. These pairs are incorporated into normal sheathed cable, often with regular telephone pairs. This construction gives a cable–pair impedance that can be held to close tolerances, giving minimal echoes from manufacturing irregularities. Because of the balanced design and effective shielding, there is no crosstalk limitation on the number of circuits within a cable and the noise level is low. The characteristic impedance of video pairs is almost a pure resistance of 124 ohms at high frequencies. Attenuation at 4.5 MHz is 18.6 dB per mile at 75°F. Since there is some variation in attenuation due to temperature changes (approximately one tenth of one percent per degree Fahrenheit), these cables are normally placed underground.

The 16–gauge video pairs are terminated on shielded–cable terminals for connection with office–type cables. Various solid–dielectric coaxial and shielded–pair lines are used for office cabling. Office cables for connections to cable terminals, patching jacks, and amplifiers are kept short to minimize noise susceptibility. Where cables longer than a few hundred feet cannot be avoided, connecting–cable equalizers are often necessary.

Gain and equalization on wire facilities are provided via two general families of equipment. One, the A4 system, is used for one–link, nonrepeatered circuits of less than 0.5 mile. A more elaborate family made by several manufacturers is capable of multilink circuits with repeater sections up to 82.5 dB of loss at 4.5 MHz, i.e., 4.5 miles of 16–gauge video cable. Under good conditions, multiple–repeater circuits of up to 15 miles or so are feasible; more typically, the need to control differential gain and phase and to maintain good signal–to–noise ratio may limit the section length to 72.5 dB and the total length to about ten miles.

Wire video transmission involves the amplification of weak signals at the ends of long cable sections where the high frequencies have been attenuated to as low as 67 dB below 1 volt. Amplifiers receiving these low–level signals are susceptible to outside disturbances. Noise is avoided by installing the equipment so that it is not subject to physical or electrical disturbance. Precautions are required to bond and ground equipment frames properly and to eliminate differences of potential between the video system ground and the ground used for broadcaster equipment, particularly at remote pickup sites.

It is useful to measure the video signal in terms of the peak–to–peak voltage including the synchronizing pulse and to express the amplitude in dBV. One volt peak–to–peak is the reference and is defined as a level of 0 dBV. Other voltages are compared to this reference by the relation $(20 \log E/1.0)$ dBV. With the reference level established, the signal amplitude at any point in a circuit is used to identify the transmission level at that point. Circuit gain or loss is the difference in dB between these levels. (Note the difference from CATV practice, in which one millivolt is the reference and levels are expressed in dBmV.)

It is convenient to specify levels at two frequencies (e.g., −10/+5 dBV) to give the slope of the attenuation/frequency characteristic. The first number (−10 in this example) refers to the transmission level near zero frequency, while the second number (+5) refers to the level at 4.5 MHz. The numbers specify the amplitudes of sinewave test signals at that point resulting from 0–dBV signals introduced at a 0/0 point. Thus, at any point, the numerator of the level fraction is the voltage that would be measured if a low–frequency test signal of 0 dBV were applied to the transmitting terminal. The denominator of the level fraction is the voltage that would be measured if a 0–dBV signal at 4.5 MHz were applied at the transmitting terminal.

This level designation is convenient for expressing the slope of the transmission characteristic. For example, assume that a 0/0 signal is applied to a transmitter to feed two miles of cable having 36 dB of slope between near–zero and 4.5 MHz. If there is 15 dB of preequalization, the output of the transmitter is −10/+5 dBV; with the 36 dB of cable slope, the level at the receiver input would be −10/−31 dBV. As another example, an equalizer with 20 dB of slope and 4 dB of flat loss would be referred to as having a transmission characteristic of −24/−4 dB; the terminology used above for levels is used here to define loss. Thus, at near–zero frequency, the loss of this equalizer is 24 dB; at 4.5 MHz, the loss is 4 dB.

The design of the cable system provides high–frequency preemphasis of 0 to 32.5 dB at the transmitting terminal. The low–frequency transmission level on the video pair is maintained at −10 dBV for all system layouts. This is possible because the cable loss is essentially zero at zero frequency. Thus, the maximum high–frequency level on the line leaving the terminal or repeater is +22.5 dBV.

At repeaters and receiving terminals, the low–frequency input level is −10 dBV. The 4.5–MHz input level depends on the loss of the preceding line section and the output level of the preceding amplifier. For noise control, the 4.5–MHz level at the input to an amplifier should be no lower than −60 dBV.

The output of the receiving terminal is normally 0 dBV for either a 75–ohm unbalanced or a 124–ohm balanced output, matching the standard customer interface level. In a television center equipped with attenuators for minor level adjustments, the 124–ohm balanced output may operate at +2 dBV.

Equalization

The equalization of a system requires control of the attenuation and delay characteristics over the transmission band.

Manufacturers use various approaches to equalization. One method is to use fixed plug–in cable equalizers in values from 2.5 to 20 dB in 2.5 dB steps, each having an attenuation/frequency characteristic inverse to that of average video cable. Combinations of these fixed units equalize the system to within ±1.25 dB. Variable equalizers at the receiving terminal supplement the fixed equalizers to provide final adjustment of the attenuation/ frequency characteristic and to compensate for miscellaneous variations. A series of delay equalizers, also installed in the receiving terminal, is used to correct the residual envelope delay distortion not compensated by the fixed equalizers. They introduce a small amount of slope, which is corrected with the variable equalizers.

Since video signals are highly sensitive to echoes (ghosts) resulting from mismatched cable terminations, this family of equipment includes complex–impedance terminating networks that simulate the characteristic impedance of the cable as it changes with frequency. This is considerably different practice from other services, which use fixed resistive impedances (135, 150, 600 ohms, etc.) to terminate the cable.

Analog Fiber Systems

Conventional video pairs are in declining use because they are not readily usable for services other than television and, if combined with ordinary pairs in the same sheath, cannot be removed independently. Thus, they tend to lead to "stranded investment" when a video circuit is moved or disconnected.

Fiber facilities are much more flexible and consistent with present–day plant growth. As a result, there is fairly wide application for analog video transmission over fiber via pulse frequency modulation techniques. This modulation method avoids the non-linearities that are inherent in direct analog modulation of a light–emitting diode (LED) or laser diode.

One typical fiber link of this type offers either LED or laser light sources operating at wavelengths of 0.83 or 1.30 μm. It provides a 10–MHz video bandwidth, with both balanced and unbalanced inputs/outputs, and can equalize the office cabling at the sending and receiving ends. Audio subchannels are available by using carriers above the video, specifically the 5.8 and 6.4 MHz frequencies that are the usual standard for diplexing. A non-demodulating repeater is available for use over extended distances.

Cable section lengths obtainable with this system vary with the options chosen. Typical attainable distances are shown in Table 16-5.

Table 16-5. Typical Fiber Section Lengths

Wavelength	Transmitter	Cable	Distance (mi)
0.83 μm	LED	Multimode	2
	Laser	Multimode	6
1.30 μm	LED	Multimode	6
	Laser	Multimode	16
	Laser	Single mode	22

These fiber systems offer the bandwidth to handle high-definition video using, for example, 1125 scan lines. Some of them provide passbands up to 25 MHz wide. Others offer the option of a 1.55-μm laser transmitter and an operating range of 40 miles on single-mode fiber.

Microwave Radio

Analog microwave radio systems have been used for television transmission since they first became available. Almost any system in the 4-GHz or higher bands is suitable for a video signal plus audio carriers. However, most such operation is in the 11-GHz area for reasons of relative ease of obtaining frequencies in crowded areas and of small antenna size. These systems, of course, are one-way, single-antenna installations. They usually involve short radio paths not needing protection, but use a frequency spectrum shared with satellite services.

For remote pickup "TV pool" use, a variety of portable 6-, 11-, 18-, and 23-GHz microwave radio systems is available. These are generally tripod-mounted units with the transmitter or receiver built on the rear of a small parabolic antenna. They are "frequency–agile" designs capable of quickly switching among the channels in the licensed band. They are licensed on a blanket basis covering an area and require only relatively simple frequency coordination before use.

Digital Systems

A facility of growing importance for video use is the 45–Mb/s DS3 channel. Direct digital coding of an NTSC video signal, at a reasonable sampling rate (e.g., 11 megasamples per second) and with 8–bit coding, yields high quality but requires about 90 Mb/s or two DS3s. Nine– or ten–bit coding requires even more bit capacity. However, the use of any of a number of intraframe or interframe coding techniques to reduce redundancy can deliver full–motion broadcast–quality video and stereo audio via a single DS3 [14], and can provide teleconference or broadcast facilities [15] for general–purpose use. With further compression, video teleconferencing (but not broadcast video) can operate on a 1.5–Mb/s channel. Coding techniques currently in use are equipment–specific; however, it is likely that a standard compatible method will emerge in the next few years. Digital transmission can deliver essentially the same high quality across a continent as across a LATA, obsoleting transmission requirements based on distance. It is consistent with the trend toward interconnection of exchange carriers and ICs on a DS3 basis. Detailed operating measurements of equalization, noise, differential gain, etc., are replaced by simple error–rate tests.

Television Operating Center

The location containing the equipment to process analog television services in a major city is the television operating center (TOC), or "hub" in tariff terms. Operations performed in the TOC include the administration of service orders, setting up and testing television circuits, switching of circuits, and maintaining the services after they have been established. Close cooperation is required among the TOC, studios, and remote pickup sites.

Circuits are brought into the TOC where their video levels can be adjusted to standard value and where their frequency characteristics can be equalized. Testing, monitoring, and switching are performed at a reference point within the TOC video switch or at a splitting pad. A few video switches are arranged for remote control by the broadcasting user.

The TOC video layout varies with local conditions. Generally, radio terminals carrying television circuits are somewhat removed from the TOC and are connected to the video switch by connecting circuits having jack appearances in the TOC video patch bays.

The TOC test positions provide for making video tests under standard measuring conditions that eliminate uncertainties due to unequalized cable lengths. At these test positions, transmission may be evaluated to and from any point.

The video monitoring equipment at the TOC normally consists of a waveform monitor, a vectorscope, and a picture monitor. A subcarrier demodulator and modulator are added for testing the sound channels. The video monitoring circuit is looped through the picture monitor and is connected to the waveform monitor and vectorscope to yield simultaneous displays. Cable lengths in the transmitting circuits, receiving circuits, and monitoring equipment are chosen so that the test equipment and jack fields are at a flat 0/0 point.

References

1. Church, S. "Telephone Systems and Interfacing—Part 1: Interfacing to the Dial–Up Network," *NAB Engineering Handbook*, Seventh Edition (Washington, DC: National Association of Broadcasters, 1985), p. 6.4–51.

2. CCITT Rec. G.722, "7–kHz Audio Coding Within 64 kb/s," Study Group XVIII Report (Aug. 1986).

3. Petruschka, O. "International Standard for 7–kHz Audio," *TEL–COMS* (July 1988).

4. Anderson, M. M., G. W. Pearson, and O. Petruschka. "Digitized Voice for Hi–Fi Performance," Bellcore EXCHANGE, Vol. 4, Iss. 2 (Mar./Apr. 1988), pp. 8–12.

5. Federal Communications Commission. *Rules and Regulations, Title 47, Code of Federal Regulations, Part 74* (Washington, DC: U.S. Government Printing Office, 1987), Sections 74.401–74.663.

6. Sibley, L. A. "Program Transmission and the Radio Networks," *AWA Review*, Vol. 3 (Holcomb, NY: Antique Wireless Association, Inc., 1988).

7. Blackburn, R. J. "Bellcore Makes Progress in U.S. Field Trial of Land–Based Digital Network for Broadband Television Distribution/Collection," *Bellcore Digest of Technical Information*, Special Report SR–TSY–000104, Vol. 4, No. 5, Bellcore (Aug. 1988), pp. 1–2.

8. *Program Audio Special Access and Local Channel Services— Transmission Parameter Limits and Interface Combinations*, Technical Reference TR–NPL–000337, Bellcore (Iss. 1, July 1987).

9. *Television Special Access and Local Channel Services— Transmission Parameter Limits and Interface Combinations*, Technical Reference TR–NPL–000338, Bellcore (Iss. 1, Dec. 1986).

10. Federal Communications Commission. *Rules and Regulations, Title 47, Code of Federal Regulations, Part 73* (Washington, DC: U.S. Government Printing Office, 1987), Sections 73.681, 73.682, and 73.699.

11. *ANSI Standard T1.502– 1988*, "American National Standard for Telecommunication—System M—NTSC Television Signals—Network Interface Specifications and Performance Parameters" (New York: American National Standards Institute, 1988).

12. *EIA Standard EIA–250–B*, "Electrical Performance Standards for Television Relay Facilities" (Washington, DC: Electronic Industries Association, Sept. 1976).

13. Network Transmission Committee. *NTC Report No. 7*, "Video Facility Testing—Technical Performance Objectives" (Washington, DC: Public Broadcasting System, Jan. 1976).

14. Limb, J. O. et al. "Digital Coding of Color Video Signals—A Review," *IEEE Transactions on Communications Technology*, Vol. COM-25 (Nov. 1977), pp. 1349–1385.

15. Judice, C. N. "Report from the Frontier: Visual Telecommunications," Bellcore EXCHANGE, Vol. 4, Iss. 3 (May/June 1988), pp. 14–19.

Chapter 17

Digital Services

Digital services are the major growth sector in the exchange carriers' special–services offerings. This chapter covers data services provided by the Digital Data System (DDS). It also addresses newer offerings called Basic Dedicated Digital Service (BDDS), furnished by a related but simpler network plan. It describes the 56–kb/s dial–up data service known generically as Public Switched Digital Service (PSDS). Finally, it covers high–capacity digital channel services at rates of 1.5 and 45 Mb/s.

17–1 THE DIGITAL DATA SYSTEM AND THE BDDS NETWORK

The DDS is a dedicated facility *network* that provides *services* known originally as Dataphone® Digital Service, but now identified by a variety of names specific to particular exchange carriers. No explicit name attaches to the newer *network* that supplies Basic Dedicated Digital *Service*; in this chapter, the term "BDDS network" is used. Like DDS–based services, BDDS, a generic term, carries local service names. The DDS and BDDS networks share the same technology with regard to customer interfaces, service speeds, loop transmission, and synchronization. They differ in terms of interoffice multiplexing, hubbing techniques, test access, and nominal service reliability.

These networks are designed for the service needs of a growing class of customers. Digital data transmission is required widely for communication between data terminals. The DDS is a network independent of, but sharing facilities with, the switched message network. The BDDS network shares those facilities more extensively. Provision has been made in the network plan for use of

---

Dataphone is a registered trademark of AT&T.

existing and new facility designs, for a rapid or slow growth rate, and for a variety of service speeds [1,2]. Service is offered to points within the local access and transport area (LATA) and also to interexchange carriers (ICs) for LATA access use [3].

These are all–digital systems, with service objectives [3] stated in terms of digital parameters. Signal formats are digital throughout the system; the data signals are combined and separated by time–division multiplexing (TDM). Signal processing is facilitated because it involves mainly logic functions.

Initially, the telecommunications network was composed exclusively of analog transmission facilities. Customer–generated digital signals, normally generated in the form of baseband binary pulses, could not readily be transmitted over these facilities. Signal processing that was necessary to transmit digital data was provided by modems. These and other forms of terminal equipment handled digital data at speeds appropriate to the voiceband and, in addition, provided higher–speed transmission by using wider–than–voice bandwidth [4]. The DDS provided the first network for fully digital transmission as an alternate to analog technology. The BDDS network, a more diffused follow–on, made digital services affordable in smaller serving areas with fewer circuits.

These networks provide duplex point–to–point and multipoint private–line data transmission at a number of synchronous data rates called service speeds. Alternate–voice service and voice circuit coordination are not provided directly, but may be derived via suitable customer–provided terminals. Four–wire duplex operation eliminates transmission delays inherent in reverse–channel or turnaround operation on half–duplex channels. In addition to high facility utilization and consistent transmission performance offered by both networks, the DDS provides high end–to–end reliability, low average annual downtime (time out of service), and the ability to monitor and protection–switch most transmission facilities and terminals on an in–service basis.

Service objectives have been established for both networks. Signal formats and multiplexing arrangements are specified. The network configuration permits logical growth and flexibility in rendering service. The longer facilities are used efficiently, particularly in the DDS, to provide an economical system.

388

Basic differences between the two networks are structural: the DDS is a highly centralized design, with typically a single hub office per LATA; the BDDS network uses several simple "service nodes" per LATA. The DDS employs special–purpose, dedicated equipment with unusually complete service–protection features, while the BDDS network uses general–purpose facilities and equipment such as digital cross–connect systems (DCSs). The DDS includes protected interoffice facilities throughout; the BDDS network, to a lesser degree. They use compatible technology, so a BDDS circuit can actually be routed through DDS facilities where convenient.

Service Objectives

Stringent service objectives are important factors in establishing system design and administration. The three primary concerns are *quality*, *availability*, and *maintainability*.

The objective for transmission *quality* is that there should be at least 99.5–percent error–free seconds between stations in the LATA. This objective relates to the efficiency of data communications, since errors often reduce throughput (a measure of transmission efficiency) by requiring retransmission of blocks of data. The percentage of error–free seconds pertains to all service speeds and may be translated to error–ratio limits for each service speed. The 56–kb/s requirement is very stringent and is used for allocation of requirements to each portion of the two networks.

The objective for availability is that DDS circuits within a LATA should have at least 99.96–percent long–term availability or, in terms of outage, an average downtime of no more than 3–1/2 hours per year. The term *availability* is preferred to *reliability* since the latter may be construed as the percentage of time the channel is both connected–through and satisfactory error–wise. In other words, quality and availability together determine the reliability of the channel. For multipoint circuits, the objective applies to communication between the master station and any particular outlying station.

The objective for *maintainability* is that no single DDS outage should exceed two hours, an objective that recognizes the perishable nature of much data and the increasing impact on customer operations as an outage persists. This objective relies on substantial mechanization of trouble detection and location, and on emergency restoration practices to minimize the occurrences of long outages.

Signal Format

With the evolution and expansion of digital transmission facilities, new alternatives for efficient transmission of binary data signals became available. Digital carrier systems began with D–type pulse code modulation (PCM) channel banks to encode 24 voiceband signals into a TDM signal for transmission over a regenerative repeatered line. The resulting 1.544–Mb/s line speed is defined as the DS1 rate. These systems became predominant because they were almost universally more economical than analog systems for deriving voice–grade channels, while giving equivalent or better voice quality. Their adaptability to digital data transmission was demonstrated after their success in voice service.

Since the voice–frequency interface to the D–type channel bank is analog, there is no overwhelming advantage in choosing T1 systems as links for conventional data private lines. However, significant improvements in both efficiency and data performance are realizable if the input digital data signals can be received and regenerated at the telephone office in a baseband digital format. They can then be treated by logic processes and time–division multiplexed directly onto a T1 line or other DS1–rate facility without analog–to–digital conversion.

The first applications of these digital processing techniques were in a series of wideband T1 carrier terminals designed to give a number of standard bit rates for optional synchronous or asynchronous operation. The maximum rate, achieved by dedicating the entire T1 line signal to the wideband service, was 500 kb/s. In terms of bit rate per voice channel displaced, this was equivalent to about 20.8 kb/s per channel. When the theoretical 64–kb/s digital data capacity of a voice channel slot in the DS1 signal is considered, the 20.8 kb/s per channel was relatively inefficient even though it was about eight times as fast as could be

390

obtained through analog techniques at the time. The cause of the inefficiency was that, in adapting the synchronous DS1 bit stream to the transmission of asynchronous signals, three DS1 pulse positions were required to encode one data signal transition.

Much higher efficiency can be achieved if all data inputs are synchronous and if individual input bit streams are at clock speeds that are submultiples of the DS1 rate. This is the basis of the DDS and BDDS networks. The total usable bit rate is then a substantial fraction of the DS1 rate itself. Pulse slots must still be reserved for identification and demultiplexing of individual data signals (framing), for transmission of status and control signals, and for ensuring that bit pulses occur often enough in the multiplexed stream for clock recovery. Nevertheless, these networks can achieve up to 58.667 kb/s per voice channel displaced, which is 92 percent of the theoretical maximum. This figure represents 56 kb/s of data plus an optional "secondary channel" that raises the rate to 58.667 kb/s. This efficiency, together with the high quality of service achieved through tight control of errors and a coordinated maintenance plan, makes digital networks an attractive medium for data transmission.

The data signals are transmitted in specified signal formats and are limited to several choices of service speed and TDM arrangements.

Baseband digital data signals are multiplexed into the same DS1 bit stream as that formed by digital switches and D–type channel banks. This format consists of 24 sequential 8–bit bytes (channel or framing slots) plus one framing bit; the sequence repeats 8000 times a second. Thus, the multiplexed data signal is of the right format to be transmitted over any facility in the digital hierarchy. Moreover, use of a multiplexed data word consistent with the DS1 voice–channel structure permits formation of DS1 signals containing both voice and data in cases where dedication of a full DS1 facility to data transmission would be inefficient.

Digital Data Capacity and Service Speeds. Since each 8–bit byte recurs 8000 times a second, the data capacity of each channel slot is theoretically 64 kb/s. However, for actual use, this 8–bit byte requires the reservation of one of the eight bits (designated the C bit) to pass network control information and to satisfy the minimum pulse–density requirement for clock recovery on T1 lines. Thus, use of the seven bits remaining in the byte results in a maximum service speed per displaced voice channel of 7 bits/byte × 8000 bytes/second = 56 kb/s. This maximum rate is one of the standard service speeds. To add a secondary channel at 2.667 kb/s, the C bit is time–shared every third frame. Three other service speeds, called subrate speeds, are provided at 2.4, 4.8, and 9.6 kb/s with the option of secondary channels at 0.133, 0.266, and 0.533 kb/s respectively. The secondary channels are useful in multipoint customer data networks for purposes of centralized monitoring and control, comparable to the secondary–channel network management systems used on analog private lines. Some exchange carriers also offer service at 1.2 and 19.2 kb/s.

For the lower service speeds, submultiplexing (used primarily in the DDS) requires the reservation of an additional bit per byte to establish synchronization patterns, routing each byte to the proper output port of the receiving submultiplexer. Therefore, the maximum capacity of a subrate byte (in the absence of secondary channels) is 48 kb/s. This byte in a DS1 frame may be assigned to five 9.6–kb/s, ten 4.8–kb/s, or twenty 2.4–kb/s channels respectively.

Multiplex Subhierarchy. The DDS and BDDS networks may use facilities in any part of the digital multiplex hierarchy, DS1 through DS4 or higher. However, to use facilities efficiently over long distances, a subhierarchy of TDM arrangements has been provided in the DDS. Two signal formats used in the subhierarchy are ultimately combined into a DS1 signal for transmission. The relationships between these formats and the parts of plant in which they appear are shown in Figure 17–1. The customer signal, at one of the service speeds (2.4, 4.8, 9.6, 19.2, or 56.0 kb/s), is converted at the loop input to a 50–percent duty cycle return–to–zero (RZ) signal at the selected service speed for transmission over the loop to the serving central office. (Where the secondary channel is used, the rate sent into the loop is higher, comprising 3.2, 6.4, 12.8, or 72.0 kb/s for the four main service speeds respectively, because of an added framing bit.)

At the central office, the signal is converted to a 64–kb/s, bipolar, 100–percent duty cycle, non–return–to–zero (NRZ) signal

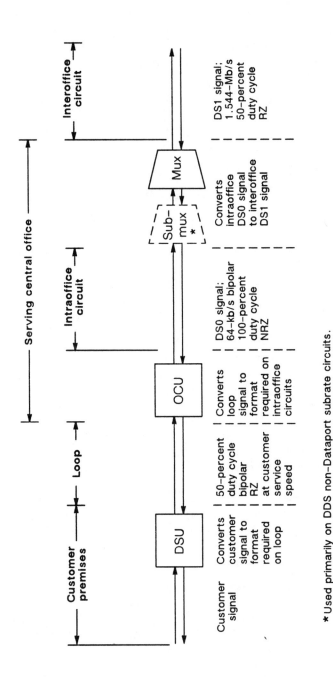

Figure 17-1. Hierarchy of data signals.

* Used primarily on DDS non-Dataport subrate circuits.

for intraoffice transmission. The format, called the DS0 signal, may contain multiplexed (packed) signals of like service speeds, particularly in larger DDS offices. It may also carry only a single data signal of the customer service speed, carried on the DS0 signal by a process called stuffing.

Finally, the DS0 signal is processed for transmission over interoffice facilities. The processing consists of stuffing and multiplexing signals for transmission in the DS1 format.

Network Configuration

These networks are flexible enough to reach almost any area that has service demand. Exchange access, described in Chapter 18, is provided to ICs for their interLATA use.

As shown in Figure 17–2(a), the hubbing plan for DDS includes three broad classes of offices, each serving as a concentration point for customer data streams. The offices are organized in a hierarchy based on their expected size in terms of circuits. The hierarchy for BDDS, given in Figure 17–2(b), is simpler, using only end offices and service nodes.

Hub Office. The typical DDS–equipped LATA contains a single hub office. This is the cross–connect and test access point for all stations in its serving area; hence, it is generally the location of the serving test center (STC). Signals from the hub area are concentrated at the hub into efficiently packed data streams for transmission to distant offices or delivery to ICs. The hub also contains the timing supply from which all timing supplies in its homing offices are synchronized, unless a universal synchronization system is available in all offices in the area. BDDS service nodes are conceptually similar, but rely on simpler equipment (D4 channel–bank mountings, DCSs, and switched–access test systems) shared with ordinary services. There is little focus in BDDS on technically efficient submultiplexing, given the declining costs of digital facilities.

In some cases, there are lower–ranking DDS offices termed collection hubs. These are placed at the center of areas of major demand and are given cross–connection facilities so as to collect circuits from outlying end offices for transmission to the main hub.

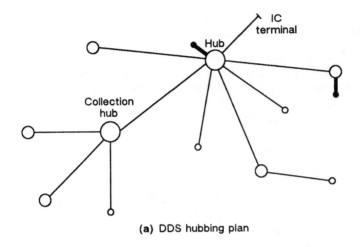

(a) DDS hubbing plan

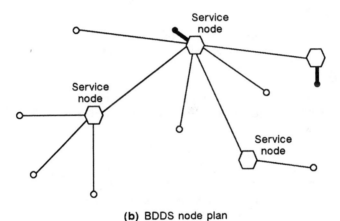

(b) BDDS node plan

○   Mux office (T1DM/T1WB5)
o   Dataport office
●   Baseband office
━━   Trunk cable
──   DS1 facility

Figure 17-2. Network configuration.

End Offices. The hub office directly serves all customer locations within its baseband loop transmission range. However, clusters of customers are often served by center–city wire centers

nearby. These offices are designated baseband offices and serve their customers via trunk cable from the hub.

Where transmission limits or economics require, other (DDS) local offices provide subrate concentration; the resultant signals are multiplexed for transmission over DS1 facilities to the hub office. Since all individual channels must connect with a central point for maintenance purposes, digital facilities must be available between each local office and its hub or service node. However, a direct T1 system is not dedicated from the hub to each local office unless the amount of data traffic warrants. Particularly in DDS, the T1 systems terminating in local offices at both ends may carry data circuits originating at the far office to the near office, where the signal may be demultiplexed and remultiplexed along with other data traffic and carried on another T1 system toward the hub. As shown in Figure 17-2, a treelike hierarchy results.

End offices may use dedicated DDS multiplexers (T1DM or T1WB5), in which case they are termed multiplex offices. Especially with BDDS services, they may simply use Dataport channel units in ordinary carrier systems, causing them to be termed Dataport offices. (They may also use a combination of Dataport DS0 facilities and DDS submultiplexing, which classes them as hybrid offices.)

Table 17-1 compares the DDS serving arrangement in one major metropolitan zone with that in a suburban and rural area in the same state. In this particular example, no collection hub is used. A comparable BDDS network might use a dozen service nodes in place of the hub.

Table 17-1. DDS Serving Arrangements

	Metro Area	Rural Area
Hubs	1	1
End Offices		
Baseband, homed on hub	5	1
homed on intermed. ofc.	23	6
Dataport, homed on hub	8	11
homed on intermed. ofc.	10	0
Multiplex (T1DM/T1WB5)	15	3

Transmission Line and Terminal Facilities

The high technical efficiency of these networks is achieved by offering a few service speeds in a synchronous format and by using facilities that carry digital data signals in flexible multiplexed combinations over long or short distances. Facility terminations, timing supplies, multiplexers and submultiplexers, automated test systems, and cross–connect systems have been developed specifically to meet data–network needs.

Loop Transmission. The transmission of signals from a remote customer location to the serving central office is usually over a four–wire loop made up of local cable pairs, by themselves or with the addition of digital loop carrier (DLC). Two processing steps are involved, one at each end of the loop. The required facilities, which include a service unit at the customer premises and a channel unit at the central office, are shown in Figure 17–3.

The signal transmitted over the loop is synchronized to the appropriate clock rate. This clock is recovered at the station from the incoming data or idle–code signal transmitted from the central office. The signal sent toward the central office also contains coded sequences containing bipolar violations to distinguish them from valid data. These special codes are inserted in place of six successive zeros (seven zeros in a 56–kb/s signal) to maintain pulse density sufficient for clock recovery at regenerative repeaters and for the transmission of status and control information. For example, circuit–idle codes to and from the station are sent in this manner, as are trouble and loop–back codes toward the station.

Customer–Premises Terminations. The data service unit (DSU) provided by the customer, shown schematically in Figure 17–4, typically accommodates the EIA–232–D interface [5] for subrate services, with a second EIA–232–D interface if a secondary channel is involved. It uses a combined EIA–CCITT–type interface for the 56–kb/s service speed. The transmission and clock leads are provided in a manner similar to CCITT interface specification V.35, while the control leads conform to EIA–232–D. In any case, the unit contains circuitry that converts the customer signal to the 50–percent duty cycle bipolar RZ

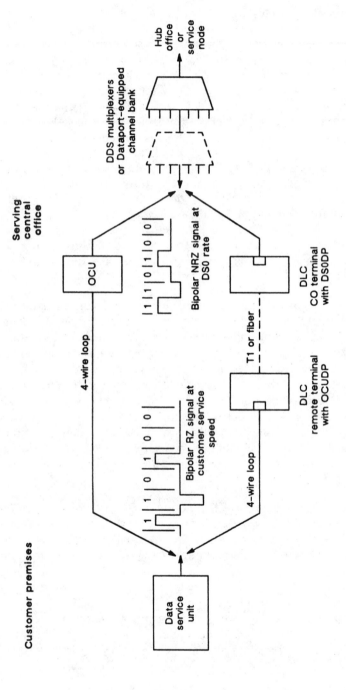

Figure 17-3. Customer loop facilities for DDS or BDDS.

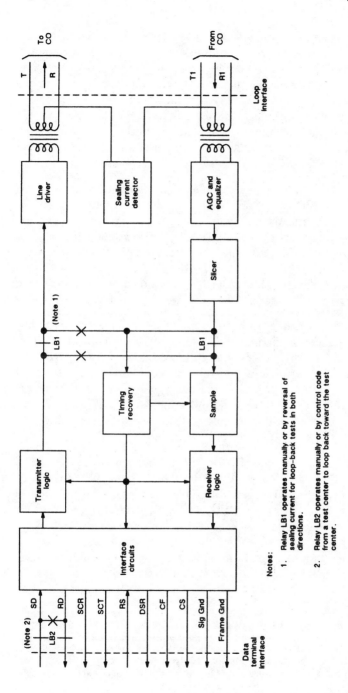

Figure 17–4. Typical data service unit without secondary channel.

Notes:

1. Relay LB1 operates manually or by reversal of sealing current for loop–back tests in both directions.

2. Relay LB2 operates manually or by control code from a test center to loop back toward the test center.

signal in the transmitting direction [6,7]. The transmitting circuits provide the logic for zero substitution and idle–code generation. The receiving circuits contain logic for recognizing idle and trouble codes; these circuits convert the incoming signal to the EIA (or CCITT) format and remove any zero–substitution codes. Clock–recovery circuits pass the received clock signal to the customer's data terminal. Where the secondary–channel option is in use, the DSU serves as a two–channel multiplexer to derive the added channel.

The DSU contains a sealing–current detector* and a line driver consisting of a transmitting amplifier and filter. Automatic gain control, equalization, and slicing circuitry recover and re-constitute the received signal. The DSU also contains circuitry to recognize and respond to loop–back commands from the serving test center. Many commercial DSUs are switchable among several service speeds. All units are powered from an ac source provided by the customer.

In the early history of these services, a simplified channel termination called a channel service unit (CSU) was available. This device, intended for cases where the exchange carrier provided a basic channel terminator and the customer furnished the rest of the circuitry, was a DSU less the clock–recovery and data–interface circuits, with a simple four–wire bipolar interface to the customer. However, with the entire DSU function now being provided by the customer, CSUs have faded from wide use.

Office Channel Unit. In the serving central office, the local loop is terminated in an OCU optioned to match the service speed [8]. The line–side (baseband) portion of the OCU functions like that of the DSU. The OCU includes provision for automatic shaped gain control (called automatic line build–out). It sends and reverses sealing current. The OCU can be looped back through a fixed pad on the loop side on command from the test center. The OCU may be a discrete equipment unit or,

---

* DC sealing current is transmitted over a pair of wires to maintain a low resistance at splices and cross–connection points by breaking down small accumulations of dirt and oxides to reduce noise and other trouble conditions. A reversal of sealing current, initiated by the office channel unit (OCU) in response to a control code received from the DDS STC or BDDS service node, causes the DSU to establish a loop–back for testing.

especially for BDDS, may be built into an OCU Dataport (OCUDP) channel unit for a carrier terminal. Such an OCUDP may also be used in the remote terminal of a DLC system, with a complementary DS0DP used in the central–office terminal.

In the direction of transmission from the loop, the OCU converts the baseband signal to a standard DS0 signal regardless of service speed. The process is reversed in the opposite direction. Figure 17–5 illustrates the byte organization of the 64–kb/s signal for each service speed. In the case of a DDS submultiplexer, the DS0 rate is maintained for subrate inputs by a process called *byte stuffing*, which places the same subrate byte in 5, 10, or 20 successive 64–kb/s bytes corresponding to the 9.6–, 4.8–, or 2.4–kb/s service speeds, respectively.

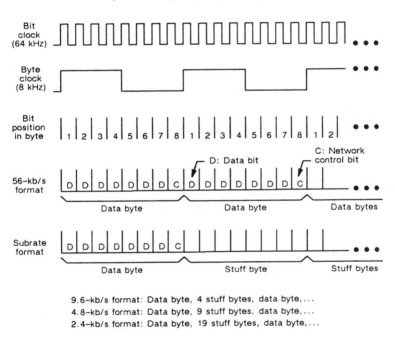

9.6–kb/s format: Data byte, 4 stuff bytes, data byte,...
4.8–kb/s format: Data byte, 9 stuff bytes, data byte,...
2.4–kb/s format: Data byte, 19 stuff bytes, data byte,...

Figure 17–5. Format of DS0 signal.

Establishment of the one fundamental (DS0) rate permits the use of common test equipment, efficient multiplexing and cross–connection, and the same multipoint junction arrangement for all service speeds. It also permits distribution of network timing signals within the office at a common rate.

Submultiplex Stage. As can be seen from Figure 17–5, the 56–kb/s data signal usually comprises different data words in successive DS1 bytes. This is not usually true for the subrate speed formats because of the redundancy inherent in byte stuffing. For example, the data content of a 2.4–kb/s subrate byte would occupy the same channel slot in the DS1 signal for twenty successive frames, although needed only once in twenty frames. Hence, direct multiplexing of the subrate OCU outputs into DS1 channel slots could result in up to a twentyfold loss in transmission efficiency. Where long protected DS1 facilities are required, that technical inefficiency may be uneconomical. Therefore, *subrate data multiplexers* (SRDMs) are available for each subrate service speed [9] for DDS use. These provide ports that accept up to five 9.6–kb/s, ten 4.8–kb/s, or twenty 2.4–kb/s DS0 signals from OCUs of corresponding service speeds. The SRDM sequentially combines the data bytes from the OCUs into a single stream, substituting the bytes from different channels for the repeated, or redundant, bytes in the OCU signals. Figure 17–6 illustrates this process for five separate DS0 signals combined by a 9.6–kb/s SRDM. The output of the SRDM is also a DS0–rate signal with successive bytes containing data from different channels that can then be multiplexed into a DS1 frame. Observation of the byte formats for the subrate speeds in Figures 17–5 and 17–6 also shows how the pulse slot for the first bit of each OCU subrate data byte is used by the near–end SRDM to insert bits (designated S) for synchronization of demultiplexing at the far–end SRDM.

It is possible to connect 4.8–kb/s or 2.4–kb/s DS0 signals to ports of a 9.6–kb/s SRDM, or 2.4–kb/s signals to the ports of a 4.8–kb/s SRDM, to defer installation of a separate multiplexer for each rate. This procedure does not fully pack the DS0 output bit stream, but provides considerable flexibility during initial growth. There is also an *integrated subrate multiplexer* (ISMX) combined with an OCU shelf that is used widely for economy in local offices. The ISMX packs either five or ten subrate signals of the same speed into a single DS0 output. At the hub office, the SRDM function can be performed three ways: in a dedicated DDS equipment bay, via special plug–in units inserted in an

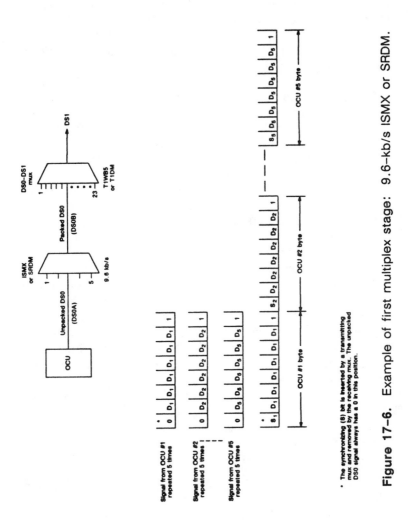

Figure 17-6. Example of first multiplex stage: 9.6-kb/s ISMX or SRDM.

ordinary channel bank mounting, or by a suitably equipped DCS. The latter two approaches are prevalent in the BDDS network.

The output of an SRDM or ISMX is routed to the input of a DS0–to–DS1 multiplex stage. With a typical mix of 2.4, 4.8, 9.6, and 56–kb/s circuits as found in practice, about 120 circuits can be handled on a single DS1, a substantial increase over the 24 circuits that would apply with analog operation. After a packed DS0 signal has been demultiplexed by a receiving SRDM, the resultant unpacked DS0 signals may be reapplied to other SRDM inputs for repacking in order to route DDS traffic in different directions efficiently.

Multiplex Stage, DS0 to DS1. Separate DS0 signals (from office channel units, submultiplexers, or both) can be multiplexed into a DS1 signal using either of two available types of multiplexer; the choice depends on such factors as the number of DS1 signals to be transmitted.

Data Multiplexer. The T1DM multiplexer is typically used where a large number of DDS signals are to be transmitted on several DS1 channels [10]. As shown in Figure 17–7, the T1DM accepts up to 23 DS0 signals and places them into the first 23 8–bit channel slots of the DS1 frame. The 24th channel slot of the frame is reserved for the T1DM to insert a special synchronizing byte as shown. Six of the synchronizing byte positions contain a unique code (10111..1) repeated in each frame. Logic circuits in the T1DM monitor both this pattern and the framing (F) bit pattern (identical to the superframe format in a D4–type bank) to maintain synchronization. This combined synchronization scheme is virtually immune to simulation by a single data signal. Moreover, because of the overall redundancy, up to 4 out of 12 successive frames may contain at least one error in the

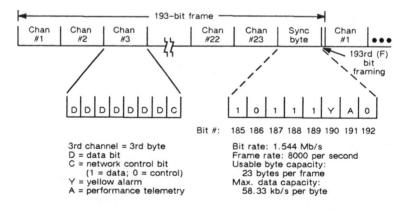

Figure 17–7. T1DM frame format.

seven synchronization bits per frame before synchronization is lost.

Bit positions 190 and 191, although part of the synchronizing byte, are not used for synchronization as such. Position 190 (the Y–bit) is normally a 1 and is used to alarm the far–end T1DM (yellow alarm) in the event the near–end T1DM loses synchronization for more than 300 ms. Position 191 (the A–bit) provides an 8–kb/s data channel for telemetering performance information based on monitoring the quality of the received sync byte.

The T1DM has automatic protection of both its common equipment and its individual ports. It requires timing from a collocated timing supply.

Local Data Multiplexers. A data–only multiplexer for DDS end offices, the T1WB5, usually operates into a T1DM at the hub end. Used for light–route applications, it has the T1DM's capacity of 23 DS0 signals. (There is also an earlier 12–DS0 model titled T1WB4.)

Unlike the T1DM, the T1WB5 synchronizes only on the F–bit sequence. While this is not as rugged as the T1DM synchronizing process, simultaneous logic comparison of the main alternating framing sequence and the interleaved subframe pattern prevents the T1WB5 from falsely locking onto a data sequence that might simulate one or the other individually. The recognition of three or more errors out of five bits in the main framing sequence starts a search for resynchronization. If the incoming signal returns to a good condition, average synchronization recovery time is about 15 ms. An out–of–synchronization code is transmitted on each of the data channels if the T1WB5 remains out of synchronization for 400 ms or more.

The T1WB5 arrangement is less expensive for small (single–digroup) DDS offices than the T1DM since it does not offer automatic port–failure restoral and has its own integrated timing supply.

Digital cross–connect systems with subrate multiplex and data bridging features may be used in DDS hubs in place of, or in addition to, the older dedicated equipment frames. Digital data bank (DDB) equipment may be used this way also.

Dataport Operation. As the cost of digital line facilities in the local area has fallen, it has become progressively less critical to make efficient use of the line capacity within the LATA. Other factors—flexibility, speed of installation, ease of reuse—make it desirable to equip ordinary carrier systems for a mix of voice and data services. Dataport channel units in D–type channel banks give this capability [11], a stimulus for the expansion of DDS serving areas and for the introduction of BDDS. For subrate services, they perform byte–stuffing similar to that in the OCU. The resulting redundant transmission forms the basis of a power-ful error–correction technique that makes prequalification of the carrier facility unnecessary. For station loops, DLC systems can be equipped with Dataport channel units, providing service quickly within a carrier serving area without any need to deload cable pairs or install loop repeaters. The major proviso in equip-ping any carrier system with Dataport is to ensure that the chan-nel bank [or central–office terminal (COT)] nearest the hub is clocked from the digital synchronization network, and that the channel bank (or remote terminal) nearest the station is also syn-chronized, usually by simply setting a looped–timing option. Back–to–back connections between carrier systems involve either DS0 Dataport units or DCSs.

Cross–Connect Facilities. Two cross–connect fields for sig-nals are used at DDS hub offices. Both utilize a universal 64–kb/s cross–connect frame. One field, designated DSX0B, intercon-nects all cross–office paths between DS1 multiplexers and SRDMs (those with multiple circuits per DS0). The second field, designated DSX0A, provides access to the unpacked side of SRDMs and to local OCUs (DS0s carrying only a single circuit). A DCS with subrate data cross–connect features [12] may also be used in place of a physical DSX0, T1DMs, and SRDMs, particularly in the BDDS network; a BDDS node may use a DSX0 panel, a conventional frame, or a digital cross–connect.

Multipoint Service. Multipoint service accommodates a du-plex multiparty system in which there are one *master station* (usually a computer) and two or more simultaneously connected remote terminals, all operating at the same service speed. Multi-point junction units (MJUs), or DCSs with digital bridging fea-tures, are employed in hub offices and service nodes to link the branches together, as shown in Figure 17–8. Incoming from the master station, the DS0 bipolar NRZ signal from the DSX0A frame is converted by the junction unit to unipolar format. The signal is regenerated and retimed, bridged to four outputs, and reconverted to bipolar NRZ for connection back through the DSX0A frame to the distant stations. Since each distant station receives the same signal, the master station provides coding for address identification.

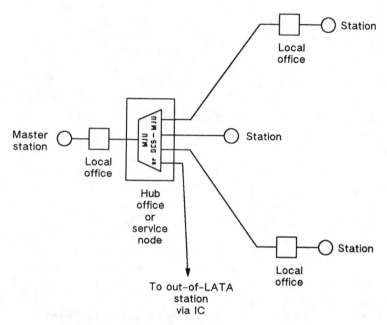

Figure 17–8. Five-station multipoint circuit.

Incoming from the branches, the MJU [13] converts the bipolar NRZ signals from the DSX0A frame to unipolar and connects them to a logic circuit that includes regenerators, control logic, and an adder. The one input signal that contains data at the moment is retimed and reconverted to bipolar NRZ for connection through the DSX0A to the master station if the remaining channels are transmitting the idle code or all ones. The logic circuit substitutes an all–ones code on a branch input to the adder when that branch transmits a trouble control byte or idle code. This feature prevents trouble on one remote branch from interrupting data transmission between the remaining distant stations and the master station since the all–ones code is blocked by the adder. Two distant stations may not transmit data simultaneously since the adder will garble such messages.

MJU hardware is of two vintages: the original DDS MJU bay using pairs of two–port plug–in units, and a later DDB version in which four–port MJUs are inserted into a D4 channel–bank mounting.

If more than four total branches require interconnection at a hub, physical MJUs are chained to provide additional branches. A multipoint circuit can include junction units at more than one hub or service node to form a tree network spreading away from the master station. The MJU, and its DCS equivalent, is thus a versatile unit for building up multipoint circuits. Since all inputs are unpacked bipolar NRZ signals at the DS0 rate, the same equipment is used regardless of the service speed.

Digital Facilities. DS1–level transmission in the DDS or BDDS network uses any of the usual facilities. The original choice for LATA–level use was T1 span lines with protection switching. The span lines, regular and protection, are qualified as to error rate via a preservice test. The regular and protection lines are usually dispersed–routed for added reliability; in no case should they share the two sides of a repeater or a common power loop. Protection is normally on a one–for–one basis, occasionally one–for–several. The exception is Dataport operation, where the maximum circuit load is only 24 voice and data circuits per DS1, rather than 100–plus data circuits. Here, protection is not usually applied but error correction is routinely used.

An early digital facility for long–haul DDS service was the 1A Radio Digital System (1A–RDS), also known as data under voice (DUV). By use of seven–level coding, the 1A–RDS added one DS1 signal below the band of the frequency–division multiplexed (FDM) message load over each radio system of a long–haul analog radio route. Thus, up to 18 DS1 signals could be added per microwave radio route without displacing voice channels. More recently, other digital facilities have superseded T1 and 1A–RDS construction. Most such facility types include one–for–several protection as part of their basic structure. DDS performance and reliability criteria are routinely met by T2 span lines, present–generation digital microwave radio, and fiber systems.

Design Requirements

A specific set of engineering rules is needed to meet service objectives. The major engineering considerations include network synchronization, maintenance and restoration, and local loop design.

Network Synchronization

A timing network is crucial to the operation of the system to ensure that all data signals at the DS1 rate and below remain synchronized. Loss of timing in any portion of the network could cause data bytes to be skipped or read twice (called *slip*); therefore, transmission between stations in that portion and stations in the rest of the network would be garbled. Special synchronization equipment for data service in the parts of the hierarchy operating above DS1 rates is not required, since the stuffing and multiplexing schemes used in the digital hierarchy allow timing recovery for all DS1 signals at the far ends of such facilities.

The original DDS timing network, shown conceptually in Figure 17–9, is a treelike structure with no closed loop. It consists of a master timing supply, nodal or secondary timing supplies (in hub offices), and local timing supplies (in local offices with T1DMs). The T1WB5 integrated timing supplies and Dataport bank clocks in local offices without T1DMs are also part of the timing hierarchy. Although not shown in the figure, the timing network extends to the customer premises since the service–speed clock rate is derived from the incoming baseband signal. This network predated other synchronization systems like the building integrated timing supply (BITS) concept [14]; as the BITS becomes widely available, the DDS timing network merges with it.

The master timing supply is physically a nodal timing supply slaved to the reference clock for the area. All other nodal timing supplies are locked to this master. Only one incoming DS1 signal is designated for synchronization, but facility protection switching ensures a reliable synchronization signal. Local and T1WB5 integrated timing supplies are slaved to hub office supplies in a similar manner except that primary and alternate working T1 lines,

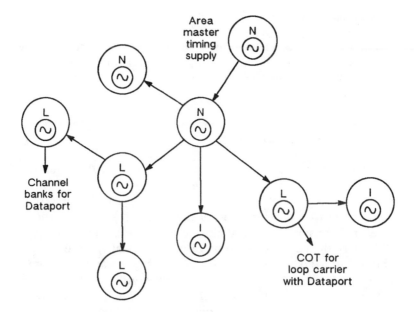

Area
master
timing
supply

N = Nodal or secondary
    timing supply

L = Local timing supply

I = Integrated timing
    supply (T1WB5)

Figure 17-9. Synchronous timing network.

over separate routes, are designated for timing purposes whenever possible. Where multiple lines exist, they must originate in
the same higher office to avoid the creation of a closed timing
loop. Where a BITS is present, the BITS timing is used for DDS
equipment.

Figure 17-10 is a block diagram of the nodal, secondary, and
local timing supplies. The timing supplies contain redundant interface units, phase-locked loops (PLLs), and output circuits.
Each interface unit extracts the framing signal from its incoming
DS1 signal ahead of any multiplexers. The working interface unit
(determined by the position of the A contacts) delivers the resulting signal to both PLLs. Each loop contains a stable oscillator

410

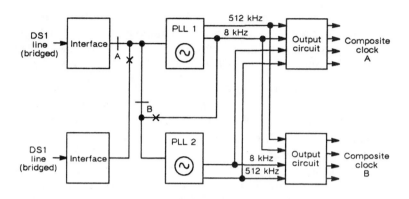

Figure 17-10. Timing supply.

that locks to the input signal from the interface unit and contains circuitry to produce 512–kHz and 8–kHz output signals. These phase–locked signals are redundantly fed to the output circuits. Each circuit produces the modified–bipolar timing signal with 62.5–percent duty cycle containing both 8–kHz (byte) and 64–kHz (bit) timing components as shown in Figure 17–11. The composite timing signal is redundantly distributed, over pairs in separate cables, to the bay clock, power, and alarm unit in each bay of DDS equipment. This unit derives separate 8– and

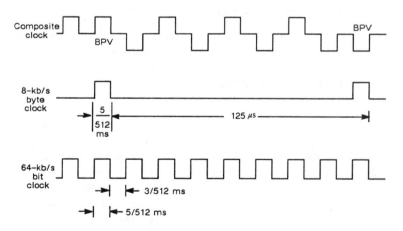

Figure 17-11. Timing waveforms in data equipment.

64–kHz signals for intrabay distribution. Nonredundant composite clock is also cabled to Dataport channel–bank and COT bays. If both interface units fail, the B transfer contacts operate. Under these conditions, PLL 1 runs free but PLL 2 is synchronized to it.

The major difference between the nodal and local timing supplies is the stability of the PLL oscillator. If the external clock source is interrupted, the oscillator in the nodal timing supply is stable enough to supply timing for at least two weeks at a worst-case slip rate of only one DS1 frame every 24 hours. It is thus considered to be of Stratum 2 quality. Therefore, the section of the network dependent on the nodal supply could communicate with stations external to that section without severe service degradation; internal stations would suffer no degradation at all. Oscillator stability in the secondary and local timing supplies is such that their dependent networks remain slip–free for only about five seconds when incoming synchronization is interrupted. The local timing supply is smaller and far less expensive than the nodal supply; it is adequate because of the protection applied to the incoming synchronization path. Each T1WB5 integrated timing supply provides timing for all office equipment associated with it without a separate timing supply. These low–ranking supplies are considered to be of Stratum 4 (channel–bank) quality.

Maintenance and Restoration

A comprehensive maintenance plan is essential to meeting overall DDS service objectives. Central–office equipment and interconnecting facilities at all levels of the hierarchy must provide high reliability and the means for prompt facility restoration as controlled by the Facility Maintenance and Administration Center.

High–Capacity Facilities. The maintenance plan assumes automatic protection switching of fiber or radio paths, the use of telemetered performance monitoring, and centralized fault location. In the event of failure of both the working and the protection channel, service should be restored by existing methods of emergency restoration within 20 minutes unless an entire route has failed. In–service monitoring and alarming are normal for all portions of the DDS carrying DS1–level signals.

T1 Span Facilities. T1 span lines used with DDS multiplexers normally require one–for–one automatic protection switching. The two lines are double–fed through a bridging repeater with monitors on both lines at the end to control a receiving–end transfer switch. Switching to the spare is based on either an excessive bipolar–violation rate or a loss of pulses on the regular line. The protection system includes the necessary alarms for status indication.

Baseband and Dataport Facilities. Automatic protection switching is not used on baseband or Dataport facilities other than DLC systems. Customer reports, received at the serving test center, initiate fault location on customer loops. Office troubles cause alarms to register at offices and test centers. A maximum time of 90 minutes is the objective for restoring DDS service that has failed due to outages on loop facilities.

Serving Test Center. The STC (or special–service center or centralized test center) controls maintenance activities for circuits served by the office and for all circuits and hierarchy equipment homing on its hub office or service node. Figure 17–12 shows typical routing and equipment layouts for a DDS hub office and a BDDS service node. The figure shows that all DDS circuits terminating in the serving area are accessible to the STC at the *unpacked* DS0 rate by routing them through a DSX0A cross–connect field or subrate DCS. Each submultiplexer at a given local office has a mirror image at the hub. The DSX0A includes a universal 64–kb/s cross–connect panel. Test access is normally via automatic byte–access techniques on a DS0 or DS1 basis. The test system provides maintenance access to each circuit. It generates DS0 test signals and up to 15 control codes for loop–back and straightaway testing. A multipoint signalling feature is available for the testing of multipoint circuits.

Faults occurring in the serving link between the hub and the termination at the customer premises are located by establishing loop–backs at intermediate points in response to coded commands. Loop–back points are found in the service units at the customers' premises and at the line sides of all office channel units and OCU Dataports. The loop–back in the DSU is established as shown in Figure 17–4. As indicated in this figure, some of the commands are actually decoded by the OCU, which then

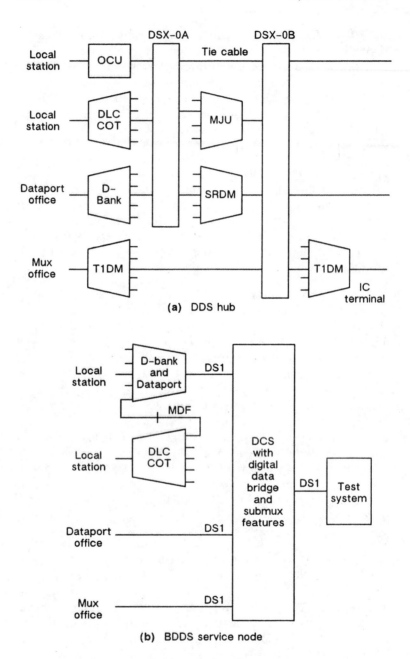

(a) DDS hub

(b) BDDS service node

Figure 17–12. Typical hub/node layouts.

reverses the sealing–current polarity on the line to cause the loop–back in the service unit.

The near–end test–access point for a circuit between stations served from different hubs or nodes can control the same set of loop–backs at the far end if required. The objective of loop–back fault location is sectionalization of a trouble condition to a major subdivision of a point–to–point link (or to a particular multipoint branch) within 15 minutes.

Additional Maintenance Features. Trouble isolation in DDS multiplexers is aided by other maintenance features. For example, all major components are monitored and alarmed at the offices in which they are located. Units like T1DMs provide an extensive set of diagnostic indications. The only exceptions are OCUs, ISMXs, individual T1WB5 ports, and individual Dataport channels. These alarm indications are telemetered to a distant point if registration at the STC is desired.

Another maintenance feature is the provision of monitoring and splitting jacks. Connections between directly connected OCUs and multiplexers in DDS local offices are normally routed through a multiplexer jack and connector panel (M–JCP). Portable test sets providing hub–style command codes and error measurement are available for testing at these points. ISMXs also have jacks for splitting or monitoring the internal connections between OCUs and the submultiplexer circuits, as do Dataport channel units.

Local Loop Design

The root mean square (rms) power of the bipolar baseband signal transmitted on customer loops varies in accordance with the peak voltage of the pulses [6,7] and with the density of ones in the customer signal. The maximum allowable power is +6 dBm. DSUs and OCUs contain automatic line build–out circuits to accept incoming signals through a range of loop insertion losses. The maximum calculated loop loss for operation within error–rate requirements depends on the specific equipment. It is 31 to 40 dB at a frequency corresponding to half the service speed (Nyquist frequency). The exact limit depends on equipment vintage: 34 dB of actual measured loss for older subrate equipment and 31 dB for older 56–kb/s equipment (more with some extended–range units). To use the higher losses, the customer–premises equipment must be capable of the longer range, and coordination with other services in the cable must be observed more closely. The maximum loop length is thus dependent on service speed, equipment vintage, and the length and gauge of available cable pairs. Mixed–gauge loops are used routinely.

A cumulative maximum of 6 kft of bridged tap is acceptable on loops for all speeds but 56 kb/s. For 56–kb/s loops, total bridged tap must be limited to 2.5 kft. No single bridged tap may exceed 2.0 kft.* In addition, load coils and build–out capacitors must be absent from pairs used for baseband data loops.

Because of the ruggedness of the signal, self–interference or interference from other signals is not a problem when loops are designed according to the previous guidelines. However, due to the relatively high power transmitted, DDS/BDDS signals do have the potential to interface with some analog and data–over–voice services in the same cable sheath. Assignment restrictions are used to prevent interference from data pairs into pairs used for 15–kHz program service or analog loop carrier systems.

If the office serving the customer is a baseband office, the interoffice cable pairs between the baseband office and the hub or multiplex office must be included in the overall loop–loss calculation. For the 56–kb/s service speed, the insertion–loss limit of 31 dB at the Nyquist frequency (28 kHz) may be unattainable on long loops. A digital repeater, or extended–range OCU and DSU, can be used for loop extension. However, DLC facilities are often available. They remove the restriction of loop loss and the need to deload pairs.

17-2 PUBLIC SWITCHED DIGITAL SERVICE

Public Switched Digital Service is a precursor to integrated services digital network (ISDN) service. It provides a switched

---

* The single-tap limitation is due to the effect of a quarter-wavelength resonant stub, which would result in an edge-of-band dip in the loss/frequency characteristic should the tap exceed 2.0 kft.

capability that allows the subscriber to send digital data at 56 kb/s over the public switched digital network. Two loop (nonloaded) systems for PSDS are discussed: circuit switched digital capability (CSDC) [15] with a 1AESS™ end–office and DATAPATH™ [16] with a DMS®–100 end office. Both systems use time–compression multiplexing (TCM) to achieve full–duplex operation over the two–wire local loop, but are not compatible with each other. A third system, USDC™, provides service via 1AESS offices. TCM operates in a "ping–pong" fashion where the digital transmission in each direction is in alternate bursts over the common two–wire connection. Appropriate guard times are provided to allow each burst to reach the far end before transmission starts in the reverse direction. The line data rate is 144 kb/s for CSDC and 160 kb/s for DATAPATH using bipolar (alternate mark inversion) pulses. Loop range limits may be due to loss, bridged taps, time delay and, for long loops, longitudinal balance. The maximum designed loop losses are 45 dB at 72 kHz or 80 kHz respectively, corresponding to about 14 kft of all–26–gauge or about 22 kft of all–24–gauge, both without bridged taps. Automatic line equalization is provided for a wide range of loop makeups such as mixed gauges and bridged taps. Knowledge of the location and tap length is highly desirable to match the loop to the requirements of this service. Operation in the same cable sheath with other systems is limited by crosstalk compatibility. Generally, PSDS is not compatible with analog subscriber carrier systems in either adjacent or nonadjacent binder groups (for single–channel carrier), or in adjacent binder groups (for multichannel analog carrier or data–over–voice). It is compatible with T1 carrier and the ISDN digital subscriber line.

All systems require digital connectivity for the network between end offices. The 1AESS end office requires a special circuit to support PSDS. In each case, network calls are dialed up via conventional process and, upon answer, the connection is cut through to circuits to provide end–to–end transparent 56–kb/s connectivity. This is provided by the seven most significant bits of

---

ESS is a trademark of AT&T.

DATAPATH is a trademark and DMS is a registered trademark of Northern Telecom.

USDC is a trademark of Integrated Network Corporation.

each 8-bit byte in the network DS1 channel. The eighth bit is used for other purposes, such as robbed-bit signalling. The integrity of the seven data bits must be preserved through the connection path, i.e., trunks, switches, cross-connects, etc. Digital processing that alters these bits must be avoided. Echo cancellers in the path are disabled and digital pads are bypassed or set to zero.

Information on generic feature requirements for the PSDS network is contained in Reference 17. These concern operations, call processing, numbering plan, signalling, transmission performance, etc. There is no accepted performance requirement for bit error ratio or for other measures of impairment. A bit error ratio of 10^{-6} or better on end-to-end connections for at least 95 percent of the calls would be a reasonable objective. Performance issues may be addressed by a standards committee.

17-3 HIGH-CAPACITY DIGITAL SERVICES

A major element in the demand for special services is high-capacity digital channels. These are generally at the DS1 speed of 1.544 Mb/s (T1, in popular terms) or the DS3 rate of 44.736 Mb/s (T3). While occasionally used as single channels for high-speed data transfer [18], their main application is as multichannel backbone facilities in private networks, as covered in Chapter 14. Individual voice or data channels can be derived from the high-speed bit stream via multiplexers provided by the customer, the exchange carrier, an IC, or a combination. Technical parameters for these services in an exchange-access environment are given in Chapter 18.

DS1 Services

These services developed from T1-carrier technology, which set the general levels, impedances, pulse format, etc. Options such as automatic protection switching to a backup channel and central-office multiplexing provide flexibility for customers' use. The low cost of these services compared to groups of voice services makes them particularly attractive. They interface directly to the trunk ports of most digital private branch exchanges (PBXs).

Signal Format. The signal transmitted from and delivered to the customer equipment is in 100–ohm balanced bipolar format. The signal must also meet two format constraints: at least three pulses in every sequence of 24 bit intervals (12.5–percent ones density), and no more than 15 consecutive zeros.

Service Objectives. Quality and availability objectives have been established for these services. The quality objective is to provide an average performance exceeding 95–percent error-free seconds. The second objective is to provide at least 99.7–percent channel availability, i.e., the fraction of time the channel is operative. Thus, an average outage of 0.3 percent (one day per year) is permissible as the average observed over a period of several years.

Typical Channel Layouts. As shown in Figure 17–13, the channel layouts for 1.544–Mb/s services use local and interoffice digital facilities most commonly provided by T1 span lines and channels on high–capacity fiber systems, but any facility capable of carrying one or more DS1 signals may be used.

The CSU is considered to be customer–premises equipment. It monitors input and output signals in the transmitting direction to ensure that signal format requirements are met. It usually contains a regenerative repeater, timing circuits, a bipolar–violation remover, and maintenance circuits. A loop–back arrangement at the network interface permits remote testing from the central office or the distant end, based on a digital code.

Since there may be an appreciable length of customer–provided cable between the network interface and the CSU, normal design practice is to use a shortened cable section between the last exchange–carrier repeater and the network interface. The loss of this section is held to 15 dB at 772 kHz, leaving at least 7.5 dB of loss for the customer's cable.

The Federal Communications Commission registration rules [19] limit the "encoded analog" content of DS1 signals applied by the customer equipment to values compatible with the switched message network, providing protection in cases where those signals are switched into the network.

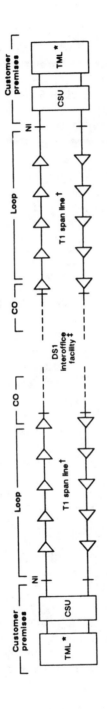

Figure 17-13. DS1 channel.

* PBX, T1 resource manager, data mux, channel bank, etc.

† May include a DS1 channel of a loop fiber system, extended to customer premises by a short T1 span.

‡ Any combination of T1 span lines and/or DS1 channels of T1C, T1D, T1G, T2, fiber, or digital-radio systems.

DS3 Services

A newer offering, these channels offer capacity equivalent to 672 voice circuits for the largest customer networks. Central–office multiplexing is possible to derive a mix of DS1 and voice signals for transmission to multiple sites.

Signal Format. The signal to and from the customer equipment is in 75–ohm unbalanced form (coaxial cable). It must meet DSX–3 voltage levels as detailed in Chapter 18.

Facilities. The high speed involved requires fiber or digital–radio facilities, generally with a multiplexer at the customer's location to derive several DS3 signals from a higher–capacity line.

Service Objectives. Performance criteria for these channels are derived from their wide usage in an exchange access context (Chapter 18). A typical quality objective is 98.75–percent error-free seconds; a normal availability goal is 99.925–percent channel availability.

References

1. Knapp, N., Jr., N. E. Snow, et al. "The Digital Data System," *Bell System Tech. J.*, Vol. 54, No. 3 (May/June 1975), pp. 811–964.

2. Sibley, L. A. "Digital Data System—Status and Plans," *Conference Record*, IEEE International Conference on Communications (1979), pp. 54.1.1–54.1.4.

3. *Digital Data Special Access Service—Transmission Parameter Limits and Interface Combinations*, Technical Reference TR–NPL–000341, Bellcore (Iss. 1, Mar. 1989).

4. Mahoney, J. J., Jr. "Transmission Plan for General Purpose Wideband Services," *IEEE Transactions on Communications Technology*, Vol. COM–14, No. 5 (Oct. 1966), pp. 641–648.

5. *EIA Standard EIA–232–D*, "Interface Between Data Terminal Equipment and Data Communication Equipment

Employing Serial Binary Data Interchange" (Washington, DC: Electronic Industries Association, Jan. 1987).

6. Bell System Technical Reference PUB 62310, *Digital Data System—Channel Interface Specification*, American Telephone and Telegraph Company (Iss. 2, Sept. 1983).

7. *Secondary Channel in the Digital Data System: Channel Interface Requirements*, Technical Reference TR-NPL-000157, Bellcore (Iss. 2, Apr. 1986).

8. *Generic Requirements for the Digital Data System (DDS) Network Office Channel Unit*, Technical Advisory TA-TSY-000083, Bellcore (Iss. 2, Apr. 1986).

9. *Generic Requirements for the Subrate Multiplexer*, Technical Advisory TA-TSY-000189, Bellcore (Iss. 1, Apr. 1986).

10. *Digital Data System (DDS)—T1 Data Multiplexer (T1DM) Requirements*, Technical Advisory TA-TSY-000278, Bellcore (Iss. 1, Nov. 1985).

11. *Digital Channel Banks—Requirements for Dataport Channel Unit Functions*, Technical Advisory TA-TSY-000077, Bellcore (Iss. 3, Apr. 1986).

12. *Digital Cross-Connect System—Requirements and Objectives for the Sub-Rate Data Cross-Connect Feature*, Technical Advisory TA-TSY-000280, Bellcore (Iss. 2, May 1986).

13. *Digital Data System (DDS)—Multipoint Junction Unit (MJU) Requirements*, Technical Advisory TA-TSY-000192, Bellcore (Iss. 2, Apr. 1986).

14. *Digital Synchronization Network Plan*, Technical Advisory TA-NPL-000436, Bellcore (Iss. 1, Nov. 1986).

15. *Circuit Switched Digital Capacity Network Access Interface Specifications*, Technical Reference TR-880-22135-84-01, Bellcore (Iss. 1, July 1984).

16. *DATAPATH Network Access Interface Specifications*, Technical Reference TR-EOP-000277, Bellcore (Iss. 1, Sept. 1985).

17. *LSSGR, LATA Switching Systems Generic Requirements*, Technical Reference TR–TSY–000064, Bellcore (Iss. 2, July 1987), Vol. 1 and Vol. 3, FSD–32–10–1000, "Public Switched Digital Service (PSDS)."

18. Lee, Y. C. and D. L. Waring. "Expanding the Horizons for Data Networks," Bellcore EXCHANGE, Vol. 4, Iss. 3 (May/June 1988), pp. 2–7.

19. Federal Communications Commission. *Rules and Regulations, Title 47, Code of Federal Regulations, Part 68* (Washington, DC: U.S. Government Printing Office, Oct. 1987), Section 68.308(h).

Chapter 18

Access Services

At divestiture in 1984, the geographic areas served by the exchange carriers were divided into local access and transport areas (LATAs). The LATAs define the boundaries within which exchange carriers may offer telecommunication networks and services. Services between LATAs or between LATAs and other national systems are offered by interexchange carriers (ICs) or international carriers. For interconnections the exchange carriers provide the ICs, by tariff, with switched–access service and special (nonswitched) access service to the LATA networks. This permits the ICs and their customers to originate and terminate interLATA or international telecommunications. Services in the interstate [Federal Communications Commission (FCC)] access tariffs are also available to any user whose requirements fall under interstate jurisdiction, including major firms operating intracompany networks, and radio or television broadcasters. This chapter covers the access services offered for voice, voiceband data, TV video, analog and digital data, and other kinds of signals.

Detailed information on access services is available from Technical References, which contain information on network arrangements, service features, interface descriptions, and signalling arrangements. They provide transmission specifications covering various bandwidths (analog) and bit rates (digital). These documents are listed in Table 18–1.

18–1 SWITCHED–ACCESS SERVICE

Provisions for voiceband switched–access service are covered in Sections 4, 6, 7, and 15 of Reference 1 and in Reference 2 in some detail. The following overview of these references describes various arrangements for access, including equal access, wide area telecommunications service (WATS) access lines (WALs), jurisdictionally interstate service (JIS), and interLATA directory assistance.

Table 18-1. Technical References on Access Services

Document No.	Title
TR-NPL-000275	Notes on the BOC IntraLATA Networks—1986
TR-NPL-000334	Voice Grade Switched Access Service
TR-NPL-000335	Voice Grade Special Access Service
TR-NPL-000336	Metallic and Telegraph Grade Special Access Services
TR-NPL-000337	Program Audio Special Access Service
TR-NPL-000338	Television Special Access and Local Channel Services
TR-NPL-000339	Wideband Analog Special Access Service
TR-NPL-000340	Wideband Data Special Access Service
TR-NPL-000341	Digital Data Special Access Service
TR-NPL-000342	High-Capacity Digital Special Access Service

Feature Groups

Several switched-access arrangements for LATA access are offered by exchange carriers to interface with an IC at a point of termination (POT). Within the LATA the connection can be routed via access tandem to an end office, or directly to an end office, and then to the subscriber. The arrangements provided are called feature groups (FGs), which differ in transmission and signalling capabilities and are chosen by the ICs to satisfy their needs. There are four feature groups: A, B, C, and D. Simplified comparisons of these groups and typical applications are shown in Table 18-2.

Table 18–2. Simplified Feature Group Comparison

	Feature Group			
	A	**B**	**C**	**D**
Access code	7–digit (NXX–XXXX)	7–digit 950–0/1XXX	None	10XXX or none with presubscription
Routing	Direct or tandem	Direct or tandem	Direct or tandem	Direct or tandem
Dialing procedure	NXX–XXXX	950–0/1XXX	(0/1) + 7/10 digits	(10XXX) + (0/1) + 7/10 digits
Type of EC termination	Line at EO	Trunk	Trunk	Trunk
Typical service use	FX, ONAL, MTS (switched–end)	MTS/WATS (switched–end)	MTS/WATS	MTS/WATS (switched–end)
Lines accessible	All lines	Lines identified by IC–designated NXX codes	All lines	Lines identified by IC–designated NXX codes
Typical standard features	Dial tone, dial pulse address signalling	Answer supervision, multifrequency signal	Answer supervision, multifrequency signal	Answer supervision, multifrequency signal, presubscription
Typical supplemental features	DTMF address signalling, dial code screening	ANI, dial-pulse, direct inward dialing	ANI, international direct distance dialing	ANI, international direct distance dialing
Available to	All	All	AT&T only (interim only)	All

Equal Access

Equal–access service is provided only by Feature Group D (FGD) arrangements, either directly to the end office or via access tandem to the end office. They offer presubscription by the subscriber to the selected IC. Equal access can be defined as providing interfaces with ICs that have the same type and quality as are offered to AT&T.

Equal access requires special arrangements that can be provided only by stored–program–control–type switching systems (conforming) or by adjunct devices added to earlier systems. This involves dialing, signalling, routing, billing, and other capabilities that such switching systems can provide. In addition, the transmission requirements for FGD are more stringent than for the other groups.

Examples of conforming end–office switching systems are No. 1 and 1AESS™, 5ESS®, DMS®–10, and DMS–100 systems. These systems can provide all feature groups, but only systems of conforming types can provide FGD for equal access. Equal–access tandem switching capabilities are limited to electronic systems such as the 1AESS, 4ESS, 5ESS or DMS–200 switching systems.

Other Switched Access

Access service (but not equal access) to other switching systems (nonconforming) is provided by Feature Groups A, B, and C. Examples of nonconforming switching systems are the older electromechanical systems such as the step–by–step and crossbar systems without adjuncts. FGC is for connection to AT&T facilities on an interim basis. In time, FGD arrangements replace FGC as offices are upgraded to conforming types.

The several switched–access routing arrangements (direct and via) for trunk–side end–office connections to an IC are shown in Figure 18–1 for conforming end offices (equal access) with FGD arrangements and for nonconforming end offices with FGB or FGC. (FGC, not shown, uses direct routing only.)

Access Transmission

For transmission purposes a more general diagram, Figure 18–2, shows the possible application of all feature groups. These differ in transmission performance and signalling capabilities to meet the varying needs of the ICs.

The transmission parameters [2] for the trunks involved in the switched LATA access services are denoted by voice–grade (VG) transmission types A, B, and C, and data–grade types DA and DB. Type A has a four–wire or equivalent transmission facility and interface. The facility is either a digital (one encode/decode step) or a selected analog carrier type. Type B is a four–wire

---

ESS is a trademark and 5ESS is a registered trademark of AT&T.
DMS is a registered trademark of Northern Telecom, Inc.

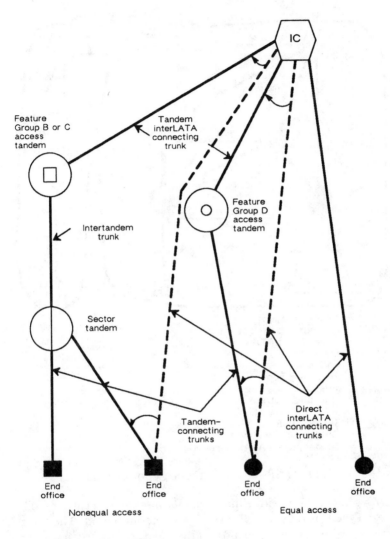

Figure 18–1. LATA access routing—FGB, FGC, and FGD.

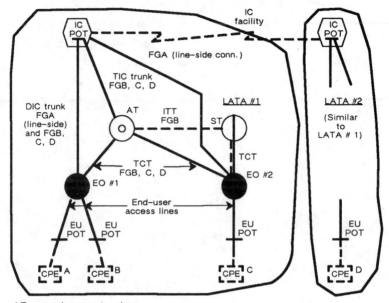

AT	=	Access tandem
CPE	=	Customer-premises equipment
DIC	=	Direct interLATA connecting trunk
EO	=	End office
EU	=	End user
FG	=	Feature group
IC	=	Interexchange carrier
ITT	=	Intertandem trunk
POT	=	Point of termination
ST	=	Sector tandem
TCT	=	Tandem-connecting trunk
TIC	=	Tandem interLATA connecting

Figure 18-2. LATA access connections.

arrangement, also, with either digital or most analog carrier facilities. Type C is only for AT&T over an interim period. The transmission type selected depends on the needs of the ICs for a particular feature group and interface type, and the availability of facilities. In practice, the predominant interface is DS1 or DS3 digital.

The feature group arrangements with the transmission types for the various trunks are shown in Figure 18-3 for the LATA switched-access service offering. Included on the figure is a diagram for WALs.

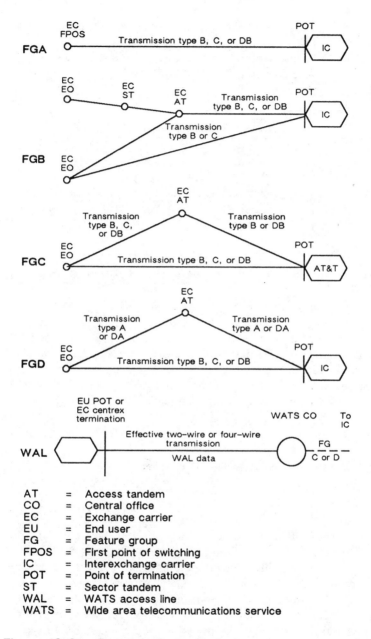

AT   =   Access tandem
CO   =   Central office
EC   =   Exchange carrier
EU   =   End user
FG   =   Feature group
FPOS   =   First point of switching
IC   =   Interexchange carrier
POT   =   Point of termination
ST   =   Sector tandem
WAL   =   WATS access line
WATS   =   Wide area telecommunications service

Figure 18–3.   Transmission and data parameter types by
switched–access service offering.

Some applications of the feature groups and transmission types are shown in the following figures, which also include insertion losses and other information. Feature Group A is illustrated in Figure 18–4, FGB in Figure 18–5, FGC in Figure 18–6, and FGD in Figure 18–7.

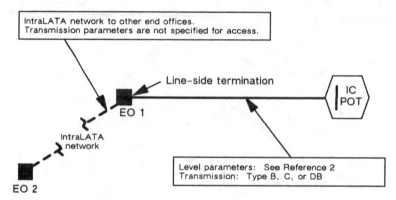

Figure 18–4. Feature Group A illustration.

The arrangements for an all–digital access network (facilities and switches) will be part of the evolving region digital switched network plan discussed in Chapter 2. With such a network, trunk loss becomes meaningless as the digital bit stream is transported through the network without modification. The desired connection loss for voiceband is then accomplished by inserting analog loss at the end digital–to–analog conversion point. Many other benefits and economies accrue with digital networks, leading to their rapidly replacing earlier networks.

Voice-Grade Transmission

The transmission parameter limits and interface combinations to cover most of the possible applications of various facilities and arrangements are included in Reference 2. For simplicity, only the parameters for FGD via a digital access tandem are given here. This example covers transmission types A, B, and C for digital facilities (one analog–to–digital and reverse conversion). The limits are stated in terms of acceptance limits (ALs) and immediate action limits (IALs). These parameter limits for VG transmission are indicated in Table 18–3. The limits are based on

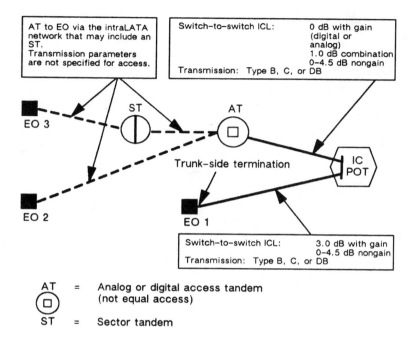

AT = Analog or digital access tandem
     (not equal access)
ST = Sector tandem

Figure 18–5. Feature Group B illustration.

meeting the predivestiture end–to–end grade–of–service performance of similar calls.

Echo Control

The access echo–control limits of Table 18–3, item 3, apply if
either the access tandem switching system is two–wire (such as
the No. 1AESS switching system) or the end office is two–wire. If
the access tandem is four–wire but the end office is two–wire,
then the limits (item 3c of Table 18–3) apply from the POT via
AT (access tandem) to a termination at the end office. Chapter 8
provides more detail on return–loss balance requirements for direct and two–wire tandem routing.

The circuit configurations to measure the limits of Table 18–3,
item 3, for FGD are shown in Figure 18–8 for a two–wire access
tandem and a two–wire end office. The equal level echo path loss
(ELEPL) limits apply. ELEPL is defined as the measured echo

433

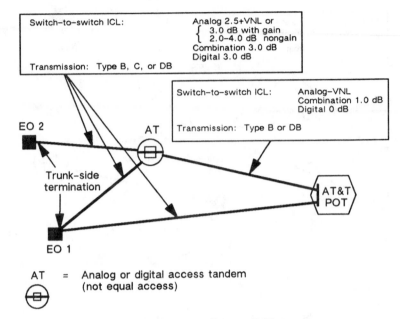

Figure 18-6. Feature Group C illustration.

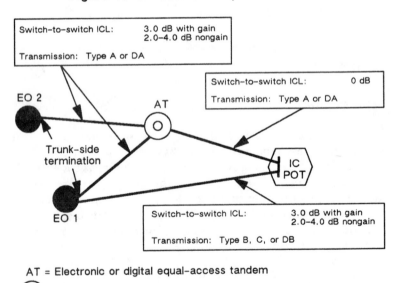

Figure 18-7. Feature Group D illustration.

Table 18-3. FGD Loss Deviation, Attenuation Distortion, Echo-Control Limits, and Noise Limits

	Transmission Type					
	A		**B**		**C**	
	4-Wire Transmission at POT Interface		4-Wire Transmission at POT Interface		2-Wire Transmission at POT Interface	
Voice Transmission Parameter	**AL**	**IAL**	**AL**	**IAL**	**AL**	**IAL**
1) Loss deviation from EML at 1004 Hz (dB)	±0.7	±2.0	±0.7	±2.5	±0.7 with gain ±1.2 w/o gain	±3.0
2) Attenuation distortion (dB) (Note 1)						
404 and 2804 Hz	-0.5 to +2.5	-1.0 to +3.0	-1.5 to +3.5	-2.0 to +4.0	-1.5 to +5.0	-2.0 to +5.5
2804 Hz (Note 2)	-0.5 to +2.5	-1.0 to +3.0	-1.0 to +2.5	-1.0 to +3.5	NA	NA
3) Echo control (Notes 3,4,5)						
(a) Measured at the POT to EO (direct):						
ERL	NA	NA	20	16	16	13
SRL	NA	NA	13	11	8	6
(b) Measured at the POT to AT:						
ERL	25	21	NA	NA	NA	NA
SRL	18	14	NA	NA	NA	NA
(c) Measured at the POT to EO (via the AT):						
ERL	20	16	NA	NA	NA	NA
SRL	13	11	NA	NA	NA	NA
4) C-message noise (Note 6)	27	32 to 42	27	35 to 45	27	38 to 45
5) C-notched noise with -16 dBm0 signal (Note 6)	40 to 42	45	40 to 42	47	40 to 42	47

AL = Acceptance limit   NA = Not applicable
EML = Expected measured loss   POT = Point of termination to IC
ERL = Echo return loss   SF = Single frequency
IAL = Immediate action limit   SRL = Singing return loss

Notes:

1. Loss deviation at 404 Hz and 2804 Hz relative to the loss at 1004 Hz. The "+" means more loss and "-" means less loss than at the reference frequency.

2. When the IC is provided an interface with SF signalling, these limits apply between the exchange carrier SF unit and the POT.

3. Echo control is specified in dB as ELEPL at four-wire interfaces (types A and B) and as return loss at two-wire interfaces (type C).

4. Both the low-band and high-band tests must meet the SRL limits specified.

5. Measured only if the AT and/or EO is a two-wire switch.

6. Limits in dBrnc0 for digital facility (0 to 1000 miles) with one A/D - D/A conversion plus switch. See Reference 2 for details on other facilities.

path loss (EPL) corrected for the transmission level (TL) values at the point of measurement. Hence, ELEPL = [EPL] – TL (transmit) + TL (receive), all in dB. For example, if a –16 dBm signal is applied in the transmit direction at the –16 TL point and the received signal from a reflection of the four–wire–to–two–wire conversion is –9 dBm at the +7 TL receive point, the ELEPL is [–16–(–9)] –(–16) +7, or 16 dB. ELEPL is discussed in Chapter 20.

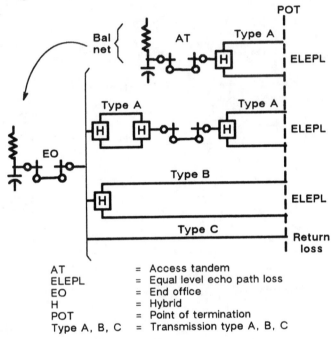

AT	= Access tandem
ELEPL	= Equal level echo path loss
EO	= End office
H	= Hybrid
POT	= Point of termination
Type A, B, C	= Transmission type A, B, C

Figure 18-8. Arrangement for testing echo control—Feature Group D.

Other configurations to cover Feature Groups A, B and C are given in Reference 1.

Voiceband Data Transmission Limits

The switched–access arrangements also provide for voiceband data transmission. The applications of data transmission types DA and DB are shown in Figures 18–4 through 18–7, respectively. Only the data transmission limits for DA and DB with FGD are shown in Table 18–4. Data limits for the circuits associated with the other feature groups are given in Reference 2.

Table 18–4. Voiceband Data Transmission Limits for FGD

Data Transmission Parameter	Type DA IAL (Note)	Type DB IAL (Note)
Signal to C–notched noise ratio with –13 dBm0 holding tone	33 dB	30 dB
Envelope delay distortion		
(a) 604 Hz to 2804 Hz:		
Less than 50 route–miles	500 μs	800 μs
50 route–miles or more	900 μs	1000 μs
(b) 1004 Hz to 2804 Hz:		
Less than 50 route–miles	200 μs	320 μs
50 route–miles or more	400 μs	500 μs
Impulse noise: 15 cts/15 min at threshold	65 dBrnc0	67 dBrnc0
Intermodulation distortion		
(a) Second order (R2)	33 dB	31 dB
(b) Third order (R3)	37 dB	34 dB
Phase jitter, degrees peak-to-peak (4 Hz to 300 Hz)	5°	7°
Frequency shift	± 2 Hz	± 2 Hz

Note: These limits are not provided by special conditioning, but are the limits that are to be met when testing is performed at the request of the customer.

American National Standards Institute Performance Standard

The performance of the switched–access networks is of consid erable importance as two access networks (originating and terminating) are involved in every interexchange connection. The American National Standards Institute (ANSI)–accredited standards committee, T1–Telecommunications, is preparing an ANSI standard [3] on these networks. The standard will apply to direct or tandem connections and has been modeled after the capabilities offered by FGC and FGD. The draft standard is based on a number of considerations such as end–user service perception, the technical capabilities of analog and digital transmission systems, and the performance of analog and digital switches. Requirements for voice and voiceband data transmission parameters are given.

WATS Access Line

The WAL service is used to connect a centrex end office or the POT of an end customer's line to an exchange carrier switch, which is capable of performing screening functions for 800 service, WATS, or similar services, and which can be connected via FGC or FGD to the POT of an IC for routing traffic in and out of the home LATA. There are three two–wire (line–side) and one four–wire (trunk–side) versions of WAL circuits, depending on customer transmission and signalling needs. The WAL arrangements are illustrated in Figure 18–9.

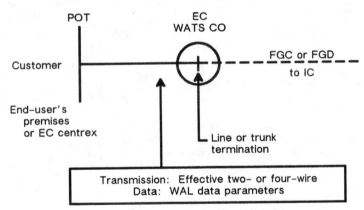

Figure 18-9. WAL illustration.

The voice transmission parameter limits are indicated in Table 18–5 for the several WAL arrangements. The limits cover loss deviation, attenuation distortion, echo control, and noise for a standard two–wire arrangement used for short distances, a two–wire arrangement with improved attenuation distortion for better

Table 18-5. WAL Loss Deviation, Attenuation Distortion, Echo Control, and Noise Limits

Voice Transmission Parameter	Standard 2-Wire		Improved 2-Wire Atten. Dist.		Improved 2-Wire		4-Wire	
	AL	IAL	AL	IAL	AL	IAL	AL	IAL
Loss deviation from EML at 1004 Hz (dB)	±1	±4	±1	±4	±1	±4	±1	±3
Attenuation distortion (dB) 404 - 2804 Hz (Note 1)	-2.5 to +8	-3 to +9	-1.5 to +5	-2 to +6	-1.5 to +5	-2 to +6	-1 to +4	-1 to +4.5
Echo control (dB) (Notes 2,3,4) ERL	5.5	5	5.5	5	14	13	19	15
SRL	3	2.5	3	2.5	7	6	12	9
C-message noise, 0-50 miles (Note 5)	32	35	32	35	32	35	32	35

AL	= Acceptance limit	ERL	= Echo return loss
CO	= Central office	IAL	= Immediate action limit
EML	= Expected measured loss	SRL	= Singing return loss

Notes:

1. All limits are the deviation from the loss at 1004 Hz. The "+" means more loss and "−" means less loss than at the reference frequency.

2. Echo control is specified in dB as return loss for effective two-wire transmission and as ELEPL for effective four-wire transmission.

3. Both the low-band and high-band tests must meet the SRL limits specified.

4. Measured only if the WATS CO furnishes a two-wire switch termination.

5. Limits in dBrnc0 for a combination of one digital facility in tandem with a short cable link and a switch (see Reference 2 for limits on other combinations).

frequency response, a two-wire arrangement with improved attenuation distortion and improved echo control for longer LATA and interLATA connections, and a four-wire arrangement with the least attenuation distortion and best echo control. Also, the four-wire circuit can provide optional features such as dialed-number identification, answer supervision, E and M signalling, etc.

The WAL data transmission parameter limits are indicated in Table 18–6.

Table 18–6. WAL Data Transmission Limits

Data Transmission Parameter	IAL (Note)
Signal to C-notched noise ratio with –13 dBm0 holding tone	30 dB
Envelope delay distortion 604 Hz to 2804 Hz 1004 Hz to 2404 Hz	1000 μs 500 μs
Impulse noise (threshold at 67 dBrnc0)	15 counts in 15 minutes
Intermodulation distortion (a) Second order (R2) (b) Third order (R3)	31 dB 34 dB
Phase jitter (4 Hz to 300 Hz)	7 deg. peak-to-peak
Frequency shift	± 2 Hz

Note: These limits are not provided by special conditioning, but are the limits that are to be met when testing is performed at the request of the customer.

18-2   INTERLATA DIRECTORY ASSISTANCE

InterLATA directory assistance (DA) service is similar to that provided prior to divestiture for answering foreign numbering plan area DA calls (area code–555–1212) by connecting to an operator at an automatic call distributor (ACD) having access to directory information. At that time, these calls were routed to the ACD over an intertoll trunk from a toll switching office of the long-haul network. With present operation, the routing of an incoming interLATA DA call via IC to an ACD is shown in Figure 18–10. The routing from the point of termination for an IC can be direct to the ACD or via an access tandem with FGC (for an

interim period) or FGD. To provide the operators with predivestiture transmission quality, transmission type A or B is recommended.

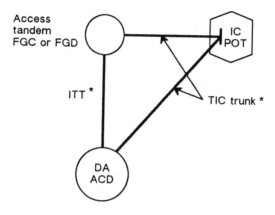

* See Chapter 8 for transmission limits (four-wire).

ITT =   Intertandem trunk
TIC =   Tandem interLATA
         connecting

Figure 18-10. InterLATA directory assistance configuration.

18-3   VOICE-GRADE SPECIAL ACCESS SERVICE

Voice-grade special LATA access services are covered in Reference 4:

(1)   13 voice-grade special access services, VG1 to VG12 and
      VGC (customized)

(2)   several jurisdictionally interstate services where the serv-
      ice is for an end-to-end user crossing a state boundary,
      for an interstate corridor (e.g, Northern New Jersey to
      Manhattan), or for an international connection.

The various service arrangements have different capabilities
and qualities. Reference 4 indicates the arrangements for two-
point, multipoint, two-wire, four-wire, voice, voiceband data, or
voice/data service. Interface and signalling considerations and

channel transmission specifications are included. Generally, the IC specifies the insertion losses that are needed; these must be within the adjustable range given in the transmission specification. In addition, the IC may select an echo control option that is needed to provide the desired quality of service.

There are many different signalling and transmission options for special access services. Table 18–7 lists the codes and selected options that define features of the services.

The 13 VG special access services with defined service features, transmission specifications, and interfaces are described in Reference 4. They are listed in Table 18–8 for typical applications. Only the transmission parameters for VG2 service will be described here. A limitation common to all services is the maximum signal allowed on subchannels (multiple data tones) to prevent overloading analog carrier facilities. The allowable maximum three–second average total power of simultaneous signals is indicated in Table 18–9.

Voice–Grade Service VG2

Voice–grade service VG2 is suitable for the access segments of two–point and multipoint voice connections for private–line and special–services circuits. Typical two–wire and four–wire configurations are shown in Figure 18–11 for two–point service and in Figure 18–12 for multipoint service. The two–wire applications are sometimes used for one–way transmission in either direction but not both simultaneously. The four–wire type can be used for either one–way or two–way simultaneous transmission.

The transmission specifications with ALs and IALs are based on providing the same or better quality of service as that provided for similar circuits before divestiture. The transmission parameters and limits for loss deviation, three–tone slope, and attenuation distortion are given in Table 18–10. The several options for echo control are given in Table 18–11. The option desired is selected by the IC for echo performance reasons (because of two–to–four–wire conversions) depending on the length (delay) of the circuit, the use or nonuse of echo cancellers, etc.

Table 18-7. Glossary of Protocol Codes and Selected Options

Code	Options	Definition
AB		Accepts and provides a nominal 20-Hz ringing signal at IC terminal[1]
AC		Accepts and provides a 20-Hz ringing signal at end-user premises
	R	Two-digit code select (≤ 10)
AH		Analog high-capacity interface
CT		Centrex tie-trunk termination
DA		Data stream in voiceband at end-user premises
	S	Sealing-current option for four-wire transmission[3]
DB		Data stream in voiceband at IC terminal[1]
DD		Dataphone Select-A-Station interface at IC terminal[1]
DE		Dataphone Select-A-Station interface at end-user premises
DS		Digital hierarchy interface
	9GO	Ground-start signalling – open end (DS1)
	0GO	Ground-start signalling – open end (DS2)
	6GO	Ground-start signalling – open end (DS3 or higher)
	9GS	Ground-start signalling – closed end (DS1)
	0GS	Ground-start signalling – closed end (DS2)
	6GS	Ground-start signalling – closed end (DS3 or higher)
	9LO	Loop-start signalling – open end (DS1)
	0LO	Loop-start signalling – open end (DS2)
	6LO	Loop-start signalling – open end (DS3 or higher)
	9LS	Loop-start signalling – closed end (DS1)
	0LS	Loop-start signalling – closed end (DS2)
	6LS	Loop-start signalling – closed end (DS3 or higher)
	9EA	E and M signalling (DS1)
	0EA	E and M signalling (DS2)
	6EA	E and M signalling (DS3 or higher)
	9NO	Transmission only – no signalling (DS1)
	0NO	Transmission only – no signalling (DS2)
	6NO	Transmission only – no signalling (DS3 or higher)
DX		Duplex signalling at either IC terminal location[1]
	X	Simplex reversal (four-wire)
EA		Type I E and M signalling at either IC terminal or end-user premises[1]
	E	IC or end user originates on E lead
	M	IC or end user originates on M lead

Table 18–7.   Glossary of Protocol Codes and Selected Options
(Continued)

Code	Options	Definition
EB		Type II E and M signalling at either IC terminal or end–user premises[1]
	E	IC or end user originates on E lead
	M	IC or end user originates on M lead
EC		Type III E and M signalling at IC terminal, IC originates on M lead[1]
EX		Back-to-back carrier arrangement with tandem signalling
	A	Exchange carrier has closed end
	B	Exchange carrier supplies dial tone
GO		Ground–start signalling – open end[1]
	X	Simplex reversal (four-wire)
GS		Ground–start signalling – closed end[1]
	C	Centrex foreign-exchange (FX) trunk termination
	M	CO answering-service concentrator
	X	Simplex reversal (four-wire)
LA		End-user loop signalling – Type A registered port, open end
LB		End-user loop signalling – Type B registered port, open end
LC		End-user loop signalling – Type C registered port, open end
LO		Loop-start signalling – open end[1]
	X	Simplex reversal (four-wire)
LR		20-Hz automatic ringdown interface at IC or end user, with exchange-carrier-provided PLAR[1,5]
	A	D4-type PLAR channel unit signalling format[5,6]
	B	D3-type PLAR channel unit signalling format[5,6]
LS		Loop-start signalling – closed end[1]
	M	CO answering-service concentrator
	X	Simplex reversal (four-wire)
NO		No signalling interface, transmission only[1]
	S	Sealing-current option for four-wire transmission[3]
PR		Protective relaying[2]
RV		Loop reverse-battery supervision[1]
	O	Battery supplied by exchange carrier, IC originates
	T	Battery supplied by end user, end user terminates
SF		Single-frequency signalling within VF band at an IC terminal[1]

Table 18-7.   Glossary of Protocol Codes and Selected Options
(Continued)

Code	Options	Definition
	AB	SF to manual ring [4]
	EA	SF to E and M signalling [4]
	GO	SF to loop signalling, ground start, open end [4]
	GS	SF to loop signalling, ground start, closed end [4]
	LO	SF to loop signalling, loop start, open end [4]
	LS	SF to loop signalling, loop start, closed end [4]
	LR	SF to automatic ring [4]
TF		Telephotograph interface [1]

Notes:

1. Where applicable, these interfaces are available with the "SX reversal option" and the "improved termination option."

2. PR is specifically intended for the transmission of audio-tone protective relaying signals used in the protection of electric power systems. Protective relaying channels carry the "trip" signals needed to protect (isolate or deenergize) power lines when faults occur. Due to the severe electromagnetic environment present during faults, and due to the simultaneous need for uninterrupted transmission of control signals for protective relaying, short-term enhanced powers are permitted at the interface. To minimize crosstalk into adjacent channels due to the enhanced signal levels, this interface is restricted for use only in protective relaying applications.

3. Sealing-current option provides a dc current flow on four-wire "dry" metallic loops to help maintain continuity. With a DA protocol at the EU POT, the two SX paths are connected together within the exchange carrier equipment to complete the sealing-current path from the CO. With a "NO" protocol at the EU POT the sealing-current path from the CO is completed by the customer. Nothing should be connected into the SX path that would cause the path to become polarity-sensitive, or that would substantially increase the resistance of the path between the two pairs. In the absence of equipment for other reasons, the sealing-current path may also be obtained by ordering the improved termination option.

4. Applicable option needed only when associated interface protocol is high capacity, i.e., "DS."

5. PLAR (private-line automatic ringdown) is a signalling method. Generally, when an LR protocol at an EU is used in conjunction with a DS or SF protocol at the IC, the exchange carrier configuration will resemble the open end of a loop-start FX service.

6. Limited to locations with embedded technology; i.e., "A" for a D4 or "B" for a D3 channel bank.

Table 18–8. Voice–Grade Special Access Services

Service	Typical Application
VG1	Two–point nonswitched voice
VG2	Two–point switched voice or end–link or mid–link of multipoint
VG3	Trunk, PBX, or centrex nondata
VG4	Voice plus control tone
VG5	Two–point or end–link or mid–link of multipoint low–speed voiceband data circuit
VG6	Two–point or end–link or mid–link of multipoint voiceband data circuit
VG7	Voiceband medium–speed data
VG8	Voiceband data trunk
VG9	Voiceband data trunk – IC to IC
VG10	Two–point or end–link or mid–link of a digital data off–net extension
VG11	Two–point or end–link or mid–link for telephoto/facsimile
VG12	Two–point or end–link or mid–link audio–tone protective relaying
VGC	Customized

The C–message noise limits in dBrnc0 for various lengths of transmission facilities are shown in Table 18–12 for various combinations of facilities.

The insertion loss for the access segment is specified by the IC. To accommodate the ICs, a range of compatible values is offered by exchange carriers. This is given in terms of transmission levels for the transmit and receive directions. Table 18–13 indicates the

Table 18-9. Maximum Transmitted Power of Simultaneous Signals (at +13 TLP)—Three-Second Average

No. of Channels	Per Channel Maximum rms Power (dBm)	No. of Channels	Per Channel Maximum rms Power (dBm)
1	0	11	−10.4
2	−3.0	12	−10.8
3	−4.8	13	−11.1
4	−6.0	14	−11.5
5	−7.0	15	−11.8
6	−7.8	16	−12.0
7	−8.5	17	−12.3
8	−9.0	18	−12.6
9	−9.5	19	−12.8
10	−10.0	20	−13.0

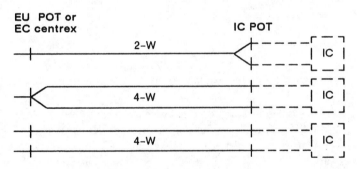

EC = Exchange carrier
EU = End user
IC = Interexchange carrier
POT = Point of termination

Figure 18-11. Two-point VG2 configurations.

ranges provided at the end–user point of termination (EU POT). Table 18-14 indicates the transmission level ranges provided at the interexchange carrier point of termination (IC POT).

Jurisdictionally Interstate Service

This special access service is for an intraLATA end–user–to–end–user, private–line voiceband application with the connection

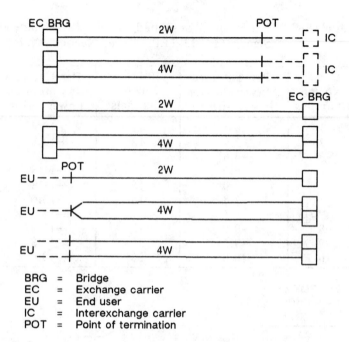

BRG = Bridge
EC = Exchange carrier
EU = End user
IC = Interexchange carrier
POT = Point of termination

Figure 18-12. Multipoint VG2 configurations.

crossing a state boundary or otherwise coming under Federal jurisdiction [4]. The JIS is similar to the service provided to intraLATA customers of the exchange carrier. The several service designations and typical applications are listed in Figure 18-13(a). A description of only one service, JIS VG7, for two-point line/trunk—voice/data, and typical configurations are shown in Figure 18-13(b).

Transmission limits for loss deviation, slope, attenuation distortion, noise, and echo control are shown in Table 18-15. For interfacing with the user, the connecting transmission levels must be within the compatible level ranges shown in Table 18-16.

18-4 "METALLIC" AND TELEGRAPH GRADE SPECIAL ACCESS SERVICES

"Metallic" and telegraph grade special access services are provided to the interexchange carriers between the EU POT and the IC POT. These services are described briefly.

Table 18-10. VG2—Specifications for Loss Deviation, Slope, and Attenuation Distortion

	Acceptance Limit	Immediate Action Limit
Loss Deviation from EML at 1004 Hz	±1.0 dB	±1.5 dB
Three-Tone Slope*	−0.5 to +3.5 dB	−1.0 to +4.0 dB
Attenuation Distortion*		
404–2804 Hz	−0.5 to +3.5 dB	−1 to +4 dB
304–3004 Hz	−0.5 to +4.5 dB	−1 to +5 dB
2804 Hz[†]	−1.0 to +2.5 dB	−1 to +3.5 dB

* Limits are the maximum deviation at 404 and 2804 Hz from the loss at 1004 Hz. The "+" limit means more loss and "−" limit means less loss.

† For SF applications, these limits apply to the 2804-Hz loss deviation from the 1004-Hz loss for that portion of the channel between the exchange carrier SF unit and the IC POT.

Metallic Access Services

Metallic special access services MT1, MT2, and MT3 are nonswitched services for alarms, protective relaying, etc. Access service MT1 can accommodate signal rates up to 30 baud on a pair. Access service MT2 can accommodate up to 15 b/s on a pair or a derived narrowband transmission path. It may or may not have dc continuity. Access service MT3 provides for dc and low-frequency control signals.

Telegraph Grade Access Services

Telegraph access services TG1 and TG2 are nonswitched services for telegraph channels. Access service TG1 provides asynchronous transitions between two current levels at rates up to 75 baud, half-duplex or duplex, on a two-point or multipoint basis. Neither dc continuity nor capability to transmit continuously varying ac is assured. Access service TG2 is the same as TG1, but for rates up to 150 baud.

Table 18-11. VG2 Echo Control

Transmission and Test Configuration		Type of Measurement	AL (dB)		IAL (dB)	
Measure @	Terminate @		ERL	SRL	ERL	SRL
Effective 2-Wire Transmission						
EU POT (2W) or CO (2W)	IC POT (4W)	Standard RL	5.5	3	5	2.5
		Improved RL	15	9	13	8
IC POT (4W)	EU POT (2W) or CO (2W)*	ELEPL-1	7.0	4.5	5.5	2.5
		ELEPL-2	18	12	16	11
4-Wire Transmission with 2-Wire Interface						
EU POT (2W) or CO (2W)*	IC POT (4W)	RL	26	20	24	18
IC POT (4W)	EU POT (2W) or CO (2W)*	ELEPL	23	16	20	14

* If a centrex central office, these values apply with 2-dB pad in.

AL	=	Acceptance limit
CO	=	Central office
ELEPL	=	Equal level echo path loss
ERL	=	Echo return loss
EU	=	End user
IAL	=	Immediate action limit
IC	=	Interexchange carrier
POT	=	Point of termination
RL	=	Return loss
SRL	=	Singing return loss

Note: Improved return loss and ELEPL-2 may require exchange carrier equipment at the two-wire interface.

The transmission parameter limits and interface combinations are given in Reference 5 for these services. There are also "custom" channels MTC and TGC with specially developed parameters.

18-5   PROGRAM AUDIO SPECIAL ACCESS AND LOCAL CHANNEL SERVICES

Program audio services are provided between an IC POT and an EU POT or between POTs of the same or different end users. One-way, full- or part-time services are offered for radio

Table 18-12. C-Message Noise Limits (dBrnc0) for Various Lengths (Miles) of Facility Combinations

Facility Combinations	0-50		51-100		101-200		201-400		401-1000	
	AL	IAL	AL	IAL	AL	IAL	AL	IAL	AL	IAL
D	27	29	27	29	27	29	27	29	27	29
D*	30	35	30	35	30	35	30	35	30	35
C + L/C	30	32	NA	NA	NA	NA	NA	NA	NA	NA
D + L/C	29	32	30	33	30	35	30	37	30	39
D* + L/C	32	38	32	39	32	40	32	37	32	39
CAC + L/C	29	32	29	33	30	35	NA	NA	NA	NA
CAC* + L/C	30	32	32	39	33	40	NA	NA	NA	NA
D + C + L/C	31	38	31	39	31	39	31	39	31	39
D* + C + L/C	33	38	33	39	33	39	33	39	33	39
CAC + C + L/C	31	38	31	39	32	40	NA	NA	NA	NA
CAC* + C + L/C	32	38	33	39	34	41	NA	NA	NA	NA
D + CAC + C + L/C	32	38	32	39	33	41	33	41	33	42
D + CAC* + C + L/C	33	38	34	39	35	41	35	41	35	42
D* + CAC + C + L/C	33	38	33	39	35	41	35	41	35	42
NCAC + C + L/C	35	38	36	39	37	41	39	43	41	45
NCAC + D + L/C	34	38	36	39	37	41	39	43	41	45
NCAC + D* + L/C	35	38	37	39	37	41	39	43	41	45
NCAC + D + C + L/C	35	38	36	39	37	41	39	43	41	45
NCAC + D* + C + L/C	36	38	37	39	38	41	39	43	42	45

```
   C  = Cable
 CAC* = Compandored analog carrier (N1 or equivalent)
 CAC  = Compandored analog carrier (N2, N3, or equivalent)
   D* = Digital carrier (D1 or equivalent)
   D  = Digital carrier (D2-D5, DLC, or equivalent)
 L/C  = Loop or cable
 NA   = Not applicable
 NCAC = Noncompandored analog carrier
```

broadcasting, audio recording, audio teleconferencing, etc. Because FCC-licensed broadcasters are considered to be in interstate business, their program and television channels fall under the FCC interstate access tariffs even if the channels are physically intrastate.

Table 18-13. VG2 Compatible Transmission Level Ranges at EU POT

2-Wire Transmission at the EU POT		
Specified Protocol Code	**Transmit**	**Receive**
AC,GO,LA,LB,LC, LO,LR,LS,NO	0	≥ -10
GO,LA,LB,LC,LO, LS,NO (with effective four-wire transmission)	0 to +5 [0]	-2 to -10 [-4]

4-Wire Transmission at the EU POT		
Specified Protocol Code	**Transmit**	**Receive**
AC,LR,LS,NO	0 to +5 [0]	0 to -16 [-4]
AC,LR,LS,NO (with improved termination option)	-16 to +7 [0]	-16 to +7 [-4]

Note: [ ] denotes recommended TLP.

Six types of program audio services of various bandwidth and quality may be offered.

(1) Service AP0 provides a nonequalized channel with a nominal passband from 300 Hz to 2500 Hz, not universally offered.

(2) Service AP1 provides a channel with a nominal passband from 200 Hz to 3500 Hz.

(3) Service AP2 provides a channel with a nominal passband from 100 Hz to 5000 Hz.

(4) Service AP3 provides a channel with a nominal passband from 50 Hz to 8000 Hz.

Table 18-14. VG2 Compatible Transmission Level Ranges at IC POT

Specified Protocol Code	Level Ranges	
	Transmit	Receive
A. With EC Equipment at IC POT		
EX	-3.7 to -2.1 [-2.1]	- 2.1
AB, GS, LO, LR, LS	-16 to +7 [0]*	-16 to +7 [-4]*
NO, SF†	-16	+7
SF physically at POT	+7	-16
B. Without EC Equipment at IC POT		
AB, GS, LO, LR, LS,	-3 to +5 [0]‡	0 to -6 [-4]§
NO, SF†	-3 to +5 [0]‡	0 to -6 [-4]§
C. Digital Signal at IC POT		
DS	DRS	DRS

*   With improved-termination option.

†   Where EC carrier is located at the IC POT.

‡   The full range of levels may not be available for all configurations.

§   In general, the receive TLP will be a function of the connecting EC cable facilities; however, there will be instances when EC apparatus on the circuit can be adjusted to provide the full range of TLPs shown.

Notes:

1.   [ ] denotes recommended TLP.

2.   DRS = Digital reference signal.

(5)   Service AP4 provides a channel with a nominal passband from 50 Hz to 15,000 Hz.

(6)   Service APC provides a channel customized to meet special requirements.

The transmission parameter limits and interface combinations are given in Reference 6 for these services. Chapter 16 discusses the construction and uses of these services in greater detail.

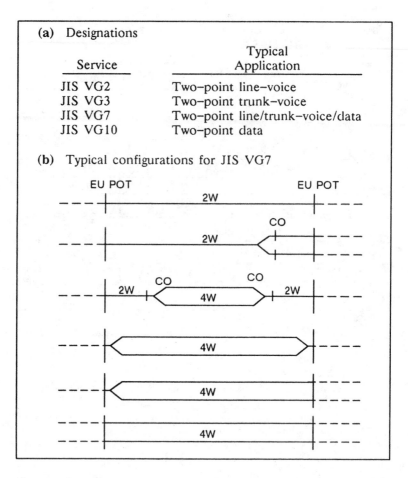

(a) Designations

Service	Typical Application
JIS VG2	Two–point line–voice
JIS VG3	Two–point trunk–voice
JIS VG7	Two–point line/trunk–voice/data
JIS VG10	Two–point data

(b) Typical configurations for JIS VG7

Figure 18–13. Jurisdictionally interstate special access service.

18–6   TELEVISION SPECIAL ACCESS AND LOCAL CHANNEL SERVICES

Television special access and local channel services are offered to the ICs, to any customer to access an IC, or for point–to–point service within a LATA. These services are nonswitched services for full–time and part–time commercial TV, noncommercial TV, closed–circuit TV, etc.

Table 18-15. Special Access Service JIS VG7 Transmission Limits

Loss Deviation from Loss at 1004 Hz
    EML $\pm$ 4.0 dB
 	
Three-Tone Slope[1]
    404 and 2804 Hz: -2 to $+5$ dB
 	
Attenuation Distortion[1]
    504 to 2504 Hz: -2 to $+5$ dB
    304 to 3004 Hz: -3 to $+12$ dB
 	
C-Notched Noise (-13 dBm0 Holding Tone)
    Maximum: 52 dBrnc0
    Minimum signal to noise: 25 dB
 	
Envelope Delay Distortion
    804 to 2604 Hz: $\leq 1250\,\mu$s
 	
Impulse Noise (-13 dBm0 Holding Tone)
    $\leq$ 15 cts/15 min over 71 dBrnc0
 	
Intermodulation Distortion
        R2 $\geq$ 28 dB
        R3 $\geq$ 35 dB
 	
Phase Jitter
        20 to 304 Hz: $\leq$ 8° peak to peak
        4 to 304 Hz: $\leq$ 12° peak to peak
 	
Frequency Shift
        $\pm$3 Hz
 	
C-Message Noise

Miles	0-200	201-400	401-1000
dBrnc0	$\leq$40	$\leq$42	$\leq$44

Table 18–15. Special Access Service JIS VG7 (Continued)

Echo Control				
		Interface Being Measured	Impedance Balance	
Facility			ERL	SRL
Effective 2–wire 2		2–wire or 4–wire	≥ 6	≥ 3
4–wire		2–wire or 4–wire	≥ 18	≥ 12

Notes:

1. Limits are the maximum deviation from the loss at 1004 Hz. The "+" limit means more loss and "–" limit means less loss.

2. An effective two–wire facility may be all two–wire or contain four–wire portions (such as carrier with two–wire extensions). A four–wire facility must be all four–wire; however, one or both interfaces may be two–wire.

Service TV1 provides a one–way circuit for a standard monochrome or color video signal and one or two associated 15–kHz audio signals. In most cases the audio is diplexed on the video.

Service TV2 is the same as TV1 except one or two audio channels of bandwidth 5 kHz are provided on separate facilities from the video. Service TVC provides a "customized" channel with special transmission parameters.

The transmission parameter limits and interface combinations are given in Reference 7 for these services. Technical specifications for the video and audio channels are included in Volume 1, Chapter 26, Part 5. Chapter 16 of this volume provides further detail on these services.

18-7 WIDEBAND ANALOG SPECIAL ACCESS SERVICES

Wideband analog special–access (nonswitched) services may be provided to the ICs. The exchange carrier provides the wideband access service from the EU POT within the LATA to the IC POT, also within the LATA.

Table 18-16. Special Access Service JIS VG7 Compatible
Transmission Level Ranges

Two-Wire Transmission at the EU POT

Specified Protocol Code	Transmit	Receive
NO	0	0 to -10* -1†
LA, LB, LC	0	0 to -4.5* -4.0†
GO, GS, LO, LS	0	0 to -4* -3.5†
EA, EB	-2*	$-(VNL + 2)$*
CT	-2	$-(VNL + 2)$

Four-Wire Transmission at the EU POT

Specified Protocol Code	Transmit	Receive
GO, GS, LS, LO	0	0 to -4* -3.5†
EA, EB	-2*	$-(VNL + 2)$*
EA, EB	-6*	$-(VNL + 2)$*
EA-M, EB-M	-16*	$+7$*
EA-E, EB-E	$+7$*	-16*
CT	0	-5
DS	DRS	DRS

*   With gain device.
†   Without gain device.

The various services that can be provided by this access, which covers transmission channels capable of group, supergroup, mastergroup, or other bandwidths, are listed in Table 18–17. The detailed transmission characteristics for these services are covered in Reference 8.

Table 18–17. Wideband Analog Special Access Services

WA1* (Channel group)

Access service WA1 provides a transmission channel with a passband from 60 kHz to 108 kHz.

WA2* (Supergroup)

Access service WA2 provides a transmission channel with a passband from 312 kHz to 552 kHz.

WA2A* (Mastergroup)

Access service WA2A provides a transmission channel with a passband from 564 kHz to 3084 kHz.

WA3

Access service WA3 allows transmission of a wideband signal at an EU premises. The frequency of the signal is approximately 300 Hz to 16 kHz.

WA4

Access service WA4 provides a transmission channel with a passband between approximately 29 kHz and 44 kHz.

---

* Optional central-office multiplexing of compatible analog hierarchy can be provided.

18-8 WIDEBAND DATA SPECIAL ACCESS SERVICES

Wideband data special access services are provided between the EU POT and the IC POT. These services are designated WD1, WD2, and WD3 and are for nonswitched circuits with the capability of transmitting sync and nonsync digital data. They require a 303–type data station at the EU POT and are available only to existing customers (as a means of maintaining circuits that existed at divestiture). A description of each service type is shown in Table 18–18. The services are covered in Reference 9.

Table 18–18. Wideband Data Special Access Services, Service Designation

---

Access Service WD1

Access service WD1 provides for the transmission of synchronous serial 19.2–kb/s data and a voice–frequency coordinating channel.

Access Service WD2

Access service WD2 provides for the transmission of synchronous serial 50–kb/s data and a VF coordinating channel.

Access Service WD3

Access service WD3 provides for the transmission of synchronous serial 230.4–kb/s data and a VF coordinating channel.

OPTIONAL FEATURES

Access Service WD1

Optional services are available for the transmission of synchronous serial 18.75–kb/s data or asynchronous serial data with a minimum signal element width of 52 μs.

Access Service WD2

Optional services are available for the transmission of synchronous serial 40.8–kb/s data or asynchronous serial data with a minimum signal element width of 20 μs.

Access Service WD3

Optional services are available for the transmission of asynchronous serial data with a minimum signal element width of 4.3 μs.

18-9   DIGITAL DATA SPECIAL ACCESS SERVICES

Digital data special access services are provided between the EU POT and the IC POT. These services are designated access services DA1, DA2, DA3, and DA4. They are for nonswitched serial synchronous data service at 2.4, 4.8, 9.6, and 56 kb/s,

respectively. 1.544–Mb/s high–capacity digital service channelized to a D4 framing format may be used to provide up to 24
56–kb/s channels. Reference 10 gives service descriptions, interface codes, technical specifications, and illustrations of some applications and their interfaces. Chapter 17 covers the Digital Data
System (DDS) in detail. Examples of the application of access
service DA4 are:

Digital Data—Provides for two or more points of private–line
digital transmission operating at a synchronous data rate of 56
kb/s, without alternate voice or voice coordination channel.

Packet Access Line—Provides 56–kb/s data access for data device or host computer to the packet network.

Packet Switch Trunk—Provides a 56–kb/s data channel between packet switches.

Packet Off–Network Access Line—Provides a 56–kb/s off–network access for packet switches. Uses digital data and analog
facilities.

Typical interfaces for access service DA4, for a single 56–kb/s
channel and for a 56–kb/s channel carried on a 1.544–Mb/s
digital service, are shown in Figures 18–14 and 18–15, respectively. The interface levels for the 56–kb/s service are shown as
voltage limits over the bit time interval in Figure 18–16 to define
a "one" (plus or negative) and a "zero."

The expected service performance is:

Offering—Monthly average equal to or greater than
99.875–percent error–free seconds.

Service Acceptance Limit—Acceptable for service turn–up
when no more than two errored seconds are observed during a
15–minute loop–back test.

Immediate Action Limit—When, as a result of a customer
trouble report, the performance of a 15–second channel loop–
back test results in the observation of any errored seconds.

Application   56-kb/s digital data

Physical Description   Jack or connector

Access Service   DA4

Electrical Features

   Impedance   135 Ω

   Levels   See Figures 18–17 and 18–18

   Speed   56-kb/s bipolar return to zero,
                        timing on signal from EC to
                        customer.

Illustration

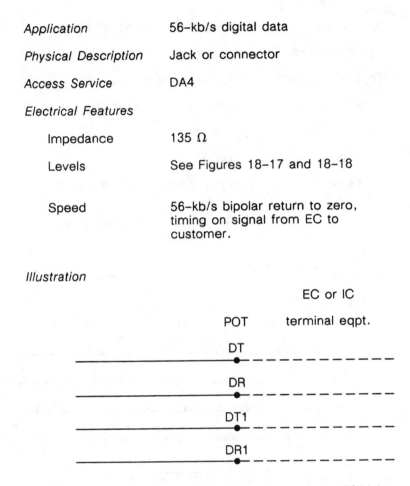

Figure 18–14. Digital data special access interface—56 kb/s,
access service DA4.

18–10 HIGH–CAPACITY DIGITAL SPECIAL ACCESS SERVICES

High–capacity digital special access services are provided between the EU POT and the IC POT, or between one of these
sites and an EC office where multiplexing is offered. These services are designated HC1, HC1C, HC2, HC3, and HC4, and are

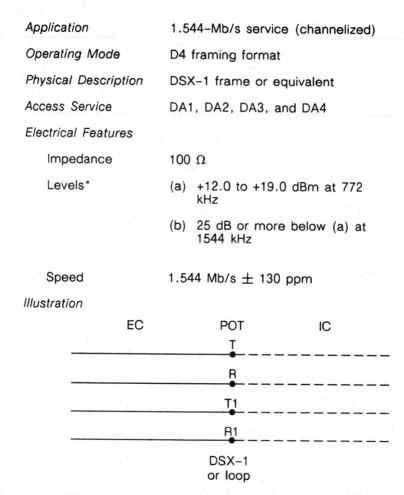

Application	1.544-Mb/s service (channelized)
Operating Mode	D4 framing format
Physical Description	DSX-1 frame or equivalent
Access Service	DA1, DA2, DA3, and DA4
Electrical Features	
Impedance	100 Ω
Levels*	(a) +12.0 to +19.0 dBm at 772 kHz
	(b) 25 dB or more below (a) at 1544 kHz
Speed	1.544 Mb/s ± 130 ppm
Illustration	

EC   POT   IC

T

R

T1

R1

DSX-1
or loop

* *CCITT Recommendation G.703* for 3-kHz band around 772 kHz, all 1s transmitted. For a 2-kHz band, (a) becomes 12.6 to 17.9 dBm and (b) becomes 29 dB or more.

Figure 18-15. Digital data special access interface—1.544 Mb/s, access service DA1-DA4.

for isochronous serial data service at 1.544, 3.152, 6.312, 44.736, and 274.176 Mb/s, respectively. Optional features that can be provided include automatic protection switching, central-office multiplexing, and a transfer arrangement. The multiplexing

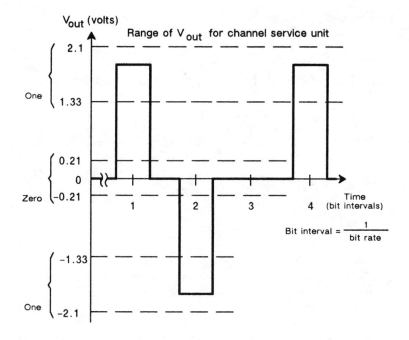

Note: V_{out} is ac voltage
across a 135-ohm resistive termination.

Figure 18-16. Voltage ranges—digital data special access
interface.

may be between digital signal levels (e.g., DS3 to DS1) or from
DS1 to voice or digital data.

Reference 11 gives a description of the access services, interface codes, technical specifications, illustrative examples, and
performance objectives for all of these services. The newer *ANSI
Standard T1.102-1987* on digital hierarchy interfaces [12]
covers the characteristics of DS1, DS1C, DS2, and DS3 signals
appearing at the applicable digital cross-connections. The
DSX-1 interconnection specification is given in Table 18-19
[12]. The associated pulse templates for new and older equipment are given in Figures 18-17 and 18-18, respectively [12].
(As with other digital services, the POT may also be at the end of
a cable loop where DSX-type signals do not pertain.) The

performance objectives are listed in Table 18–20 for all of the access services when measured at a DS1 rate.

Table 18–19. DSX–1 Interconnection Specification (ANSI)

Line rate:	1.544 Mb/s
Tolerance:	Source timing for self–timed DS1 bit streams shall not exceed ±32 ppm with respect to the basic rate. DS1 sinks should be capable of accepting a rate deviation of ±130 ppm.
Line code:	Either of the following:
	(1) Bipolar[1] with at least 12.5–percent average ones density and no more than 15 consecutive zeros
	(2) Bipolar with eight–zero substitution (B8ZS). See Volume 2, Chapter 13, Part 1.
Signal format:	DS1 formats (See Volume 2, Chapter 13, Part 1).
Termination:	One balanced twisted pair for each direction of transmission. The distribution frame jack connected to a pair bringing signals to the frame is an outjack; the jack connected to a pair carrying signals away is termed the injack.[2]
Impedance:	A test load of 100 ohms resistive ± 5 percent shall be used at the interface for the evaluation of pulse shape and the electrical parameters specified below. This test-load requirement applies only to the DSX–1 appearance of a source at its outjack.

Table 18–19. DSX–1 Interconnection Specification (ANSI) (Continued)

Pulse shape:	An isolated pulse from newly designed equipment shall fit the template shown in Figure 18–17. The template of Figure 18–18 shall be used to represent pulses generated by older equipment.
Power level:	(1) New equipment: for an all–ones transmitted pattern, the power in a band no wider than 3 kHz centered at 772 kHz shall be between 12.6 and 17.9 dBm.
	(2) Older equipment: the power in a band no wider than 3 kHz centered at 772 kHz may be between 12.4 and 18.0 dBm.
	(3) The power in a band no wider than 3 kHz centered at 1544 kHz shall be at least 29 dB below that at 772 kHz.
Pulse imbalance:	There shall be less than 0.5 dB difference between the total power of the positive pulses and of the negative pulses.
Signal–to–noise ratio:	Not available. The effects of noise at DS1 have been controlled by using a short repeater spacing for the first span leaving an office.
Jitter:	See Volume 2, Chapter 15, Part 4.
Cable characteristics:	Reference cable for DSX–1 specifications is 655 feet (200 meters) of 22 AWG multipair, ABAM or equivalent (PIC construction with overall outer shield).

Table 18–19. DSX–1 Interconnection Specification (ANSI) (Continued)

Test access:	High impedance bridging monitor access should be provided across the outjacks. This bridging circuit consists of 432–ohm ± 5 percent resistors connected in series with tip and ring.

Notes:

1. Bipolar is also known as alternate mark inversion. Successive logical "ones" are coded as pulses of alternating polarity. Zeros are coded zero. Two or more successive pulses of the same polarity are termed bipolar violations.

2. The North American convention for naming jacks is opposite to that in *CCITT Recommendation G.703*.

Of these services, HC1 and HC3 are receiving strong attention for use in both customers' private networks and ICs' facilities. These 24– and 672–channel bundles are replacing a great many individual voice–grade circuits for reasons of economy and convenience to the user.

18-11 CELLULAR MOBILE CARRIER INTERCONNECTION

Cellular mobile carriers (CMCs) provide end–link telephone service to their customers via radio links to a CMC mobile telephone switching office (MTSO). The MTSO is interconnected to the network of exchange carriers to reach telephone subscribers [13]. This section describes some transmission guidelines for these interconnections, which are noted in Figure 18–19. The guidelines were a combined effort of the cellular industry, exchange carriers, and Bellcore. Other circuits, like the MTSO–to–cell–site link, may be furnished by the exchange carrier from its normal private–line and high–capacity offerings.

The CMC concept provides many contiguous "cells" of assigned radio coverage, configured so that many mobile telephone customers can be served within a cell or when they move from cell to cell, and the radio frequencies can be reused efficiently.

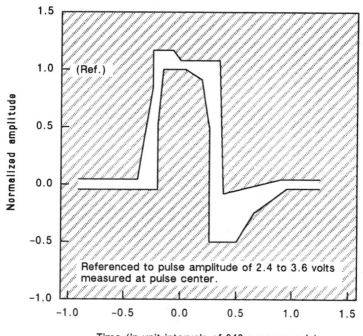

Figure 18-17. DSX-1 isolated-pulse template and corner points (new equipment).

Maximum Curve

Unit intervals	-.77	-.39	-.27	-.27	-.12	0.0	.27	.35	.93	1.16
Normalized amplitude	.05	.05	.8	1.15	1.15	1.05	1.05	-.07	.05	.05

Minimum Curve

Unit intervals	-.77	-.23	-.23	-.15	0.0	.15	.23	.23	.46	.66	.93	1.16
Normalized amplitude	-.05	-.05	.5	.95	.95	.9	.5	-.45	-.45	-.2	-.05	-.05

Calls made by mobile customers are connected via a cell to an MTSO, which is interconnected to the exchange carrier network for connections outside the MTSO. Further details are available in Volume 2, Chapter 22.

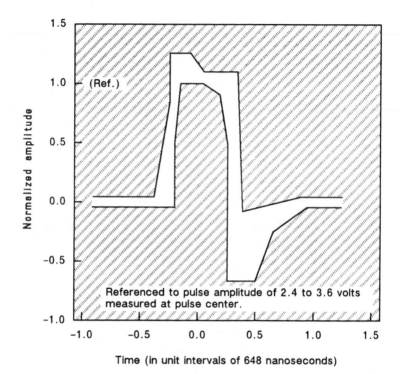

Time (in unit intervals of 648 nanoseconds)

Maximum Curve

Unit intervals	-.77	-.39	-.27	-.27	-.12	0.0	.27	.34	.58	1.16
Normalized amplitude	.05	.05	.8	1.22	1.22	1.05	1.05	.08	.05	.05

Minimum Curve

Unit intervals	-.77	-.23	-.23	-.15	-.04	.15	.23	.23	.42	.66	.93	1.16
Normalized amplitude	-.05	-.05	.5	.95	.95	.9	.5	-.62	-.62	-.2	-.05	-.05

Figure 18-18. DSX-1 isolated-pulse template and corner points (older equipment).

Three interconnection arrangements have been defined between the CMC-designated points of termination and the exchange carrier. Each has been assigned 3 dB loss (0 to 4 dB for two-wire nongain circuits). These are as follows.

Table 18-20. High–Capacity   Digital   Special   Access
Services—HC1, HC1C, HC2, HC3, and HC4
Performance Objectives at DS1 Rate

Error–Free Seconds

Performance is a measure of the error performance while
the circuit is available. This parameter is measured by counting
the number of errored seconds* (ES) that occur within a given
period of time. The objective is a performance level of
98.75–percent error–free seconds over a continuous 24–hour
period.

Acceptance limits. Three 15–minute runs, spanning the
EC's switching busy hour at the wire center of measurement,
should be made and ES should be counted and recorded.
None of the three 15–minute periods should exceed 20 ES. If
one of the measurements exceeds 20 ES, repeat one 15–minute run. If this gives 20 ES or under, accept the facility. If all
three tests exceed 20 ES, repair the facility or change assignments. If two tests exceed 20 ES, repeat three 15–minute tests
and follow the above procedures.

Immediate action limit. Short durations of ES may be expected and will come clear without repair action. A 5–minute
interval with more than 60 ES should be reported.

Availability

The availability objective over one year is 99.925 percent,
where an unavailable circuit is defined per the following
CCITT definition:

Unavailable = bit error ratio (BER) greater than 1 per
1000 bits in each of 10 consecutive 1–second intervals.

Restoration = BER less than 1 per 1000 bits in each of
10 consecutive 1–second intervals.

---

*   An errored second is any second in which one or more bit errors are
received.

Table 18–20. High–Capacity Digital Special Access Services—HC1, HC1C, HC2, HC3, and HC4 Performance Objectives at DS1 Rate (Continued)

Availability (Continued)

Outage = accumulated intervals between circuit unavailable and circuit restoration.

$$Availability = \frac{(total\ time) - (outage\ time)}{(total\ time)}$$

Jitter

Max peak–to–peak timing jitter = 14 DS1 time slots (introduced by the EC network on the HC1 service).

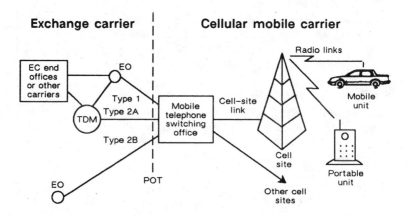

Figure 18–19. Cellular mobile carrier to exchange carrier switched interconnection (types 1, 2A, and 2B).

Type 1 Interconnection to an Exchange Carrier End Office

This interconnection provides the MTSO with access to an exchange carrier end office that can establish paths to telephone customers served by the same and other offices within the

LATA, to other MTSOs within the LATA, to an interchange carrier, to an operator, and to directory assistance systems.

Type 2A Interconnection to an Exchange Carrier Tandem Office

This arrangement gives the MTSO unrestricted access to the same services as for type 1, but with improved transmission performance as losses on tandemed calls are reduced.

Type 2B Interconnection to an Exchange Carrier End Office

This interconnection provides the MTSO with access to an end office, but connections are restricted to dialable numbers (line appearances) served by that end office.

Transmission Loss

The overall transmission objective is to provide a grade of service to the cellular end user that is comparable to that provided to telephone subscribers. For this purpose and to provide a basis to establish transmission parameter limits for the interconnections, it was assumed that transmission performance and speech volumes of the cellular end user referenced to the MTSO office should be similar to those of an exchange carrier end office for average loop/telephone set users. Average end–office speech volumes are -20.7 dBm outgoing from the loop, -20.7 dBm into the loop from local end–office calls, and -26.7 dBm from remote calls (assuming 6–dB connection loss).

From the viewpoint of loss, a 3–dB type 2A interconnection from an MTSO to the exchange carrier tandem office provides the nominal loss for a tandem–connecting trunk (to an exchange carrier end office) in the exchange carrier tandem network. This type of cellular interconnection should provide a grade of service for the cellular calls similar to that for exchange carrier metro calls.

The 3–dB loss type 2B interconnection from an MTSO to an exchange carrier end office is the same as for an interoffice trunk in the LATA network; hence, the same grade of service would be provided for cellular–to–end–office calls as for calls between end offices. It is presumed that a type 2B interconnection is desirable when the community of interest is covered by only that end office and other features are not wanted.

The 3–dB loss type 1 interconnection from an MTSO to an exchange carrier end office may provide the same service features as type 2A, but at a poorer transmission performance for connections other than terminations of the connecting office. This is so because of the extra links and loss introduced on multilink connections. Table 18–21 indicates the expected losses from the MTSO to exchange carrier end offices. It may be noted that for a multilink MTSO interconnection, type 2A would provide lower–loss links between the MTSOs within the tandem network.

Table 18–21. Interconnection Losses—MTSO to End Offices

CMC - EC Interconnection Type	To Connected End Office	To Interoff Trunk to End Office	To End Office Via Tandem	To AT or IC Direct or Via Tandem
Type 1 MTSO to end office	3 dB	6 dB	9 dB	6 dB
Type 2A MTSO to tandem	–	–	6 dB	3 dB
Type 2B (local only) MTSO to end office	3 dB	–	–	–

Note: Assumes all trunks are 3–dB loss (no two-wire, 0-to-4-dB loss).

When the CMC requires call forwarding or three–way calling type services with type 1 interconnection, values of loss lower than the 3 dB normally assigned may be desired. Reference 13 discusses approaches to reduce the loss to 1 dB for such calls.

Other Transmission Parameters and Limits

The major transmission parameters affecting the quality of service for voice are loss (see the preceding section), loss deviation, three–tone slope, C–message noise, C–notched noise for digital or compandored systems, and return–loss balance. Reference 13 includes these factors and, for satisfactory voice performance, gives limits in terms of ALs and IALs for various facilities that may be used for the three types of interconnection. Only a few noise and balance values taken from the reference are presented. These are applicable to all interconnection types. Loss deviation and three–tone slope limits for four–wire and two–wire connections are given in Table 18–22. Noise limits in dBrnc0 for C–message and C–notched measurement are given in Table 18–23 for four–wire and two–wire terminations at the CMC POT. Balance guidelines are given in Table 18–24 for four–wire or two–wire physical connections. The responsibility for providing and maintaining the balance is also noted in Table 18–24.

A few examples of typical configurations that may be used between the exchange carrier switch, tandem or end office, and the MTSO are shown in Figure 18–20(a), (b), and (c).

Table 18–22. CMC   Interconnection—Loss   Deviation   and Three–Tone Slope

	4-Wire Transmission at POT Interface		2-Wire Transmission at POT Interface	
Voice Transmission Parameter	AL	IAL	AL	IAL
Loss deviation from EML at 1004 Hz (dB)	0.7	2.5	0.7 with gain 1.2 w/o gain	3.0
3–tone slope (dB) (Note 1) 404 and 2804 Hz 2804 Hz (Note 2)	-1.5 to +3.5 -1.0 to +2.5	-2.0 to +4.0 -1.0 to +3.5	-1.5 to +5.0 NA	-2.0 to +5.5 NA

Notes:

1. Level deviation at 404 Hz and 2804 Hz relative to the measured level at 1004 Hz.

2. These limits apply between the exchange carrier SF unit and the POT when the CMC is provided an interface with SF signalling.

Table 18–23. Interconnection Types 1, 2A, and 2B Noise Limits*

Facility EC to CMC POT	C-Message Noise (dBrnc0)			C-Notched Noise (dBrnc0)		
	AL	IAL		AL	IAL	
		4-Wire	2-Wire		4-Wire	2-Wire
Digital and cable and EC switch	30	35	38	40	47	47

* For 0 to 50 miles. See Reference 13 for other facilities.

Table 18–24. Interconnection Balance Guidelines

Type of Inter-connect	EC SW	POT	CMC MTSO	POT Toward EC (EC Responsibility)		POT Toward EC (CMC Responsibility)	
				ERL AL/IAL	SRL AL/IAL	ERL AL/IAL	SRL AL/IAL
1/EO	2W	4W	4W	20/16	13/11	NA	NA
(Note 2)	2W	4W	2W	20/16	13/11	20/16	13/11
	4W	4W	2W	NA	NA	20/16	13/11
	4W	4W	4W	NA	NA	NA	NA
				EC Toward POT (Note 1)		POT Toward MTSO	
	2W	2W	2W	16/13	8/6	NS	NS
	2W	2W	4W	16/13	8/6	NS	NS
				POT Toward EC		POT Toward MTSO	
2A/TDM	2W	4W	4W	20/16	13/11	NA	NA
(Note 2)	2W	4W	2W	20/16	13/11	20/16	13/11
	4W	4W	2W	NA	NA	20/16	13/11
	4W	4W	4W	NA	NA	NA	NA
2B/EO	Any	Any	Any	NR	NR	NR	NR

NA = Not applicable (no two-wire-to-four-wire conversion unit)
NR = Not required (equivalent to EC inter-end-office trunks)
NS = Not specified

Notes:

1. These values are stated as return loss; all others are stated as equal level echo path loss.

2. ELEPL values are predicated on the assumption that MTSOs and EC EOs operate at TP0 while EC tandems operate at TP2.

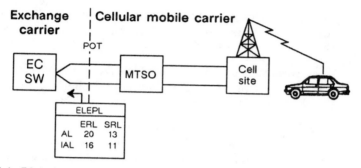

(a) EC is two-wire switch, service is type 1 or 2A, MTSO is four-wire, POT is four-wire.

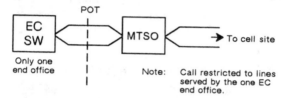

(b) EC is two-wire switch, service is type 2B, MTSO is two-wire, POT is four-wire, no balance required.

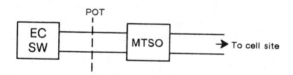

(c) All four-wire, no balance required.

Figure 18–20. Typical interconnections to cellular mobile carriers—EC SW to MTSO.

References

1. *Notes on the BOC IntraLATA Networks—1986*, Technical Reference TR–NPL–000275, Bellcore (Iss. 1, Apr. 1986).

2. *Voice Grade Switched Access Service—Transmission Parameter Limits and Interface Combinations*, Technical Reference TR–NPL–000334, Bellcore (Iss. 2, Dec. 1987).

3. Draft of *ANSI T1.506–1989*, "American National Standard for Telecommunications—Network Performance Standards—

Switched Exchange Access Network Transmission Specifications" (New York: American National Standards Institute, 1989).

4. *Voice Grade Special Access Service—Transmission Parameter Limits and Interface Combinations*, Technical Reference TR–NPL–000335, Bellcore (Iss. 2, Dec. 1987).

5. *Metallic and Telegraph Grade Special Access Services— Transmission Parameter Limits and Interface Combinations*, Technical Reference TR–NPL–000336, Bellcore (Iss. 1, Oct. 1987).

6. *Program Audio Special Access and Local Channel Services— Transmission Parameter Limits and Interface Combinations*, Technical Reference TR–NPL–000337, Bellcore (Iss. 1, July 1987).

7. *Television Special Access and Local Channel Services— Transmission Parameter Limits and Interface Combinations*, Technical Reference TR–NPL–000338, Bellcore (Iss. 1, Dec. 1986).

8. *Wideband Analog Special Access Service—Transmission Parameter Limits and Interface Combinations*, Technical Reference TR–NPL–000339, Bellcore (Iss. 1, Oct. 1987).

9. *Wideband Data Special Access Service—Transmission Parameter Limits and Interface Combinations*, Technical Reference TR–NPL–000340, Bellcore (Iss. 1, Oct. 1987).

10. *Digital Data Special Access Service—Transmission Parameter Limits and Interface Combinations*, Technical Reference TR–NPL–000341, Bellcore (Iss. 1, Mar. 1989).

11. *High–Capacity Digital Special Access Service—Transmission Parameter Limits and Interface Combinations*, Technical Reference TR–NPL–000342, Bellcore (Iss. 1, June 1989).

12. *ANSI T1.102–1987*, "American National Standard for Telecommunications—Digital Hierarchy—Electrical Interfaces" (New York: American National Standards Institute, 1987).

13. *Cellular Mobile Carrier, Interconnection Transmission Plans*, Technical Reference TR–EOP–000352, Bellcore (Iss. 1, May 1986).

Telecommunications Transmission Engineering

Section 5

Transmission Quality

The continuing provision of a public switched network transmission quality that meets customer expectations requires that all of the interconnected carriers properly plan, design, install, and maintain network transmission facilities and circuits. In addition, each of the carriers must carry out a quality control program to provide for the timely correction of transmission deficiencies as they arise.

Chapter 19 describes transmission facility planning in terms of an integrated process including both strategic and tactical elements. Exploratory studies and service planning along with documentation and coordination are all required to provide efficient, flexible network transmission facilities. The chapter explains how transmission input to the process is vital to the generation of plans that meet grade–of–service objectives.

With facilities in service, a quality–assurance function is required to determine whether facilities meet performance objectives and to provide for correction so that transmission quality will satisfy customers. Chapter 20 describes measurements that may be used to assess the quality of service being provided by the various interconnecting carriers. These include indications of customer reactions to service as well as internal measurements of transmission quality parameters. The chapter then describes how the measurements are analyzed to provide feedback to various organizations so that corrections can be made in completing the quality process. Finally, Chapter 20 discusses the importance of incorporating these activities into a company transmission quality program in order to get proper management attention in today's competitive telecommunications environment.

Chapter 21 describes the principles on which maintenance is based and discusses the strong trends toward automated facility and circuit maintenance. The chapter relates maintenance to circuit continuity and discusses some specific ways in which the basically reliable facilities are supplemented by automatic means of protection against service outage. Other considerations in maintaining service continuity and restoring service after catastrophic failure are also discussed.

Chapter 22 provides a summation and overview of the management processes and techniques that should be employed to maintain control of transmission activities in the face of continuous and accelerating rates of technological, sociological, and environmental change. The chapter concludes with a glimpse of the management challenge in implementing emerging telecommunications technologies and services as the changes continue to accelerate.

Chapter 19

Transmission Facility Planning

Efficient management of the internodal/interoffice facilities that carry the public switched network, special services, and overlay digital networks can be achieved only through careful planning. Such planning is carried out in an organized and structured manner by organizations dedicated to ensuring the deployment of network facility and equipment resources in a service/cost–efficient manner.

An efficient transmission facility planning process must include both the strategic and tactical functions and be integrated with the node–access (loop) and switch planning processes. Data requirements for planning include technical parameters on space, power, and transmission requirements, as well as compatibility and cost. Data are also required on forecasted demands and type and quantity of embedded facilities. The process is carried out in various ways to meet the specific requirements of metropolitan, rural, and intercity networks.

The data required to produce a facilities plan is supplied by various sources; forecasted demand is collected from the traffic and marketing organizations, technical information is provided by the transmission engineering group, and information about embedded facilities is extracted from inventory data bases. Thus, while facilities planning is normally carried out in organizations dedicated to that purpose, information must be supplied by the transmission engineering, provisioning, forecasting, and marketing organizations.

When additional trunks or special–services circuits are needed, they are usually installed on existing facilities where spare capacity is available. Where spare capacity does not exist, where it cannot be made available by rearrangements, or where older facilities cannot meet the requirements of advanced services, new facilities must be constructed in the form of optical fiber, digital radio, or other cost–justified serving arrangements. The construction of new base facilities (i.e., high–capacity transmission paths multiplexed into channels of lower capacity) is normally preceded by a study that compares alternative means of furnishing relief over a specified period of time. To perform these studies, geographical areas of study must be defined and planning models constructed. These models incorporate information relevant to the node–access and switching networks of the study area. Consideration is also given to interactions between the plans for the selected study area and plans made by the same or other companies for adjacent or overlapping areas. When developing planning models, the planning organization relies on the transmission engineering group to provide technical guidance regarding the feasibility of activities such as upgrading a fiber facility, splicing in a new fiber cable from a different manufacturer, putting in a new digital radio route, performing an analog–to–digital modernization, etc. Once the planning models are established and the studies are completed, the results must be evaluated and documented for management approval and ultimate implementation.

19–1 TRANSMISSION FACILITY PLANNING FUNCTIONS

Transmission facility planning can best be performed as part of a totally integrated network planning process. Integration is desirable to ensure that the resulting network plan and its implementation provide to the network the greatest efficiency and flexibility possible. Integration is preferably horizontal (combining node–access and switch planning with internodal) and vertical (joining strategic and tactical planning). Transmission facility planning, while part of the total network planning process, consists of three interrelated functions: exploratory studies, service planning, and documentation and coordination.

The *exploratory studies* function does much of what traditionally has been called fundamental or long–range planning. The necessary economic studies, developed in an atmosphere removed from day–to–day decision making and budget preparation demands and reflecting corporate goals based on business decisions, are provided by this function. Outputs of the exploratory studies function include technical planning guidelines (e.g., applicability of a new type of transmission system), and long–range direction to the evolution of the facilities network. These guidelines and studies serve to assist those responsible for the service planning function in the development of implementation plans. The exploratory studies and service planning functions must, of necessity, be coordinated closely.

The development and maintenance of unit cost estimates for facilities, equipment, and certain "special assemblies" is an important part of the planning process. These unit costs are based on detailed estimates of material prices and the associated costs for engineering, installation, and maintenance, as well as the costs of related resources such as power, cable ducts, and floor space. Unit costs are utilized within exploratory studies and tracked through service planning and actual implementation of the construction program.

Service planning of central–office transmission equipment and internodal facilities results in the establishment of near–term requirements by type and quantity, and proper timing of specific equipment and facility augments. Because of the lead time required to develop and fund a construction program, the service-planning function focuses on the first and second years of the program. The current year is reviewed to identify unanticipated requirements and to modify existing plans according to changing conditions. The third and fourth program years are included for construction budget purposes and for the planning of items such as conduit and radio, which have long lead times.

The *documentation and coordination* function serves to summarize the planned construction program in a form useful to other work groups and to issue specific project requests to the various provisioning areas such as outside plant, central–office equipment engineering, and the circuit provision center. This function also provides assignment rules, either on paper or in mechanized data files, to the circuit provision center to guide it in the selection of lowest–cost facility designs for new circuits. Additionally, the documentation and coordination function monitors actual construction and assignment activities to ensure that the network evolves in a manner consistent with the plan.

These three functions are neither separate nor discrete; each supports and is supported by the others. The development, in recent years, of mechanization tools to assist the transmission facility planner in the performance of the planning functions has caused the three functions to merge into a single continuous process. Today, there is no easily distinguishable boundary separating them. Still, while recognizing the inseparability of these functions in practice, the separate identification and discussion of each is valuable for analysis of the planning process.

19-2 FACILITIES NETWORK ENVIRONMENTS

Planning studies cover specific geographical areas and reflect interrelationships with adjoining networks of the same or other companies. As planning problems tend to be different in metropolitan, rural, and intercity environments, a major consideration in the determination of boundaries of a study area is the nature of that area.

A metropolitan area consists of one or more large business districts and associated suburbs. Metropolitan areas tend to contain many central offices located in close proximity with large numbers of relatively short trunks. Outside plant facilities are predominantly underground (in conduit) and highly developed, perhaps even congested. The circuits are transported over a mix of voice–frequency (VF) cable and digital carrier. As a result, the facility network interconnecting these offices is quite complex and possibilities for routing a circuit between two points are numerous. While these multiple routing possibilities afford a high degree of flexibility to the network, they also make the job of planning for network growth difficult. On a long–term basis, there is usually one optimum routing for any particular circuit group. For the short term, the growth pattern for all circuit groups, when considered in an existing facilities network, often requires the selection of expedient circuit routings that differ from the long–term optimum. Due to the high capacity of optical fiber cable, it has become the base facility of choice for digital carrier systems, and provides the incentive for hubbed facility layouts that simplify the choice of circuit routes.

Interexchange carriers (ICs) have a significant presence in metropolitan areas. While the ICs are major purchasers of high–capacity special services and access services, their presence, together with the high concentration of large businesses, creates an environment suited to bypass of the exchange carrier. This competitive environment places on the planning organization a need for added emphasis on flexibility and responsiveness to the customer. Local fiber carriers provide further competitive pressure.

By contrast, the wire centers in rural areas are relatively small, generally isolated, and almost always unattended. Some are remote switch units in towns where the population is growing slowly or even shrinking. Since population clusters are often more than ten miles apart, there is rarely more than one wire center per cluster. Existing base facilities are often metallic cable or microwave radio. Where cable is used, the facilities are either buried (without conduit) or aerial. In new construction, current emphasis is on replacing copper cable with fiber cable. While aerial placement and placement within plastic duct is common, properly sheathed fiber can be buried without the benefit of duct.

While the rural environment generally contains several other exchange carriers with which coordination must take place, ICs typically have a minor physical presence. Customer locations are relatively far apart and few industrial parks of significant size exist. Thus, the general volume of interLATA (local access and transport area) traffic is insufficient to warrant ICs' construction of facilities deep into rural areas. Access services are typically provided through the use of access tandem switches.

The networks of predominantly high–capacity transmission facilities that interconnect metropolitan and rural areas make up the intercity environment. These facilities normally terminate in major toll centers or, in some cases, large but isolated end offices. The routes tend to be long and consist of large and rapidly growing cross sections. The possibilities for diverse routing in such networks tend to be greater than in the rural networks but less than in major metropolitan areas. The interLATA networks of the ICs have most of these same characteristics with even greater lengths, capacities, and circuit growth.

19-3 INTERACTING ORGANIZATIONS

Transmission facility planning interacts with several organizations. These organizations provide technical assistance and information to the planning process and, at times, receive information or direction from the planning organization.

Budget

The function of a budget organization is the analysis of planned capital investment costs and anticipated expenses to be incurred in current and future years. The goal is to assist the planner in devising a facility plan that fits the needs and constraints of the business. The creation of a facility plan involves the allotment of scarce capital resources among those growth projects that are needed most urgently, those discretionary programs that promise the highest return, and those market positioning projects that will maintain or enhance current and future revenues. The facilities plan covers the current year plus four additional program years and attempts to balance augments that are capital–intensive with rearrangements that are expense–intensive.

Circuit Provision

Circuit Provision makes specific assignments to facilities and equipment, administers inventory records, prepares work orders, and tracks the progress of message–trunk, special–services, and rearrangement orders. This function is particularly sensitive to customer service orders.

Distribution Services

Distribution Services handles the details of cable installation and removal, including the actual installation, splicing, and acceptance of cables, administration of cable ducts, coordination of splicing–in of apparatus cases, and other items related to outside facilities. The planning of node access facilities, particularly if they are to be used for integrated services, is carefully coordinated with the service planning function of internodal planning in order that augments of related internodal facilities and equipment may be properly sized and timed.

Equipment Engineering

Equipment Engineering orders central–office equipment, consolidates equipment requirements for switching, interoffice/internodal facilities, and loop facilities, coordinates installations with the supplier, performs acceptance tests upon completion, and handles related activities.

Network Operations

Network Operations performs activation and testing of facilities, turnup of trunks, equipment rearrangements, and similar activities.

Plug–In Administration

Plug–In Administration provides advance ordering of plug–in equipment, stocking it in a pool, and shipping it to the necessary destination. It also coordinates the repair and modification of plug–in equipment.

Transmission Engineering

Transmission Engineering provides technical assistance on design aspects of particular circuits and facilities, availability of radio frequencies, reasonableness of upgrading existing facilities, and similar matters.

19-4 EXTERIOR INFLUENCES

The planning of transmission facilities is heavily impacted by many exterior influences. These influences derive from a wide variety of situations and occurrences. The divestiture of the Bell operating companies from AT&T transformed telecommunications within the United States into a fast–moving and highly competitive market. This competition is enhanced by changing technologies, slowed by regulatory lag, and confused by the changing attitudes and needs of customers. New services, uneven growth of existing services, force reductions, migration of special services to high–capacity offerings, evolving regulation of emergency restoration and provision of priority services, new network architectures, increased network flexibility, added network capability, and increasingly complex technologies all influence the development of a transmission facility plan.

To respond effectively to these types of influence, the planning organization relies on technical assistance from the transmission engineering group. Such assistance may be required when the handoff to another company involves a "mid–span meet" or when a customer requests a service that involves special assembly. Technical assistance may also be required for dark–fiber or special–construction projects, for new services such as public packet–switched network (PPSN) and Basic Dedicated Digital Service (BDDS) with secondary channel, and for new network architectures such as the integrated services digital network (ISDN) and integrated network access.

19-5 EXPLORATORY STUDIES

A specific exploratory study can be stimulated in many ways. The most familiar mechanism is the impending exhaust of spares in existing facilities. In this case, the study is sometimes referred to as "exhaust–triggered." Other stimuli include the approval of new technical planning guidelines or the adoption of new corporate policies. For example, a study might be initiated as a result of a decision to extend digital connectivity (or equal access) to all offices in a LATA. The addition of new wire centers or additional switching capabilities can also lead to facilities planning studies, as can exhaustion of facilities in adjacent areas or in interacting routes.

Data Requirements

Several different kinds of data are needed to support the exploratory studies function of transmission facility planning. These include (1) data on existing facilities, equipment, and circuits, (2) forecasts of future demands on the network, (3) planned construction, (4) cost estimates for installing new facilities and equipment and for rearranging the existing network, and (5) knowledge of corporate goals as determined by corporate officers and executive staff.

Existing Network. In order to provide continuity in the planning process, statistics on the availability and utilization of currently installed equipment and facilities must be extracted from inventory data bases. But because exploratory studies do not require the amount of detail required to design, install, and maintain individual circuits, this data is compressed to minimize the number of classifications. Compression of data is important because it reduces the number of alternatives to be considered to a manageable number. For example, two coterminous T1 cable facilities of differing lengths that use regenerators with differing power requirements would be considered as a single facility within an exploratory study.

Until recently, the number of equipment entities that had to be considered within an exploratory study has been low because of the trend toward plug-in derived functionality from universal mountings. As the number of equipment manufacturers has increased and as each manufacturer has attempted to individualize its products, equipment has become available in varying size, functionality, and cost, thereby increasing the total number of equipment entities to be considered within a study. This has increased the complexity of exploratory studies by enlarging both the volume and technical detail that must be evaluated.

Forecasted Demand. Future demand on the network is an important element in all planning models. Prior to the 1980s, demand was, for the most part, the result of either message trunk traffic or VF (or DS0) level special–services circuits. In recent years, significant demand has developed as a result of new service offerings such as high–capacity (1.544 Mb/s or greater) channels, PPSNs, and subrate digital–data services. As a result, demand must now be forecasted in greater detail. Where once it was sufficient to forecast net point–to–point growth of VF and trunk circuits, a detailed breakdown by type, source, multiplexing and bandwidth requirements, and type of termination is now required. Volume and rate of circuit churn (the amount of add/disconnect activity within a network or route that results in zero net additions) are additional elements that must be forecasted and used as inputs to exploratory studies.

Planned Construction. Exploratory studies build upon the results of previous studies and their implementation plans as established by the service–planning function. These major construction activities, as well as more modest activities such as carrier, central–office equipment, and digital loop carrier turnup, when combined with information on the existing network, provide the network configuration of the near future to the exploratory study. The network must further evolve from this near–future configuration in response to changing demands.

Cost Estimates. The purpose of exploratory studies is the economic analysis of alternative solutions to such network problems as exhausted facilities, the need for a new wire center, or a corporate program to replace old technology. All alternatives studied must have certain elements in common. These elements include (1) a timely resolution of the problem, (2) approved and obtainable materials, (3) realistic personnel requirements, and (4) consistency with corporate goals. The economic analysis depends on cost characterizations developed and maintained for facility, equipment, and personnel resources detailed in the alternatives studied. These unit costs are based on estimates of material prices, engineering, installation, and maintenance costs and the costs of related resources such as power, floor space, or cable ducts.

Corporate Goals. The corporate goals of a company are established by its executive officers. They typically provide broad (although at times very specific) direction to the management of resources. The most obvious example of a universally accepted corporate goal is, perhaps, the transition from a part–analog to an all–digital network. It is important to note that corporate goals such as the development of ISDN do not, by themselves, provide justification to a construction project. Rather, they provide assistance in the choice of alternatives to be studied and can be the deciding factor between two alternatives of nearly equal cost.

Selection of Alternatives

In specific planning studies, careful consideration must always be given to the various ways in which relief can be provided to a congested route or to an exhausted facility. The basic alternatives are reroutes and new construction. Each has subclasses that must be evaluated separately. In some cases, both alternatives may be used.

Circuit rerouting is often an attractive means of providing for growth, especially when there is apparent spare capacity in the alternative route. Often rerouting is a short–term solution that defers the need for new construction for a few years. Construction deferrals of this type are highly desirable in that they not only delay capital expenditures but also delay the commitment of those expenditures until additional information (i.e., actual growth figures) is available.

In general, it is within the service–planning function that exhausted facilities are identified. Because one of the subfunctions of the service–planning function is the routing of circuits, obvious circuit reroutes are often planned and are sometimes in effect prior to the request for an exploratory study. When this occurs, the fact that circuits are being rerouted ("misrouted") due to an exhausted or congested facility should be communicated to those responsible for exploratory studies. The information may impact current or future studies and certainly identifies a future problem area that requires careful surveillance.

When circuit rerouting is a less obvious solution, the service planning function passes a request for a study to the exploratory studies function of internodal planning. Circuit reroutes of a more complex nature are evaluated and compared with various alternatives involving new construction within the exploratory study. This often occurs when the rerouting of circuits to avoid exhausting one facility has a significant adverse impact on a second facility. Service planning may assist in the identification of alternatives, but the evaluation and final decision will be made within the exploratory studies function.

The second major alternative is new construction. An example of this option is the establishment of a SONET (synchronous optical network) ring for service protection. Other forms of new construction include overbuilding (i.e., establishment of a second facility, usually of a different type, on an existing route), upgrading (e.g., changing the terminal equipment on an optical fiber

489

cable to increase the bit rate of transmission), and installation of a completely new facility.

The problems associated with the engineering of a route for a new cable system are a mixture of technical, economic, and sociopolitical considerations. A satisfactory route must be selected and the right of way must be procured. This may occasionally require the exercise of the right of eminent domain. Environmental and ecological effects must always be considered.

Many of the factors to be considered in the selection of a new route can be evaluated in terms of cost, reliability, or service quality. For example, it may be economical to add several miles to a route in order to avoid a natural or manmade obstacle such as a swamp, lake, river, rocky terrain, or major highway. In such a situation, the effect on performance of the added mileage would mostly likely be negligible while the cost reduction could be significant.

Where it is necessary to cross a body of water, such as a river, lake, or bay, study must be given to optimizing the crossing site. The width and depth of water, the strength of current flow, protection against anchor and ice damage, and concurrence of governmental agencies must all be taken into consideration. This requires analysis of the costs of trenching ducts into the riverbed versus adding conduits to an existing bridge versus boring under the body of water. In some cases, special engineering may be applied to permit the placement of a specially built cable or, more commonly, a longer repeater section than is normally permitted by spacing rules.

Another important factor in route selection is that of access to the cable and to repeater sites for economical maintenance and repairs. There may be a choice between selecting a route on one side or another of a range of mountains. Prevailing winds and weather patterns may strongly favor one of these choices because, for the other, there are long periods of time when the route is covered by deep snow. Another example might involve a route through a swamp that is feasible insofar as construction is concerned; however, access to repeater points located in the swamp may be difficult. Fortunately, the longer repeater spacings provided by modern fiber systems relieve this problem to a great extent. Generally, an attempt is made to select a route that parallels a highway so that relatively inexpensive access roads can be provided from the highway to the repeater sites.

Acquiring the right to place cable and repeater equipment is a significant part of the cost of a new system, especially in highly developed or natural areas. Costs, measured in terms of time, money, and public opinion, can be materially affected by environmental considerations. The need to clear timber, fill in swamplands, etc., to allow access by construction and maintenance crews is subject to environmental concerns. The reestablishment of natural habitats upon the completion of construction can also add significant costs to a project. Construction costs in highly developed areas are often high because of limited access, congestion, and the need to limit the disruption of street traffic in the area of construction.

Other factors that must be considered before a final choice is made from the available alternatives include the possible adverse effects of rerouting and rearranging circuits (e.g., operations labor, longer circuits, and costly terminal equipment), the capacity required and the capacity of the systems being considered, interactions with other study areas, relationships between proposed solutions, and capacity and design of existing buildings. A well-organized planning study usually includes the evaluation of three to six alternative solutions to a given problem. All of these alternatives should possess a degree of flexibility that allows for upward or downward changes in forecasted demand.

Just as the growth of integrated services has increased the need for coordination and planning among the internodal, node access, and switch planning organizations (i.e., horizontal integration), the rapid deployment of new technology is creating a need to tie the exploratory studies and service-planning functions closely together. No longer can long-range studies, based on 5-, 10-, or 20-year forecasts, enjoy freedom from financial and personnel constraints. The fast pace of today's ever-changing network requires that exploratory studies make decisions to be implemented only 18 to 24 months into the future. Data for these studies are often supplied through the service planning mechanization tools and contain detail never before required. Demands such as these are quickly establishing the need for vertical integration of the planning functions.

19-6 SERVICE PLANNING

The objective of the service–planning function is to produce detailed timed and sized deployment plans that minimize operating costs while maintaining an acceptably low risk of: (1) inability to provide service or (2) underusage of equipment. To accomplish this, consideration must be given to capital, expense, and other economic constraints, analysis of demand forecasts, facility and equipment utilizations, and facility and circuit routing alternatives. The plans developed must be consistent with the fundamental plan and contain the facility and equipment augments, retirements, modifications, multiplex plans, circuit routings, carrier routings, and rearrangements required to meet future service and marketing objectives. The creation of such plans is achieved through careful application of the following subfunctions:

(1) Circuit planning

(2) Carrier planning

(3) Facility planning

(4) Equipment planning.

The documentation and coordination of the implementation of these plans, once established, is the third and final function of the transmission facility planning process.

Circuit Planning

Many activities occur within the circuit planning subfunction. These activities include the processing and validation of approved circuit forecasts together with the initiation of changes to correct obvious discrepancies, the development of preferred circuit routes, the allocation of forecasted circuit growth to those preferred routes, and the creation of circuit rearrangement plans including the disconnect of the existing assignment and the establishment of the new one. Further effort includes the identification of special cutover and grooming requirements, the transmission of circuit rearrangement requests to the circuit provision center, and the response to orders that are unassignable due to the lack of facilities or equipment upon request from the circuit provision center.

Of these numerous activities, the allocation of forecasted circuit growth to preferred routing is the key to all other planning subfunctions. Proper allocation is vital. Forecasted growth must be allocated to proper route, facility type, and equipment if other planning activities are to be of value.

The activities performed within circuit planning result in the following inputs to other planning subfunctions:

(1) Carrier planning receives a chronological accumulation of the resulting carrier channel requirements by carrier type and subpath (point–to–point set of assignable facilities, regardless of their routing).

(2) Facility planning receives a chronological accumulation of the resulting VF cable pair requirements by facility type and subpath.

(3) Equipment planning receives a chronological accumulation of the resulting circuit equipment requirements by type and location.

(4) Documentation and coordination receives the plans for rearrangement of existing circuits. Included are the number of circuits, their completion dates, and associated project numbers.

Carrier Planning

The carrier planning subfunction develops carrier designs for each route and facility type. These carrier designs specify terminal type, span line, multiplex, and other carrier equipment requirements. This equipment is uniquely identified to an extent that enables proper planning. At times, this requires that similar equipment items be identified by manufacturer. The carrier planning subfunction also determines the timing and sizing of carrier system adds, disconnects, and rearrangements according to carrier channel demand as provided by the circuit planning subfunction. System additions are according to preferred equipment type and routings as identified in the carrier designs.

Additional activities that occur within the carrier planning subfunction are the development of plans for carrier system rearrangements and resulting requirements for circuit rearrangements, the initiation and monitoring of activation plans for carrier systems, the identification of base–facility exhaustion, and the identification of base facilities whose construction requires Federal Communications Commission approval because of their intended use for "interstate" service.

The activities performed within carrier planning result in the following inputs to other planning subfunctions:

(1)   Circuit planning receives circuit rearrangement requirements that result from carrier rearrangement plans. These circuit rearrangements will reappear as modified carrier channel demand in the carrier planning subfunction.

(2)   Facility planning receives chronological requirements for span lines and high–capacity base line facilities grouped by facility type for each subpath.

(3)   Equipment planning receives chronological accumulation of the resulting carrier equipment requirements (e.g., terminals, office repeater bays, multiplex equipment) by location.

(4)   Documentation and coordination receives information for each planned job, including the size, dates, facility type, subpath, activity type, and associated project numbers.

Facility Planning

The facility planning subfunction reviews the cable supply and demand (pairs and fibers) in light of utilization and retirement

494

goals, schedules, and retirement of surplus cable sheaths; it identifies circuit and carrier rearrangement requirements caused by these planned retirements. This subfunction also identifies needed cable splices, cable conversion activities, and base–facility exhaustion situations.

When necessary to the installation of base facilities (e.g., microwave radio, fiber) that have been identified by studies and approved for implementation, the facility planning subfunction coordinates activities with other companies.

The activities performed within facility planning result in the following inputs to other planning subfunctions:

(1) Carrier planning receives requests for carrier systems to be rearranged due to changes in cable plant. Carrier system rearrangements are passed to the carrier planning subfunction where they appear as demand. This demand is eventually reflected as modified demand in the facility planning subfunction. The carrier system rearrangements will also result in additional circuit rearrangement requirements.

(2) Circuit planning receives circuit identifications requiring rearrangement or redesign caused by changes in cable plant. Circuit rearrangement requirements are passed to the circuit planning subfunction where they appear as demand. This demand will generate (or modify) the carrier channel demand as passed to the carrier planning subfunction and will lead eventually to modified cable requirements.

(3) Exploratory studies receives notification of facility exhaust situations and requests for study.

(4) Documentation and coordination receives information for each planned job, including size, dates, facility type, subpath, and activity.

Equipment Planning

The equipment planning subfunction chooses the timing and sizing of circuit equipment mounting requirements for trunk relay units (in analog switching machines), transmission and signalling equipment, carrier banks, office repeater bays, multiplex bays, digital cross–connect systems, and similar types of equipment. This subfunction also forecasts all circuit and carrier plug–in requirements involving carrier channel, trunk relay, and transmission or signalling needs, and such carrier–related plug–in requirements as common boards, multiplex plug–ins, and office repeater plug–ins. Additionally, carrier systems requiring rearrangement due to upgrades of terminals are identified.

The activities performed within facility planning result in the following inputs to other planning subfunctions:

(1) Circuit planning receives identifications of circuits that are to be rearranged or redesigned due to changes in circuit equipment mountings.

(2) Carrier planning receives identifications of the carrier systems, span groups, or derived facilities that are to be rearranged or redesigned due to changes in carrier equipment mountings.

(3) Documentation and coordination receives information for each planned job including the location, size, dates, equipment type, and the activity of the job.

19-7 DOCUMENTATION AND COORDINATION

Implementation of a facilities plan requires that planning results be documented for use by organizations involved in the implementation process and monitored for continued service and cost applicability, as well as for proper execution. The appropriate data must be extracted from the facility plan, arranged into useful format, and distributed as required. Informational needs fall into three categories: (1) construction project documentation, (2) instructions for circuit layout and assignment, and (3) coordination. These categories are established to emphasize the following responsibilities of the documentation and coordination function:

(1) Document and submit for approval the serving plan as established within the service–planning function.

Documentation should include capital, expense, material and other resource requirements, risk analysis, and possible alternatives.

(2) Respond to requests for special data, studies, or actions. Such requests may include inquiries from marketing or other organizations as to resource availability and projection.

(3) Initiate plan implementation. This is accomplished through the transmittal of facility and equipment job actions to the equipment engineering organization and the appropriate distribution service centers, and through the transmittal of carrier activation schedules and preferred routings to the circuit provision center.

(4) Monitor and coordinate implementation of plan components including timing, interrelated dependencies, and activities affecting the execution of the plan. Monitoring the plan also includes the review of forecasts, actual resource usage, related projects, new technologies, etc., to determine continued applicability in terms of service and cost.

Construction Project Documentation

The ultimate purpose of the planning process is to determine how, when, and where additions to the network are to be made by conversion, overbuilding, or new construction. While each authorization for new construction must stand on its own merits, the total construction program is reviewed to ensure that each individual project is both consistent with and an integral part of the total program.

The total construction program is analyzed through the construction−budget process. Through this process, upper management is assured that corporate expenditures are in line with and will accomplish corporate goals. The process also provides a procedure through which management may approve or disapprove the proposed plan.

A construction program is well designed when demand is met by appropriate capacity at the proper time and at reasonable cost, and when construction of physical plant moves the company closer to accepted goals while meeting the financial and regulatory constraints placed on it. As a result, the documentation required for plan approval is necessarily global in nature. Approval of the construction program is based on the accomplishments of the program as determined through analysis of demand, capacity, and expenditure.

Demand analysis entails the identification of sources of demand and the level of risk associated with each source. Demands generally derive from forecast data (general trunk forecast or special–services forecast), churn, or circuit rearrangements. *Capacity analysis* entails review of timing and sizing of planned equipment and facility jobs. Charts providing utilization at time of relief, distribution of relief interval, and distribution of exhaust interval information are useful in capacity analysis. *Expenditure analysis* consists of review of material purchased as related to cost.

The documentation required for a specific project may include:

(1) Equipment by type, quantity, and location

(2) Base–facility additions (e.g., number of fibers) by type and location

(3) Required circuit and facility rearrangements

(4) Equipment and facility retirements by type, quantity, and location

(5) Nondeferrable plug–ins (i.e., plug–ins required for the turnup of the basic facility as opposed to the activation of individual channels)

(6) Scheduling information (e.g., due dates).

Information of this type is normally distributed to organizations such as equipment engineering, circuit provision, and operations. These work groups use the information as input to their normal activities (estimate preparation, workload planning, etc.).

Instructions for Circuit Layout and Assignment

For each circuit group there are usually a number of possible ways of installing circuits. These different possibilities may involve varying types of facilities and equipment, as well as different facility routes. For this reason, ensuring that circuits are assigned to equipment and facilities in a manner consistent with the facilities plan is an important part of the implementation of that plan. Improper assignment of circuits can cause exhaustion of facilities planned for other purposes while leaving the facilities planned for the provisioned circuits underutilized.

Modern mechanization systems for the planning and provisioning of circuits provide a mechanized interface for information flow from planning to the circuit provision work group. This interface can be used to ensure that actual circuit growth is provided on those facilities that were planned to support it.

Coordination

The timely implementation of a facilities plan may depend on the coordination of several construction projects, carrier system installations, and circuit rearrangements. This coordination is typically documented within the mechanized planning tool used by the planning organization. The documentation indicates interdependent dates for construction of facilities, equipment, carrier systems, etc., upon which each project depends.

It is necessary, within the planning process, to assume the completion of certain work in order to complete later activities. These dependencies, while known, must be tracked during implementation. The dependency chain should include service dates of all related activities, even if an activity is planned by a different work group. For example, the service date of a base–facility addition that is to terminate at the site of a new digital switch must be related to the service date of that switch even though the switch is not planned by the work group planning the facility addition. Carrier systems assigned to the base facility should be related to the service dates of both the facility addition and the switch, as well as to the service date of any other related activity, such as an augment of multiplex equipment.

The actual construction and installation work necessary for the implementation of the program is the responsibility of many organizations. However, each of these organizations is primarily concerned with discrete activities. The planning organization is responsible for integrating the monitoring of these activities and for ensuring that all dependencies are recognized. Certainly, all construction activities must be complete before circuits can be placed in service, and some of these activities must be completed before others can begin. Thus, the coordination required can be complex.

An additional element within the coordination category is that of monitoring the actual demand on the network. The construction program is based on forecasted demands. These forecasts may prove to be inaccurate for numerous reasons. If actual demand does not materialize in the quantities forecasted, the construction program may need to be modified. This modification may take the form of either increased or reduced equipment and facility requirements. The earlier the need for modification is recognized, the easier it is to rectify the situation. Thus, the planning organization continually monitors the demands placed on the network and, when necessary, takes action to modify the construction program.

19-8 SUMMARY

Transmission facility planning is part of a vertically and horizontally integrated total–network planning process. It consists of three interrelated functions: exploratory studies, service planning, and documentation and coordination.

The exploratory studies function performs what traditionally has been called fundamental or long–range planning but is now expanded to reflect current business decisions and budget demands. Data requirements for this function include technical data on transmission capabilities provided by the transmission engineering organization, forecasted demand provided by the marketing and forecasting groups, and data on the embedded network provided from inventory data bases. Completed studies provide direction to the service–planning function and guidance to the long–term development of the facility network.

The service–planning function develops a construction program that is consistent with the fundamental plan, and schedules the implementation of decisions based on exploratory studies and the facility and equipment augments, retirements, and modifications that are needed to meet forecasted service demands. This function is composed of four subfunctions: circuit planning, carrier planning, facility planning, and equipment planning. Together these subfunctions establish a construction program that considers capital and expense expenditures, economic constraints, future demands, facility and equipment utilizations, and circuit and facility routing alternatives.

The documentation and coordination function provides a process through which upper management is assured that corporate expenditures are in line with, and will support and accomplish, corporate goals. This process also provides a procedure through which management may approve or disapprove a proposed construction program, and a process through which implementation and monitoring of that approved program may take place.

Additional Reading

Construction Plans Department, American Telephone and Telegraph Company. *Current Planning and the Construction Budget* (Winston–Salem, NC: Western Electric Company, Inc., 1974).

Mills, J. "The Where, What, and How of Software–System Architecture," Bellcore EXCHANGE, Vol. 2, Iss. 4 (July/Aug. 1986), pp. 18–22.

Chapter 20

Measurements

The public message telecommunications service (MTS) network described in Chapter 1 incorporates an architecture that can provide transmission quality to meet current grade–of–service objectives. Each company providing MTS network services should have a transmission quality program incorporating a number of measurement plans to ensure that its portion of the network is indeed meeting network objectives and to provide essential data for correcting transmission deficiencies. This program should include reports that give indications of customer reaction to the quality of service rendered as well as plans that measure transmission performance on the elements and links that are interconnected in completing a call.

Evaluation of MTS network transmission quality is very complex; there are a large number of possible connections through a combination of analog and digital facilities involving customer–owned terminal equipment, exchange carriers, and interexchange carriers (ICs). Moreover, the various transmission impairments described throughout the three volumes of *Telecommunications Transmission Engineering* accumulate as a call progresses from origination to termination. Quality of data transmission as well as voice must be considered. This chapter describes a number of measurements and techniques that can be incorporated into a company's transmission quality program in today's rapidly changing environment.

Prior to 1984, the Bell operating companies (BOCs) and AT&T, with the full cooperation of the other telephone companies, held responsibility for end–to–end customer transmission quality on the MTS network. Before 1978, the program included the subscriber equipment. A standardized, comprehensive transmission quality program was maintained by the BOCs to provide effective transmission quality control.

Various types of measurement as well as customer comments and complaints were generally accumulated by the BOCs, which took overall responsibility in locating trouble sources anywhere in the network and instigating corrective action and improvement. Divestiture and several other changes in the network environment have dictated significant changes to each company's transmission quality program.

In the present environment, as many as five different ownership entities can contribute to transmission degradation on a single call. Figure 20-1 shows these entities on a typical interLATA (local access and transport area) call. These entities include the subscribers on each end of the call, two different exchange carriers, and possibly two different ICs, depending on which subscriber originates the call. Even this is a simplified situation; there may also be a wide area telecommunications service (WATS) reseller, an alternate operator services (AOS) vendor, a cellular mobile carrier (CMC), or a private network involved in the call. All of the carriers involved probably have numerous alternative routings and different mixes of analog and digital facilities that may be interconnected.

In the final analysis, the customer is now responsible for end-to-end transmission quality. Large users may have sophisticated computerized test systems to show when the carriers are not meeting transmission requirements. On the other hand, residential customers may now be bewildered whenever there is poor transmission. They will likely complain to the exchange carrier, which can take no action if the trouble is with the IC or station equipment. The customer generally has no way to check telephone set quality. Thus, it is vital that each telecommunications carrier maintain an effective transmission quality program. This need will become even more important as new data services are provided on the MTS network [1].

20-1 TRANSMISSION QUALITY PROGRAM

The number and scope of measurements included in a company's transmission quality program vary depending on such factors as number of customer complaints, LATA size, relative penetration of digital facilities, and the type of available mechanized

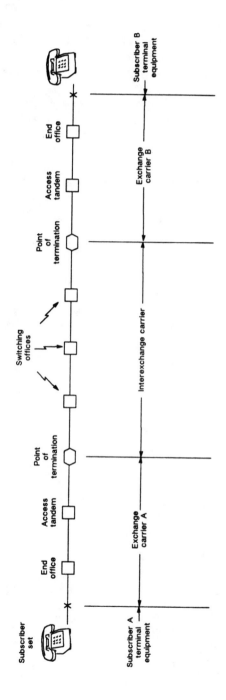

Figure 20–1. Division of ownership on a typical interLATA call through the MTS network.

measurement systems. In today's competitive environment, the cost of implementation and administration is an important factor in deciding which measurements to include in the company's program. Whenever a network segment becomes predominantly digital, it may be more economical to use measures of transmission quality on the basis of performance monitoring of the digital bit stream rather than the more cumbersome analog measurements of loss, noise, and balance. The availability of centralized, computerized network testing systems can lead to the implementation of effective economic measurement plans that are integrated with other network surveillance functions. All companies also need to consider incorporating measurements of overall network transmission quality through the portion of the network they provide.

The program should include both external and internal measurements. The difference between them is defined here on the basis of external measures reflecting customer opinions and internal measures relating to company performance measurements. This follows naturally from the fact that opinions about service originate from sources external to the company while performance measurements originate and are carried out within the company.

External Measurements

Customers expect the various interconnecting carriers to establish connections promptly on the first attempt, to provide assistance courteously and with dispatch, and to furnish accurate billing. However, they are likely to be dissatisfied with service if transmission is poor regardless of the excellence of other service features. Customer reports of trouble and service difficulties provide ongoing external indicators of the quality of service being provided.

Various types of service attitude surveys also may be made to provide insight periodically to customer opinions on the quality of transmission service. Unfortunately, it is often difficult to relate these opinions to specific carriers or plant elements.

Internal Measurements

Internal measurements consist of planned measurements of specified transmission parameters of the MTS network. Measurements for specific plant elements may be incorporated into a formalized measurement plan with standardized measurement intervals and reports. Some of these results can be converted to indices that provide the basis for evaluating telephone transmission in various administrative units of the plant on a comparison basis.

When measurements show poor or deteriorating performance, the data must be analyzed to determine the cause of trouble and to guide the course of corrective action. Comparison of measurements and the analysis of results are tools that assist in determining whether overall objectives are being met. Transmission objectives for the network are derived by procedures that, in total and by individual steps, involve compromises necessary to meet grade–of–service goals economically. The procedures may be regarded as beginning with the translation of grade–of–service objectives into transmission objectives applicable to various segments of the plant such as loops, local trunks, tandem trunks, intertandem trunks, etc. The transmission objectives must then be allocated to various entities that interconnect to provide service. Finally, these objectives must be approved by appropriate standards bodies.

In order to relate the measurements to the established transmission objectives, at least some of these internal measurements of impairments must be made in such a way that performance and objectives can be compared. The comparison process is based on measurements and tests that are practicable and economical. These measurements are valuable for assessing the performance of various administrative units relative to one another and relative to standard objectives. Additional transmission surveys of specific parts of the plant may be conducted to establish mathematical models of the plant and its performance and to provide input to the continuing process of evaluating objectives. The surveys and models permit the study and evaluation of the overall network made up of the interconnected entities. Significant studies and models of transmission performance on the pre-divestiture network [2,3] have become the basis for current performance evaluation models [4,5]. These studies and models can be used as benchmarks to compare with current measurement summaries.

Measurement Coordination

A transmission quality program will become an effective quality-assurance mechanism only through the coordination and correlation of the various external and internal measurements that have been selected for a company's transmission program. Figure 20–2 illustrates various internal and external measurements that have been used to control transmission impairments on intraLATA MTS traffic. Trunk maintenance, connection appraisal, trunk–design audit, and operator trouble–report measurements have been used to evaluate the performance of the interoffice plant. These same measurements directed to different termination points also evaluate performance on interLATA calls.

Note that these interoffice measurements omit the loops. On calls between the same two stations, the same two loops are always involved but trunks are selected more or less at random. Thus, it is desirable that loop evaluations be kept separate from interoffice evaluations so that separate control can be exercised. In this context, the loop plant transmission quality measurements illustrated in Figure 20–2 (that also apply in the same manner to interLATA traffic) include loop noise, loop survey, outside plant job quality audit, and customer trouble report.

Measurements related to transmission on the overall connections include customer–attitude surveys and in–service nonintrusive monitoring results. Typical types of external and internal measurement that are candidates for a company transmission quality program and are used to evaluate loop plant, interoffice plant, and overall transmission are reviewed in some detail later in this chapter.

Some of the transmission quality measurements can become part of established index plans that provide for directly relating results to established objectives. Any one of the transmission indices could be selected as a key service indicator (KSI). KSIs are used to focus upper management attention to important objectives.

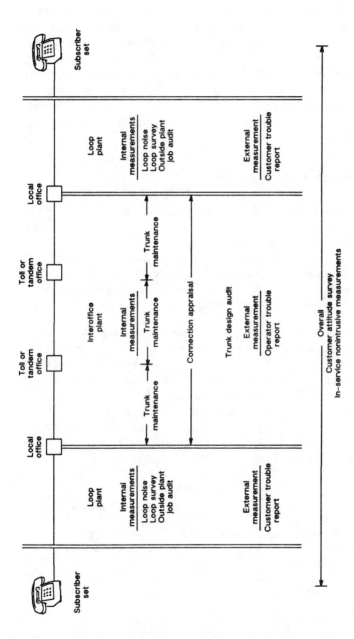

Figure 20-2. Measurements and indicators that have been used to evaluate transmission quality in intraLATA MTS traffic.

The transmission quality program should include a means for the measurements, indices, and indicators to be analyzed, correlated, and reviewed to provide guidance to one of several lines of action. The correction of performance deficiencies and possible changes in transmission objectives must be considered carefully. When performance, cost, and customer opinion are in good balance, no action may be warranted. In other cases, the expediting of digital conversions may be the best solution. Some of the indicators need to be reviewed in concert with other network performance results.

20-2 TRANSMISSION INDICES

Transmission measurement data can be used in two ways. First, since the data are too numerous to be manipulated easily, they may be simplified and combined into indices that can be used to evaluate plant performance and to compare the performance of administrative units. When these indices show poor or deteriorating performance, the data again become useful in identifying trouble and pointing toward corrective action.

Indices represent simplified summaries of large amounts of data that would be unwieldy and even useless if unprocessed. The development of indices usually involves statistical analysis of data, comparison of results with defined reference values, and weighting of results to account for indirect effects. Finally, the processed results are translated into a designation (the index) that bears a relationship to transmission grade-of-service objectives. This designation is generally designed to reflect the ratings shown in Table 20-1.

Table 20-1. Index Ratings

Numerical Designation	Band Designation	Meaning
95-100	O	Objective
90-95	L	Low
Below 90	U	Unsatisfactory

Components of indices are often given individual index ratings. The index concept is such that if all components have the same value, the overall index will have that value. If the components differ, their relationships to the overall index depend on the applied weighting factors.

The data collected for index calculation are usually sequentially reported before being tabulated and summarized. A reference value for each parameter (loss in dB, noise in dBrnc, etc.) has previously been established on the basis of fundamental studies and analysis or design. Detailed processing of raw data can be carried out, if necessary, but indices are determined only from such factors as the percentage of observations made, the percentage of observations exceeding a given reference value, or the percentage of observations outside double-ended reference limits.

Indices provide broad general evaluations of performance, show trends of performance, and permit performance comparisons within or among administrative units. Except on rare occasions, they cannot be used to isolate and identify specific troubles, but the data from which the indices are derived are often useful for these purposes. Indices can be a powerful management tool when properly interpreted and used to assist in identifying weak spots and making judgments in assigning resources.

The field analysis of results is simplified by the use of indices; this analysis focuses on the extremes of parameter distributions where some of the most significant contributions to poor transmission are found. Even where transmission indices are not used, the determination of the distribution function for any parameter and the concentration of effort to eliminate extreme values can lead to significant performance improvement.

Derivation of Transmission Indices

The calculation of transmission indices is a process too lengthy and complex to treat here in detail; however, the general principles can be reviewed.

Figure 20–3 illustrates two normal density functions that might represent data derived from internal measurements. Curve A, labeled "poor" because it has a large standard deviation, indicates that there is a large percentage of calls that may be rated poor. Curve B is labeled "good" because it has a relatively small standard deviation (i.e., has fewer calls exceeding a given limit).

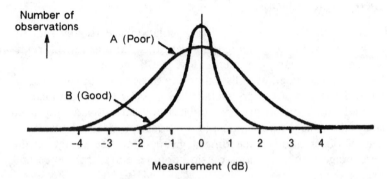

Figure 20-3. Typical transmission parameter density functions.

In the past, performance was judged by indices derived from mathematical treatment of the data to determine mean values (bias) and standard deviations (distribution grade). This approach proved unsatisfactory for field use where the assumption of normal distribution functions was not always valid and the training of field personnel to process the results proved to be impractical.

Transmission indices are now generally based on measurements that exceed specified values. For example, Figure 20-4 represents the density function for measured loss deviations on a group of circuits. Loss that is too high results in low volume on connections; loss that is too low results in uncomfortably high

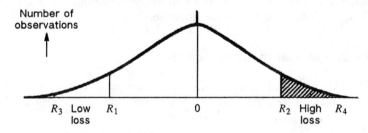

Figure 20-4. Density function for loss measurements.

512

volume, excessive echo, or circuit instability. Thus, limits R_1 and R_2 are indicated as values that should not be exceeded. The percentage of measurements outside these limits is used as a basis for determining the index. Where the density function representing facility performance has a wide distribution, it is sometimes necessary to select two additional reference points, R_3 and R_4; observations outside these limits may be given a heavier rating.

Another density function to be considered is that for noise observations as shown in Figure 20–5. In this case, there is no lower limit on observed noise values. High noise values are undesirable, so the number of measurements in excess of limit R is used as a basis for determining the noise index.

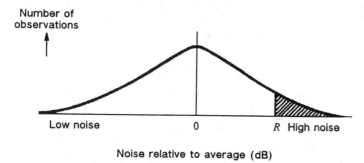

Figure 20–5. Density function for noise measurements.

Even where measurement plans are not used, these principles can be used to control any measurable parameter of any component of the plan; i.e., the identification of facilities requiring corrective action is made possible through the use of reference values. These reference values must be derived by careful analysis of data and by careful comparison with transmission objectives.

20-3 INTEROFFICE MEASUREMENTS

Transmission impairments on interconnected interoffice facilities can degrade transmission to a great number of customers who share these facilities on interoffice calls. Fortunately, these facilities terminate at central offices with relatively easy test access. Appropriate interoffice measurement plans are thus a vital part of each company's transmission quality program.

Trunk Transmission Maintenance Measurements

Much of the ability to identify trunks or connections having poor transmission was lost with the introduction of direct–distance dialing. There is usually no operator to verify satisfactory transmission, and the trunks may be called upon to carry high-speed data signals. With automatic switching, connections are set up, regardless of transmission quality, provided that address and supervisory signals are satisfactorily received by switching machines. For these reasons, it is necessary to have statistical knowledge of the quality of transmission and, at the same time, to identify those trunks that perform poorly. To acquire the necessary data, all trunks should be tested periodically even though performance could be determined statistically by measuring only a sample; the accumulated data may be used to determine an index. Such a performance indicator can provide a reliable measure of the transmission quality of message trunks; it also enables operating personnel to maintain trunks near or at design values.

Network Trunk Transmission Measurements Plan. Virtually all of the exchange carriers use the network trunk transmission measurements plan (NTTMP) to provide a performance indicator that derives indices from the test results by office, administrative area, and company [6]. While the plan can be implemented on a manual basis, almost all users employ mechanized systems that automatically dial the end–office termination, make the measurements, compute the indices, and provide printouts of trunks that are out of limits. Appropriate far–end–office terminations are used to provide two–way measurements of the following transmission parameters.

Loss: Measurement of the transmission loss of a reference 1004–Hz test tone.

C–message noise: Measurement of circuit noise with a C-weighting network.

Balance: Measurements of echo return loss and singing return loss.

Gain–slope: Measurements of the losses of 404–Hz, 1004–Hz, and 2804–Hz reference test tones.

C-notched noise: Measurement of circuit noise with a 1004-Hz test tone inserted on the facility (tone is filtered out prior to noise measurement).

NTTMP measurements of the above parameters are compared with the design values of each trunk. Measurements that are outside the specified maintenance limit (called Q1) and the immediate action limit (called Q2) are identified. Loss measurements can be Q1 or Q2 by being a specified amount either below or above design values as shown in Table 20-2. Noise and balance Q1s and Q2s are based only on being above a limit. Table 20-3 gives Q1 and Q2 limits by type of facility and circuit length for C-notched noise. Office printouts of Q1 circuits for each parameter allow scheduled maintenance to correct problems. In a similar manner, Q2 printouts list circuits to be removed from service until repairs can be made.

Table 20-2. 1004-Hz Maintenance and Immediate Action Test Limits for IntraLATA Trunks

		Multifacility Trunks	Other Types of Facilities
Actual measured loss must be within this range of design loss	Maintenance limit (Q1)	± 1.5 dB	± 1.0 dB
	Immediate action limit (Q2)	± 3.7 dB	± 3.7 dB

Indices for NTTMP are based on the percent of scheduled tests completed and the number of measurements exceeding the Q1 and Q2 limits. Index assignments of bands "O," "L," or "U" are assigned for each measured parameter and printed out monthly for each central office. Figure 20-6 illustrates a monthly central-office index report. Once the bands for each office have been calculated, the results can be formatted in various modes for other administrative units such as manager group, district, division, and company. Figure 20-7 illustrates a typical NTTMP company report printout.

Table 20–3. C–Notched Noise Limits (in dBrnc0) by Type of Facility and Circuit Length

Type of Facility	Type of Channel Bank	Circuit Length (Miles)			
		0–100	101–200	201–400	401–1000
Repeatered or non-repeatered cable	NA	B=38, A=32			
N carrier	N1	B=48, A=45			
	All other	B=41, A=38	B=42, A=39		
Digital carrier	D1	B=47, A=45	B=47, A=45	B=47, A=45	B=47, A=45
	All other	B=39, A=38	B=40, A=39	B=40, A=39	B=42, A=40
Broadband analog carrier	All	B=39, A=38	B=40, A=39	B=40, A=39	B=42, A=40

Note: Values are given in dBrnc0 with –16 dBm0 holding tone. To obtain the limit for a particular trunk, subtract the 1000–Hz expected measured loss from the requirement listed.

Noise limits presented as follows:
A = Maintenance limit (Q1)
B = Immediate action limit (Q2)

In realism, it is necessary to recognize that existing test systems do not detect digital slips or echo impairment. Slips are monitored and reported in near–real–time, but not as part of trunk tests. Echo delays, as increased by the ongoing digital conversion of the network, require special surveys for assessment.

Connection–Appraisal Measurements

The trunk transmission maintenance indicators described above can be used to maintain network trunks within design

OFFICE:	*****************************	FORM NTTMP
MANAGER:	*NETWORK TRUNK TRANSMISSION*	MONTH
DIVISION:	* *	YEAR
GEN MGR:	* *	
REGION:	* OFFICE REPORT *	
DEPT:	*****************************	AS OF
COMPANY:		

	A	B	C	D
	EXCEPTION MEASUREMENTS/ TESTS	COMPLETED MEASUREMENTS/ TESTS	%	BAND
****LOSS****				
1A Q1 (1.7 dB)	32		2.0	
1B Q2	24		1.5	
1C TOTAL		1600		O
****C-MESSAGE NOISE****				
2A Q1	131		8.2	
2B Q2	197		12.3	
2C TOTAL		1600		U
****BALANCE****				
3A Q1	87		15.0	
3B Q2	79		13.6	
3C TOTAL		580		U
****GAIN SLOPE****				
4A Q1	14	810	1.7	
4B Q2	11		1.4	
4C TOTAL				O
****C-NOTCHED NOISE****				
5A Q1	38		5.3	
5B Q2	33		4.6	
5C TOTAL		720		L

Figure 20-6. Example of a completed office NTTMP report.

objectives; however, other measurements are required to assure that customers are receiving an objective grade of service when these trunks are interconnected in establishing calls. Connection-appraisal measurement plans are used by some companies to evaluate the overall quality of interoffice transmission. This testing simulates the end-office-to-end-office calling patterns of subscribers and evaluates the individual results based on both transmission quality and occurrence of blockage. Figure 20-1 illustrates how connection-appraisal measures several trunks

517

COMPANY:		*****************************		FORM NTTMP
		NETWORK TRUNK TRANSMISSION		
		* MEASUREMENT PLAN *		MONTH
		* *		YEAR
		* COMPANY REPORT *		
		* *		
		*****************************		AS OF

	MEASURED CHARACTERISTICS	NBR OFCS	OFCxMON REPORTS	OFCxMON REPORTS BY BAND		
				--O--	--L--	--U--
1	LOSS	487	487	253	170	64
2			% OF TOTAL	52.0	35.0	13.1
3	C-MESSAGE	487	487	268	195	24
4	NOISE		% OF TOTAL	55.0	40.0	4.9
5	BALANCE	157	157	43	25	89
6			% OF TOTAL	27.4	15.9	56.7
7	GAIN SLOPE	464	464	425	30	9
8			% OF TOTAL	91.6	6.5	1.9
9	C-NOTCHED	460	460	329	110	21
10	NOISE		% OF TOTAL	7.15	23.9	4.6

Figure 20-7. Example of a completed company NTTMP report.

connected in tandem from the originating office to the terminating office (without including the loop) in a built-up connection. The measurements are made via test calls dialed to an assigned end-office telephone number providing appropriate test terminations. The loops are omitted from the tests since it would be very expensive to access the subscriber locations.

Connection-appraisal tests may involve loss, noise, balance, and gain-slope measurements; arrangements can also be made to measure other network performance parameters (e.g., high and dry, reorder) in addition to transmission. Each dialed-up connection for test includes the transmission paths (with associated transmission impairments) through the switching machines. These connections may be established over any number of trunks permitted by the switching plan; therefore, the measurements are oriented toward evaluation of overall connections rather than of individual trunks or types of trunk. Since the paths being appraised are subject to design control, performance can be predicted and compared with measured results. While connection–appraisal tests were originally performed manually, they are now almost always done via mechanized systems that make the measurements as well as compute and print out results.

Central offices are selected periodically for connection–appraisal tests. Each office is usually tested once per year. The call sample is selected on the basis of actual traffic patterns. An office connection–appraisal survey requires at least 50 connections in order to give reasonable statistical accuracy. The results of such a survey give only an approximate evaluation of performance for a single originating entity. Results are quite accurate, however, when used for comparison of entities in a large operating area.

Data accumulated during a connection–appraisal survey are usually examined immediately after completion of the survey. Completed calls are individually rated for transmission quality according to a loss/noise grade–of–service model, which combines the parameters into a single performance criterion. In addition, circuit mileage can be incorporated into the model to reflect local conditions such as predominantly long or short analog trunks. If the data show a pattern indicating trouble or design deficiencies common to the originating office, a review with operating personnel may be warranted. The measurements are sometimes useful in pointing out individual trunk transmission conditions not found by trunk testing procedures. However, trouble isolation may not always be possible because routing is not usually controlled or identified. System weak spots caused by circuit design deficiencies, improper routing, or substandard installation and maintenance may also be identified.

Computer programs that were developed to compare pre-divestiture measured results with grade–of–service models for some of the transmission parameters (assuming an average loop on each end of the measured connection) can be modified to accommodate current network conditions. These programs combine performance on the measured parameters to derive a performance index of "O," "L," or "U" for each office. Results can be summarized for originating offices and various administrative areas. Summaries are usually published either monthly or quarterly; changes in index results are a good indication of the effectiveness of correction activities. Figure 20–8 illustrates a typical quarterly company summary that focuses on the number of "L" and "U" offices.

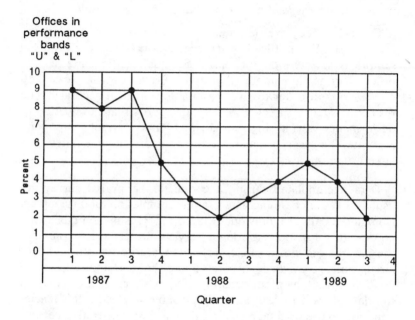

Figure 20–8. Typical quarterly company management summary of connection–appraisal results.

Digital Facility Quality Measurements

The trunk transmission maintenance and connection–appraisal indicators already discussed measure transmission quality on the MTS network, which combines digital and analog facilities. As portions of the network totally comprise digital facilities interconnected by digital switches, it will be feasible to measure transmission quality on these sections on the basis of digital performance. Available new technologies have already resulted in the development of economical digital–error monitoring systems [7]. Considerable activity is underway to establish standardized digital network performance criteria [8]. Such bit–oriented digital performance measurements may already be the most practical technique at the digital interface between exchange carriers and ICs where

520

the provision of analog test terminations required for conventional trunk measuring systems would be very costly.

Whenever these digital–error indicators are used, it is important to monitor results and trends.

Operator Trouble–Report Summaries

Customers that experience difficulties on interexchange calls will frequently call the operator to receive billing credit. Studies have shown that the number of transmission trouble reports identified by the operators has a direct relationship with network transmission quality. Figure 20–9 shows the relationship between operator trouble reports per 1000 calls and the loss/noise/echo grade of service.

Operator trouble reports are generally analyzed to identify weak–spot areas. Mechanized systems are frequently employed in this analysis and used to follow up on correction activity [9]. Individual company participation in these correction activities, along with the observation of trends, is an essential part of all transmission quality programs.

Trunk–Design Audit Indicator

An audit of a selected sample of the trunks designed during a specified time period is made by some companies to ensure that trunk designs and routes meet proper criteria. An indicator based on the number of errors on a circuit layout record provides a quality measurement of the design process. In some cases, the audit includes a physical check to ensure that the trunk circuits are actually installed as designed.

20–4 LOOP MEASUREMENTS

Loops that use digital carrier facilities can, if designed and maintained properly, have significantly better performance than common voice–frequency transmission over copper pairs. However, the large imbedded plant investment dictates that in the immediate future the vast majority of loops will remain on copper. These analog loops in a hostile environment will be called on to carry new digital services with exacting transmission requirements. Hence, loop measurements should not only be a vital part of a company transmission quality program, but will become even more important in the future.

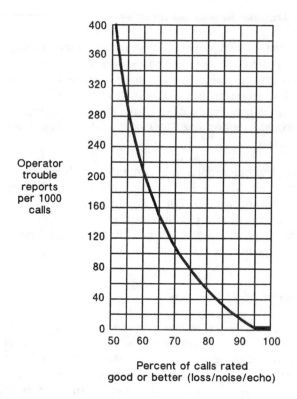

Figure 20–9. Relationship of transmission quality to operator trouble reports.

Transmission measurements on these analog loops are complicated by the difficulty and high cost of obtaining access to the station ends of loops. This makes customer trouble reports to repair service a primary indicator of loop transmission quality and thus an essential measurement in a company's transmission quality program. Fortunately, a number of computerized programs are available from several suppliers to use these reports effectively to pinpoint weak spots and emerging trouble areas.

Customer Trouble-Report Summaries

When service deteriorates to the extent that the customer experiences difficulty, the trouble is most likely on the loop or the customer-owned station equipment. Customer calls to repair service generally go to an automated repair service bureau where they are classified and summarized on a computerized system. Those reports that are found to be station-set problems are referred back to the customer. Information on reports that indicate company problems is recorded in some detail.

Significant data on these reports, from the transmission standpoint, are the codes that classify the type of report. These codes, assigned on the basis of the description of the problem given by the customer, are defined in Table 20–4.

Table 20–4. Customer Trouble–Report Codes

Code	Report Type
1	Can't call, no dial tone
2	Can't call, other
3	Transmission noise
4	Can't be called
5	Memory service failure
6	Data failure
7	Physical condition
8	Miscellaneous

Among these report types, code 3 and code 6 are most likely to reflect transmission–type troubles; they account for almost 15 percent of the total. Code 3 reports include complaints such as "can't hear," "can't be heard," "distortion," "cutoffs," "momentary interruptions," and "noise." Code 6 includes reports from customers who cannot receive or send data. Code 2 and code 4 reports may also have transmission–related signalling problems. The coded report categories are summarized as "customer trouble reports per 100 stations." By monitoring successive reports, trends of deterioration or improvement can be observed. Furthermore, the transmission trouble rates can be observed for specific sections of a territory.

Available computer programs can generate printouts of information from the transmission coded reports that can provide data that is useful in uncovering problem areas. Additional programs are available that can be used to pinpoint routes that have significantly higher trouble rates over a time period, such as six months or a year. Of particular value are printouts of repeat reports. Outside plant rehabilitation activities undertaken as a result of the analysis of this data can not only improve transmission quality, but also significantly reduce plant maintenance expense. Engineering personnel should be aware of the capabilities of these programs; they may need to participate in additional studies with operating personnel for the mutual solution of transmission–related problems.

Historically, almost two thirds of the loop transmission trouble reports relate to noise; hence, a noise mitigation program may be required in those areas where a high incidence of noise is reported. High noise in subscriber plant has three common causes that must be considered in a noise mitigation program: (1) cable sheath discontinuities, (2) unbalanced pairs, and (3) maintenance or housekeeping problems that affect pairs at random. Additional measurements and surveys may be used to determine the magnitude of the noise mitigation program needed.

Outside Plant Job Quality Audit Measurements

Some companies check a sample of outside jobs to measure compliance with transmission design objectives. These audits usually check the quality of other facets in addition to transmission. They often involve a field inspection and may include a check of the associated transmission tests that ensure new or rearranged facilities conform to design objectives. With the large amount of outside plant rearrangement activity, this type of audit plan is an excellent candidate for a company's transmission quality program.

Loss and Noise Measurements

Because of the high costs of making station visits, routine two-way transmission measurements on loops are not viable. Surveys are occasionally made of selected offices by making two-way loss and noise measurements on a sample of the loops.

Sample measurements of steady-state loop noise made from the central office are sometimes used to derive a loop-noise quality index. This index gives a loop-noise quality rating to a central office; however, the measurements do not fully reflect noise experienced by the customer since they are made with the subscriber set on-hook. Since the data thus cannot be used to pinpoint troubles, the measurement plan is not a good candidate for a transmission quality program.

20-5 OVERALL EVALUATION

The interoffice quality measurements and indicators that have been discussed are based on simulations of the talking state and have assumed average customer loops. Loop indicators do not generally consider the effects of interconnecting with interoffice facilities. Overall transmission quality measures are needed to assess the overall transmission as experienced by the customer.

Customer-Attitude Surveys

Customer-attitude surveys are used by many exchange carriers to provide an evaluation of customer perception of service and operations. These surveys use telephone interviews on a sampling basis with questions to generate opinions on a number of service criteria.

Prior to divestiture, the surveys included a number of questions about the quality of transmission on recent telephone calls. The questions related to the ability to hear and be heard, noise on the line, fading, clipping, echo, and distortion. Monthly results were summarized by various administrative organization units and distributed to upper management. While the results seldom correlated with other quality results measurements, they did indicate weak areas and generally responded to total program improvement activity.

In the present environment, customer answers to transmission questions can reflect opinions on any of several carriers; therefore, the transmission questions have generally been removed from exchange carrier customer–attitude surveys. Nevertheless, reintroducing transmission questions into these surveys remains a viable option.

In-Service Nonintrusive Measurements

Using various new digital signal processing technologies (e.g., the digital echo canceller discussed in Volume 1, Chapter 20, Part 4), in–service nonintrusive measurement devices (INMDs) have been developed to provide transmission indications on live network calls, encompassing the customer, station sets, loops, and interoffice facilities. INMDs used at any point in the network can provide a measure of near–end and far–end speech levels as well as noise levels. They will also derive return loss and delay measurements. Sophisticated INMDs can interpret automatic number identification digits to assist in pinpointing impairments to the subscriber level. At least one exchange carrier has already installed a number of measuring systems based on an INMD [10]. Activity is underway to develop appropriate standards for INMD measurements. Since the measurements are usually made in an automated mode, the relatively long time (about two minutes) required per measurement should not limit INMD application.

As measurement techniques are refined and appropriate objectives standardized, INMD may not only provide effective overall network transmission evaluation systems, but also economically replace a number of the other transmission program indicators.

References

1. Takahashi, K. "Transmission Quality of Evolving Telephone Services," *IEEE Communications Magazine*, Vol. 26, No. 10 (Oct. 1988), pp. 24–35.

2. Cavanaugh, J. R., R. W. Hatch, and J. L. Sullivan. "Models for the Subjective Effects of Loss, Noise, and Talker Echo on Telephone Connections," *Bell System Tech. J.*, Vol. 55 (Nov. 1976), pp. 1319–1371.

3. *1983 Exchange Access Study: Analog Voice and Voiceband Data Performance Characterization of the Exchange Access Plant*, Technical Reference TR–NPL–000037, Bellcore (Iss. 1, Sept. 1984).

4. Manseur, B., H. S. Merrill, T. C. Spang, and M. E. Vitella. "Estimated Voice Transmission Performance of Equal Access Service," *Conference Record*, IEEE International Conference on Communications (1985), 12.3, pp. 347–353.

5. DiBiaso, L. S. "User Reaction to Transmission Quality—A Cost Model," *Conference Record*, IEEE International Conference on Communications (1987), 34.6, pp. 1206–1210.

6. Padula, L. M. "Monitoring the Monitors," *Bell Laboratories Record*, Vol. 61, No. 9 (Nov. 1983), pp. 18–24.

7. Bergmann, H. J. "Taking the Pulse of Digital Facilities," Bellcore EXCHANGE, Vol. 2, Iss. 3 (May/June 1986), pp. 13–18.

8. Wright, T. and Y. Yamamoto. "Error Performance in Evolving Digital Networks Including ISDNs," *IEEE Communications Magazine*, Vol. 27, No. 4 (Apr. 1989), pp. 12–18.

9. Kort, B. W. "Transmission Quality Assurance in the AT&T Communications Network," *Conference Record*, IEEE International Conference on Communications (1985), 12.2, pp. 344–346.

10. Evans, K. W., D. M. Hornbuckle, and R. A. Sutton. "Tracing Quality with Live Calls," *Telephone Engineer and Management* (Nov. 15, 1986).

Chapter 21

Maintenance and Service Continuity

Facilities, circuits, and equipment must be properly installed and maintained in order to provide telecommunications services that meet transmission quality objectives. Transmission facilities are defined to include the media, the equipment making up transmission systems, and the channels derived from these systems. A circuit is composed of transmission facilities and ancillary equipment including gain, signalling, and terminating units. In order to meet the grade–of–service objectives, limits are set for allowable departures of actual parameter measurements from design values. The limits are specified in terms useful for day–to–day maintenance activities as well as management control.

This chapter describes the comprehensive maintenance considerations involved in applying the principles of maintenance and reliability covered in Volume 1, Chapter 22. These maintenance considerations include a comprehensive program in which data on the performance parameters of facilities and circuits are collected and evaluated. Testing is done immediately after installation (initial testing) and then on a routine, demand, or surveillance basis. The need to make maintenance activities more economical has led to the wide application of mechanized test, surveillance, and administrative systems.

Service continuity is maintained for many transmission services by the use of protection facilities to which service may be transferred when failure occurs. When a major facility failure occurs, service outage time is minimized by using emergency restoration or disaster recovery procedures.

21–1 MAINTENANCE PRINCIPLES

Transmission objectives for message telephone service and most other telecommunication services are generally derived on the basis of a balance between cost and grade of service. Thus it seems reasonable to conclude that if facilities and circuits are designed to meet these objectives, the services they provide will be satisfactory; however, facilities and circuits may be installed incorrectly and are subject to impairments that cause transmission quality to vary with time. Telephone operators once detected and reported these impairments, but conversion to direct dialing has long since eliminated the need for operator assistance on most connections; therefore, variations must be detected in other ways. Although trouble on nonswitched special–services circuits is reported promptly, detection of such trouble before it affects service is required in a competitive environment. Therefore, the primary function of transmission maintenance is to detect and correct substandard transmission performance. Another function is to test facilities and circuits as they are installed or rearranged to ensure that initial service objectives are met.

Transmission and Signalling Measurements

Transmission quality for individual connections, circuits, or facilities is evaluated by measurement of transmission characteristics and comparison of the results with standards that are ultimately based on subjective laboratory appraisals. Analog message trunk transmission measurements include insertion loss at 404, 1004, and 2804 Hz (the 404– and 2804–Hz losses are used to determine slope), echo return loss, singing return loss, C–message noise, C–notched noise, and impulse noise. For analog carrier and wideband facilities, loss measurements are made at frequencies standardized for each type of system or facility; average noise and impulse noise are also measured on analog carrier systems. Transmission measurements on digital facilities include synchronization, bit error ratio, and bipolar violations.

In addition to transmitting voice and data signals, trunks and most special–services circuits need to transmit control signals consisting of alerting, address, and supervisory signals for use by switching systems or station equipment. Signalling tests include dial–pulse tests where the intervals between pulses and the lengths of pulses are measured to ensure satisfactory switching–system operation, and supervision tests where the satisfactory transmission of on–hook and off–hook conditions can be verified.

Transmission Parameter Variations

Differences between actual and design characteristics of a transmission facility or circuit are caused by a variety of factors. To detect such differences, performance measurements must be made at the time of and subsequent to installation.

At installation, the transmission characteristics of new facilities may differ from the expected or design values because of design or installation errors, manufacturing or installation tolerances, or differences between actual and assumed environmental conditions. For example, computational errors may occur during design, or circuit gains may be improperly set at the time of installation. Cable conductor diameter varies slightly and, even within manufacturing tolerances, may produce a resistance value significantly different from the nominal value. Load–coil spacing tolerances may result in measurable differences between design and actual values of circuit impedances. Irregular terrain can produce unanticipated effects on radio paths. Fiber splices may introduce unwanted optical reflections.

For these reasons, transmission characteristics should be measured whenever a new facility or circuit is installed; such tests are called preservice tests. The measured characteristics are compared with expected values; if the difference exceeds initial test limits, corrective action must be taken before releasing the facility or circuit for service. Loss variations in terms of differences between calculated values [expected measured loss (EML)] and measured values [actual measured loss (AML)] represent the major use of this concept

Transmission characteristics vary after installation for several reasons. Electronic components may change under the influence of heat and time to cause changes in transmission performance. Switch contacts deteriorate with use and can introduce both loss and noise. Temperature and humidity variations, most notably in outside plant where environmental conditions are not controllable, also affect the transmission characteristics. Variations may result from installation errors; in addition, maintenance activities on adjacent facilities and equipment may cause trouble.

Depending on the type of facility and testing techniques employed, variations in transmission parameters are detected by continuous monitoring, periodic testing, or customer reports of trouble. The interoffice transmission maintenance index plan covered in Chapter 20, Part 3 calls for periodic analog measurements that allow moderate variations in transmission parameters before either a maintenance limit or an immediate action limit is reached.

In addition to the variations mentioned above, customer opinions of transmission quality, as evaluated by subjective appraisals, vary. That is, rather than there being discrete values, there are distributions of received talker volume, received noise, video echo ratings, etc., that are subjectively rated good or better (GoB). The number of trunks in switched connections and the number of transmission facilities or links in private–line circuits are also variable. These variabilities in facility and connection characteristics require that objectives be set by grade–of–service techniques that combine their effects.

Cost and Revenue

Maintenance control of transmission quality variations requires a balance between service and cost. The high cost of manual testing would lead to testing intervals too long to maintain good transmission quality. Therefore, the need to economically provide an acceptable grade of service has led to the wide application of automated surveillance and testing systems.

Trunk and special–services circuit designs that have low cost, yet meet design objectives, may require frequent testing and re-aligning, and therefore have high maintenance costs. On the other hand, better designs with higher initial investment may require little maintenance; therefore, long–term costs may be lower. Thus, operating costs in addition to capital investment must be considered in all designs.

A Comprehensive Maintenance Plan

Maintenance activities constitute a vital part of a company's transmission quality program. A comprehensive maintenance plan is required to provide methods for the operation and coordination of the various centers established to carry out the total maintenance process. This plan should provide for the orderly integration of improved automated systems into the maintenance operation.

21-2 FACILITY MAINTENANCE

Initial tests are performed on transmission media (including paired cable, coaxial cable units, radio paths, and fiber strands). Tests are also made on the various analog and digital carrier systems that are transported on these media.

Initial Testing

Subscriber cable pairs are usually tested by construction forces for open, short–circuited, and grounded conductors when splicing is completed. Initial tests for coaxial cable units consist of center conductor and insulation resistance measurements and a corona survey. Fiber is tested initially for optical insertion loss and return loss.

On digital carrier systems, initial tests include a check for digital errors using a quasi–random signal source and an error detector. The quasi–random signal is a repetitive code word that is more likely to cause digital line errors than the signals normally transmitted.

Initial tests for radio systems include measurements of received signal level, interfering carriers, frequency, radio–frequency power, radio channel gain, noise, error margin, and bit error performance.

Surveillance and Routine Testing

Continuous surveillance of overall facility integrity and performance is used to ensure that major defects and performance

533

deterioration are promptly recognized. Digital error rates, pilot amplitudes, carrier amplitudes, and noise power are continuously monitored for this purpose. This surveillance capability is generally included as part of the transmission system, and normally includes a number of service protection, alarm, and control options. In order to provide for economical operations, these surveillance measurements are almost always relayed to a Facility Maintenance and Administration Center (FMAC) which centralizes facility maintenance for a large geographic area. FMACs typically use several of the various automated maintenance support and alarm systems that are covered in detail in Volume 2, Chapter 23, Part 2. In addition to providing centralized information on the status of continuity and performance parameters at all of the stations on the various transmission systems and routes, these maintenance systems allow the FMAC to make inquiries, tests, and perform operations in response to alarm indications.

Alarm and surveillance information on digital carrier systems installed in the loop plant may also appear in the FMAC. In some cases, a separate centralized maintenance center is used to monitor loop systems.

The FMAC dispatches maintenance personnel in response to alarms and performance indicators. The FMAC also schedules other maintenance activity. In the past, the schedule was generally on a routine basis; however, the trend is now toward demand maintenance, which is scheduled on the basis of deteriorating nonservice–affecting measurements. This demand maintenance can be coordinated with such things as battery checks and tower light inspections at remote stations.

In almost all geographic areas, a computer–controlled gas–pressure monitoring system continuously analyzes the status of transducers that measure the air pressure in various cable sections. Status reports and gas–pressure alarms appear at a centralized location, usually in the same center that controls digital loop carrier maintenance.

21-3 CIRCUIT MAINTENANCE

In the past, a large amount of circuit maintenance was performed on a manual basis at testboards equipped to connect test equipment to individual loops, trunks, or special services via test jacks associated with each circuit. These manual operations have rapidly been supplanted by mechanized maintenance arrangements, resulting in improved transmission quality as well as significant cost savings.

Loops

Although transmission characteristics of customer loops are seldom measured, talking tests and tests for short circuits, crosses, and grounds are performed whenever a loop is installed or when trouble is reported. If a loop is properly designed and passes these tests, it usually meets transmission objectives.

A local test desk, equipped with test trunks for loop access, was for many years the principal vehicle for loop maintenance and testing. However, virtually all exchange carriers now have mechanized loop maintenance and testing arrangements concentrated in automated repair service bureaus (ARSBs). While a customer reporting trouble waits on the line, these mechanized systems allow an ARSB technician to call up computerized service records, control a variety of transmission tests, and generally determine whether the trouble is in the loop or the station equipment [1]. A maintenance person dispatched to clear a loop trouble can use a number of dial–up central–office test lines in making measurements, including controlling the mechanized test system itself. The field technician has a portable terminal that displays the trouble report, gives test results, accepts the report of trouble clearance, and dispatches the tester to the next case of trouble.

The automatic line insulation test equipment installed in most central offices sequentially tests each loop terminated in the office for short circuits, grounds, and foreign potential in order to detect faults before they become service–affecting. The in–service nonintrusive measurement devices discussed in Chapter 20, Part 5 may also be used for detecting loop problems before they are reported by customers.

Network Trunks

Trunk transmission maintenance is intended to keep trunks working within objective parameters. When a failure is detected by surveillance, routine testing, or trouble reports, affected trunks are removed from service and maintenance personnel are notified. The trouble is then sectionalized to the near–end office, far–end office, or intermediate facility and the appropriate repair force is notified. Once the trouble is repaired, the trunks are retested to assure proper performance and then restored to service. Similar tests are performed prior to placing trunks in service.

Message network trunk maintenance has evolved from manual methods to mechanized testing, switched (dialed) access, and continuous monitoring of transmission performance. These methods allow maintenance effort to be devoted to clearing troubles rather than to performing repetitive tests to detect defective trunks.

Manual Testing. Two persons, one at each end of a trunk, originally made transmission tests. After gaining access to the trunk, one person transmitted the signal. The person at the other end gained access to the same trunk and measured received amplitude, noise, or other characteristics. This method was expensive and slow, and required clerical effort to analyze measurement data.

One–person manual testing became possible as dial conversion progressed and dial central offices were equipped with dial–up test lines. The person making the test dials an assigned number to access the test line at the far end of the trunk, which applies an appropriate termination. When the test call is established, a transmission measuring set may be connected to the circuit under test. Tests that can be performed using dial–up test lines include the measurement of loss, noise, echo return loss, and even peak–to–average ratio (PAR). Trunk supervision features can be tested by dial–up test lines that send alternating on–hook and off–hook supervisory signals.

As previously mentioned, manual testing has a number of disadvantages. High costs tend to extend routine testing intervals so that recognition of defective trunks is delayed; thus, service is degraded for longer periods than if routine tests are made frequently. Furthermore, measurement data lags actual performance changes, which makes analysis inconclusive. Even with timely data, manual analysis is expensive. Manual routine tests are repetitive. Also, if trouble is found, sectionalization tests require the coordinated efforts of several persons. These factors lead to even higher costs, which has led to the mechanization of virtually all routine trunk testing.

Mechanized Testing. Among the most important reasons for frequent, mechanized testing of network trunks is the identification and repair of so-called killer trunks. A killer trunk is one that is incapable of completing a connection or over which there is no transmission. When such a trunk exists, it is seized frequently and then released after only a short holding time. Having been frustrated by the failure on the first attempt, the dialer is likely to disconnect and then immediately make further attempts during which the same trunk may be seized several times. Such a condition results in undesirable customer reaction to poor service. Mechanized trunk testing makes possible the early identification of a killer trunk to remove it from service until the trouble can be repaired. Without mechanized testing, the existence of such a trunk could only be recognized by multiple customer complaints or traffic observations of many calls with very short holding times.

Computer–controlled measurement systems have been used for some years to measure voice–frequency (VF) loss and noise on trunks automatically; the measurements are on an out–of–service basis at night when traffic loads are light. Later versions of these systems make terminal balance, gain–slope, C–notched noise, and operational tests in addition to loss and noise [2]. The measurement systems are interconnected via data links with other operational support systems; along with stored EML and AML data, they automatically compute the transmission indices associated with the network trunk transmission maintenance plan described in Chapter 20, Part 3. In addition to identifying killer trunks, these systems have improved transmission quality by identifying emerging transmission problems before they become service–affecting.

As the public switched network (PSN) uses digital transport facilities in conjunction with digital switching, transmission quality is improving through the elimination of intermediate digital–to–analog conversions. Digital performance and synchronization are monitored in the digital switching machines and digital cross–connect systems (DCSs) as well as on the transmission facilities. Digital performance criteria and monitoring are covered in detail in Volume 2 and elsewhere in Volume 3.

Special Services

In many respects, the maintenance techniques employed for special–services circuits are different from those employed for switched message network circuits. Special–services circuits may be switched or nonswitched and circuit lengths may vary from local intraoffice to long–haul intercontinental. Signal formats vary from very infrequent changes in direct current flow, such as is found on some alarm circuits, through the more complex signals of voice and data channels to video signals. These circuits may interconnect to a whole array of special station equipment such as private branch exchanges, key telephone sets, data sets, and sophisticated networks. A few require special transmission equalization and conditioning. In addition, special services experience an extraordinary amount of inward, outward, and rearrangement activity. The maintenance of such a wide variety of services also complicates personnel training.

Other factors that complicate the task of furnishing special services may be found by examining the basic differences in the makeup of PSN and special–services circuits. In the evolution of the PSN, performance requirements have been defined in relation to a hierarchy of switching offices, transmission circuits, and interfaces with other carriers. The only components of the network that an individual customer permanently depends on are the telephone set and the loop. Failure of any other network element is not likely to affect an individual customer because of alternate routing and the multiple switched paths of the PSN. On the other hand, the structure of most special services does not allow for exploiting these alternate–routing capabilities. For example, a foreign exchange line may traverse loop and interoffice facilities before being connected to the PSN at an end office. A major exception is special services that use a customer–controlled DCS, in which the customer may perform network management and rerouting at will.

Special gain and signalling range extension equipment may be required to provide adequate transmission and signalling performance. The service may traverse a number of administrative areas and interconnecting carriers such that ambiguities may often arise in accountability and responsibility for service.

In order to illustrate how the maintenance task is handled, it is convenient to divide special services into two categories: designed and nondesigned. Special services that use only metallic cable plant out of the local central office and require no exchange carrier equipment are usually designated as nondesigned special services. Maintenance responsibilities for these nondesigned services are generally handled out of the ARSBs in the same manner as local subscriber loops.

Designed Special Services. Special services that require any type of gain, signalling, equalization, or other equipment devices are generally classified as designed services. Designed special services must be initially laid out, installed, and maintained so that they meet tariff–dictated transmission values at the interface points with customers and other carriers. After the circuit provision center provides a design, installation control and maintenance are generally provided through a centralized special–services center (SSC). SSCs make extensive use of computerized data bases and the mechanized systems covered in Volume 2, Chapter 23, Part 3. These allow a technician in the SSC to make circuit tests from dialed–up access points in remote offices. Loop–back terminations may be provided at customer locations that allow the SSC to complete tests without dispatching a technician.

Test access for designed special–services circuits at offices not equipped with switched access terminals must still be made on a manual basis. This is usually made via mainframe connecting devices (called shoes), but in some offices may be made through VF patch jacks, carrier facility jacks, or private–line test boards.

With the proliferation of digital facilities in the the network, interoffice special services (particularly data services) are becoming more easily installed, monitored, and maintained. Nevertheless, interconnecting analog loop distribution facilities still results in time–consuming, expensive installation and maintenance. Automated digital termination systems (ADTSs) installed on the digital carrier systems, however, can significantly simplify these installation and maintenance operations as covered in Volume 2, Chapter 15, Part 9. After the placement of a portable responder at the customer location, the ADTS will automatically make analog loop measurements and gain settings. The ADTS can be remotely controlled from the SSC if desired.

21-4 SERVICE CONTINUITY

Transmission facilities and circuits must be designed, installed, and maintained to meet transmission quality objectives. In addition, they must include features that ensure that circuit continuity performance meets customer expectations. (In the final analysis, service outage is the ultimate transmission impairment.) System designs use many devices and components that must satisfy reliability criteria and, in addition, include operating margins against overload and environmental changes. Some systems also include equipment and facilities that are operated as maintenance protection facilities so that service can be transferred in the event of failure or during maintenance activities on the regular equipment. The transfer may be effected automatically by appropriate switching arrangements or manually by patching.

In the event of a major transmission route failure, emergency restoration plans should be available. Emergency repair equipment should be stored ready for use. Some systems on major routes are "hardened" to withstand natural or manmade catastrophes. Circuit routings are dispersed so that a failure on one route does not necessarily disrupt all service to a community or to some important point in the communications network.

Transmission systems are composed of three major categories of components: the transmission medium, line equipment, and terminal equipment. The first two are often considered together and referred to simply as the line. Terminal equipment includes gain adjusting, modulating, multiplexing, signalling, and interconnection components as well as common equipment such as clock sources and power components. Service protection on such systems is thus considered in terms of line, terminal, or common-equipment protection.

Automatic Protection Facilities

Individual loops, trunks, and special–services circuits are rarely provided with protection facilities because the cost would be prohibitive. (Some exchange carriers offer customers dual–routed service from different central offices at extra cost, an attractive feature for truly critical user sites.) However, automatic protection is provided on most transmission systems. This automatic protection may be arranged for either revertive or nonrevertive switching. After the failure of a normal working channel, service automatically switches to the protection channel. With a revertive arrangement, service will automatically switch from the protection channel back to the normal channel after the trouble is removed. With a nonrevertive arrangement, service stays on the protection channel for as long as that channel is usable. The provision of protection facilities is normally based on the assumption that the larger the system capacity, the more the need for protection since a greater number of circuits may be affected by a failure. The availability of a protection channel also allows the working channel to be manually switched to the protection channel for maintenance testing.

Automatic protection arrangements for the major transmission system categories are covered along with their respective system descriptions throughout Volume 2. Since propagation outage is a major concern in microwave radio systems, Volume 2, Chapters 13, 14, 15, and 16 cover in detail the considerations in engineering and designing microwave systems to meet outage objectives.

Transmission Effects. The provision of protection facilities results in exceptionally high reliability for the systems involved, especially where automatic switching arrangements are used. However, in some cases, there may be slight transmission penalties.

When systems protected by automatic switching arrangements are manually switched or gradually fail, the protection channel on most types of systems switches at a threshold level that allows service to go uninterrupted. Sudden failure of a working system, however, may cause momentary opens or momentary changes in the phase of the circuits involved. These effects, called hits, are minimal except for their production of data errors.

Common Equipment. The operation of transmission systems depends on common equipment that is shared by a number of transmission panels, bays, complete systems, or even by an entire office. Among these are power sources, carrier supplies, pilot supplies, and synchronizing equipment.

Commercial sources of ac power are used in all offices as the primary power supply for the telecommunications equipment. The circuits are arranged so that the ac supply (converted to dc) is used to maintain the office battery at full charge while simultaneously operating the equipment. In the event of ac failure, the battery alone carries the load until ac service is restored or until the office emergency alternators are switched into service. In addition to these emergency arrangements, power distribution and fusing are designed so that failure in one part of an office is limited to specific circuits or systems while other parts of the office continue to function normally.

Nearly all synchronizing signal equipment is duplicated and the signal supply circuits are arranged so that the protection equipment is automatically switched into service when failure occurs. The local sources of synchronizing signals are controlled by a master digital signal, dual–fed from external sources. If both master signals fail or if the distribution circuits are disrupted, the local sources continue to operate in a free–running mode. Telecommunications services are maintained but may experience digital slips if synchronization is not quickly restored.

Emergency Restoration

Major failures on systems that serve large cross–sections of circuits, such as those caused by the cutting of an optical cable or the destruction of a microwave tower, can produce massive service disruptions not only on the directly affected route but on many interconnected and interrelated routes, primarily because of alternate traffic routing. In order to restore as much service as possible in the shortest time, the protection facilities of operating interconnecting systems are used. The transfer to the protection facilities should be under the control of the FMAC, which normally has indications of remote protection channel availability.

In controlling restoration activities, the FMAC should maintain restoration plans that show the location of compatible, potentially usable protection channels for each major facility route. Such plans should be updated as facilities are added or removed from service so that the plans properly reflect the field situation. The plans may incorporate the use of appropriate restoration patch bays at terminal locations. The use of DCSs that can cross–connect at either the DS1 or DS3 line rate, or of SONET (synchronous optical network) add–drop multiplexers that perform a similar function, can be particularly helpful at hub locations. The computerized surveillance systems that are used in some of the FMACs have the capability to perform these restoration functions on a mechanized basis. Analog systems may experience some slight transmission degradation after they are patched for restoration; however, there is generally no difference in digital system performance.

Restoration usually results in shorter service outage time than repair procedures, but it is always desirable to release the the restoration facilities to the protection function as soon as possible. To accomplish this and to effect restoration as quickly as possible where protection facilities are not available for restoration, a wide variety of emergency repair facilities are usually available at predesignated locations. These emergency facilities include lengths of cable, repeaters, and apparatus cases. Some locations may have portable terminals, microwave equipment, and towers.

Other Service Continuity Considerations

Service continuity also depends on a number of activities outside the province of transmission engineering. Deficiencies in switching maintenance and outside–plant craft performance can lead to service outages. Appropriate network architectures can moderate the overall impact of the failure of high–capacity fiber routes and network gateways. Implementation of disaster recovery plans and network management actions can expedite the restoration of service.

Network Plans and Architectures. High–capacity digital transmission systems economically concentrate and funnel increasingly larger numbers of circuits into single cross–sections and gateways. Network plans must include appropriate diversity and strategies to minimize the effect of failure of one of these essential high–capacity network elements. These plans must consider the provision of duplicate, geographically separated gateway hubs or possibly a ring–type network architecture. A ring network can link nodes in a continuous two–way loop so that if one connecting link or node fails, the traffic can be routed in the opposite direction. The principal transmission effect of such a configuration is the mileage added around the ring in normal operation, which tends to increase the echo–delay time of the associated circuits.

Plans must also be developed with private–network customers to allow emergency alternate access to the PSN. As customers are allowed to input information to provide network control functions, strategies must be available to prevent misuse, fraud, or service interruption.

Disaster Recovery. Even with appropriate network architectures, catastrophic disasters (e.g., earthquakes, floods, hurricanes, or severe central–office fires) can cause severe service interruptions that require an extraordinary restoration effort. Restoration for these disasters is usually coordinated by a disaster recovery team, which implements a plan that had been previously prepared and periodically updated. This plan includes the application of temporary service elements such as trailer–mounted switching machines, transmission terminals, emergency generators, coin telephone booths, cellular radios, and possibly satellite terminals.

Traffic Network Management. As the telecommunications network grows, it is designed for greater efficiency and carries a larger volume of traffic. Its vulnerability to overload and breakdown also increases. Since the dependence on communications extends into every phase of life, the control and management of the network by the various interconnected carriers assume increasing importance.

A carrier's network traffic control is administered from designated points called network management centers. At these centers, switching machine traffic is continually monitored for possible overload conditions. Overload may occur as a result of unanticipated events, such as the catastrophic failures discussed above. When these events cause switching machine overload, modes of operation are altered so that alternate routing is reduced or eliminated. Under overload conditions, attempts to find an alternate route through a blocked point compound the overload by permitting added attempts that cannot be successfully completed. Other modifications of network operations can be controlled from the traffic management centers. For example, bulk traffic between two cities may be rerouted via an office not normally used for connections between the two cities.

Electromagnetic Disturbances. Electromagnetic disturbances from such sources as lightning, high–voltage power lines, high-power radio transmitters, and "magnetic storms" have long been a potential source of telecommunication circuit outage. New technologies that have been introduced into the network are more susceptible to these disturbances, which has led to new cable designs and installation techniques in order to provide acceptable service reliability [3].

With the introduction of new solid–state electronics into telephone plant, a new less dramatic but more troublesome hazard has been introduced in the form of electrostatic discharge (ESD) [4]. ESD occurs when a static charge carried by a craftsperson arcs over to a solid–state circuit element. Such an arc, annoying but harmless to the craftsperson, can render the electronic device inoperative. Procedures such as attaching a grounding cable between the craftsperson and a grounding terminal on the equipment being serviced reduce this service hazard.

21-5 INTEGRATED OPERATIONS SYSTEMS

The rapid deployment of computer–controlled centralized maintenance for circuits and facilities has resulted in improved transmission quality at significantly lower cost. In a similar way, mechanized systems in associated work centers are being used for network switching maintenance and traffic network management. Integrated operations systems (OSs) are becoming available so that all of the various network surveillance, maintenance, and control functions for transmission, switching, and traffic can be integrated into a single computer data base. Powerful work stations associated with this system will further reduce costs and improve total network performance. Generic requirements and specifications for the various interfaces of such a system are available [5]; prototype systems have been installed in several locations [6].

Integrated OSs can be used very effectively in the emerging all-digital network. They can allow a centralized assessment of total network quality with transmission as one of the components. Such assessment can provide vital information in a competitive environment. A framework that provides a matrix summary of the various network quality components has been proposed [7].

Knowledge-based, expert systems that use techniques pioneered by research into artificial intelligence are emerging to assist technicians in analyzing and pinpointing network transmission and switching problems [8]. Such expert systems will no doubt become a vital part of integrated OSs.

References

1. Fleckenstein, W. O. "Operation Support Systems: Computer Aids for the Local Exchange," *Bell Laboratories Record*, Vol. 60, No. 6 (Sept. 1982), pp. 185–193.

2. Plato, J. J. and B. S. Robb. "CAROT's New Features Expand Maintenance and Administrative Functions," *Bell Laboratories Record*, Vol. 61, No. 9 (July/Aug. 1983), pp. 18–24.

3. Parente, M. "The Energy That No One Wants on a Telephone Line," Bellcore EXCHANGE, Vol. 2, Iss. 1 (Jan./Feb. 1986), pp. 29–32.

4. Thoni, C. T. and B. A. Unger. "Take a Closer Look at Electrostatic Discharge," Bellcore EXCHANGE, Vol. 2, Iss. 2 (Mar./Apr. 1986), pp. 18–23.

5. *Operations Technology Generic Requirements*, Technical Reference TR-TSY-000439, Bellcore (Iss. 2, Nov. 1988).

6. Koblentz, M. E., M. R. Nash, and M. Seldner. "Seeing the Big Picture: Network Monitoring and Analysis," Bellcore EXCHANGE, Vol. 3, Iss. 5 (Sept./Oct. 1987), pp. 28–32.

7. Dvorak, C. A. and J. S. Richters. "A Framework for Defin-
ing the Quality of Communications Services," *IEEE Communications Magazine*, Vol. 26, No. 10 (Oct. 1988), pp.
40–47.

8. Boyd, R. C. and A. R. Johnston. "Network Operations and
Management in a Multi–Vendor Environment," *IEEE Communications Magazine*, Vol. 25, No. 7 (July 1987), pp.
40–47.

Chapter 22

Management

The foundation of an effective telecommunications network is discussed throughout the three volumes of this book. This chapter starts with an overview of the transmission management functions involved in providing transmission quality to meet customer needs at a reasonable cost. It concludes with a glimpse of the management challenge in implementing emerging telecommunications technologies and services.

Network conditions and growth, customer opinions of performance, service needs, network design and operating technology, and the external environment are changing continuously and at an ever-increasing rate. Thus, transmission management responsibilities can be fulfilled only by recognizing and responding to the dynamic nature of all these elements in addition to resolving immediate problems.

One aspect of the dynamic nature of the telecommunications network is that different impairments dominate transmission performance as the network changes. In the early days of telephony, the bandwidth limitations and nonlinear characteristics of station equipment dominated. Then, as distances increased, transmission losses had to be reduced; as a result, talker echo became a recognized limit on performance. With losses reduced by the application of electronic technology and with echo controlled by four-wire circuits and balance programs, circuit noise became the most troublesome impairment. Finally, as the network shifts from analog to digital, noise effectively disappears but echo path delay grows, while bit errors and bit-stream synchronization problems come to dominate network transmission performance. Emerging technology like 7-kHz speech coding promises to bring further changes in perceived performance. Transmission management has been a vital force in bringing about these changes.

22-1 MANAGING CURRENT ENGINEERING

The problems of maintaining high–quality transmission performance are economic and technical. While transmission management is primarily related to solving the technical problems, the solutions are acceptable only if they are economical. Transmission management functions must ensure high–quality, reliable facilities in a time frame that meets customer expectations.

Economic Factors

Customer acceptance of the grade of service provided in a competitive environment has a strong influence on revenues. Thus the transmission management control of the processes that determine the actual transmission quality is essential to a telecommunications company's successful operation. This transmission management control includes sound design methods for transmission facilities, adequate maintenance support procedures, and economical means of achieving service reliability and restoration.

While these elements of transmission management appear to be straightforward, they can be accomplished only to the extent permitted by available financial resources and within the broad limits established by overall company policy. It may be necessary to fulfill transmission management functions within the constraints of limited available funds. However, in a competitive environment, these transmission management functions include petitioning upper management for additional resources when necessary.

Technical Activities

Technical activities involved in the management of transmission quality include functions that ensure that the network is planned, engineered, installed, and maintained to meet current objectives. Quality control of actual transmission performance is exercised by making appropriate measurements and making corrections based on deviations from designs and standards. Grade–of–service calculations can ensure that new services will have transmission quality that meets customer expectations.

Coordination. Many organizations are responsible for work that may directly or indirectly affect the transmission performance of the network or its parts. In the outside plant, routes are established and cables installed in accordance with rules designed to produce satisfactory transmission performance without further engineering assistance. The establishment of new wire centers and the installation of switching machines must be monitored and coordinated if high–quality transmission performance is to be maintained. The fulfillment of traffic engineering and local access and transport area (LATA) access requirements can affect transmission performance. The introduction of new services is likely to require new sets of transmission performance objectives that may not yet be standardized. Maintaining circuits and systems is only indirectly a transmission engineering responsibility; however, transmission engineers need to be directly involved in implementing automated measuring and testing systems. Involvement in coordinating all such activities is an important transmission management responsibility.

One effective mechanism for coordinating transmission–affecting work programs is the participation by those responsible for transmission management in a variety of formal committees. In most cases, the committee should include transmission engineering membership. Some examples are the service, trunk, and facility committees. The service committees are usually involved with problems in circuit administration, installation, and operation. Trunk committees follow and coordinate trunk installations and rearrangements, while facility committees follow the progress of construction programs and make adjustments to meet emergencies and changing needs. Transmission engineering should also be represented on quality review teams that are established to monitor the performance of new installations and services. Transmission management must arrange to have technical experts available to assist field installation and maintenance forces in solving unusual problems as they arise.

Measurements and Correction. Transmission management is responsible for initiating a company's transmission quality assurance program as covered in Chapter 20, Part 1. This program should include appropriate performance measurements and customer trouble indicators that show how actual transmission performance deviates from objectives. These deviations should provide feedback to management so that appropriate corrective action can be taken.

Grade of Service. Subjective tests are used to determine the grade of service, establishing relationships between test results and value judgements of various types of impairment as covered in Volume 1, Chapter 24. Grade of service may also be used to relate transmission quality to cost [1]. A change in magnitude of an impairment can be related to a corresponding change in customer satisfaction and to cost. Thus, changes in customer satisfaction can be related directly to the cost of improving performance.

The grade–of–service concept is so flexible in application that it can be used to evaluate the performance of an entity as small as a central office on the basis of local calls only, the performance of a small area on the basis of local and intraLATA toll calls, or the performance of the entire network. To evaluate an entire switched network for loss and noise performance, average values and standard deviations on individual links that may be used by various carriers in built–up connections must be specified and combined statistically. The data is usually analyzed by a computer programmed to process grade–of–service calculations. Typical values of loss and noise are stored in memory as a part of the program. Different loss and noise values can be substituted for comparison and evaluation of their effects on the grade of service.

The proposed introduction of a new tandem switching machine may involve many changes in trunking over a wide geographic area, changing the serving home offices of many customer lines, changing end–office relationships in the hierarchy, consolidating local areas, and introducing integration of digital and packet services. Each of these changes to existing services and the integration of new services can be evaluated in terms of its effect on overall grade of service.

Other applications of the grade–of–service concept useful in transmission management are evaluating alternative solutions for a problem and establishing priorities. For example, several transmission deficiencies may exist, but resources may be available to implement only one improvement at a time. Grade–of–service calculations may be made to determine which of the alternatives is likely to produce the greatest improvement for a given budget. Thus, alternatives can be evaluated so that priorities may be assigned on a logical basis.

22-2 THE RESPONSE TO CHANGE

The optimum solution to current engineering problems as well as the fulfillment of transmission management responsibilities depends on the response to rapid regulatory, environmental, and technological change. This response must take into account the rate and interaction of the various change elements. An example of interaction is seen in intercontinental communications where the principal interactive elements have been available technologies, quality, cost, and cooperation among organizations and governments. The first circuits that linked Europe and North America were provided over radio facilities, long–wave and then short–wave, during the 1920s. Transmission was relatively poor and unreliable, and costs were high. As a result, the service was not popular and demand grew rather slowly. When repeatered submarine cable systems were first installed during the 1950s, demand expanded rapidly because transmission quality and reliability were greatly improved and costs decreased dramatically. Now, with thousands of conventional submarine cable and satellite circuits in place and optical fiber submarine cables a reality, intercontinental communication continues to grow very rapidly. With national and international standards being established, quality and reliability are still improving, costs continue to decline, and new international services are emerging.

Regulatory Changes

Divestiture of Bell exchange carriers broadened opportunities for competition in the provision of telecommunications services in the United States; however, the nature and terms of this competition are still evolving from actions of the Department of Justice, the Federal courts, and the Federal Communications Commission (FCC). For example, the FCC continues to make inquiries and rulings on the types of network interconnection to be offered vendors of enhanced services [2]. In addition, the division of jurisdiction between state regulatory agencies and the FCC is not clearly defined. Regulatory change most certainly will continue into the foreseeable future.

Environmental Changes

The regulatory environment noted above continues to evoke a host of environmental changes. Faced with increasing telecommunications needs, large business users continue to institute sophisticated private networks that use, interconnect with, or in some cases bypass the common carrier networks. These interconnections involve new and changing sets of transmission quality requirements involving both the exchange carriers and the interexchange carriers. In some cases, customers may be drawn to the economic advantages of a poorer grade of service.

Changes in population distribution affect network plans and may require the redistribution of transmission facilities. For example, there is a general population move to the suburbs. Metropolitan areas in the southern and western parts of the United States are growing rapidly while population centers in other parts of the country are relatively static; in some cases, population is actually decreasing. Ecological concerns have intensified a trend to out–of–sight plant, which has an advantage in reliability and transmission stability but may increase costs.

Innovations

Innovations, that is, the application of new technologies, the introduction of new methods, and the development of new network services, are company responses to environmental and regulatory changes. In many cases, the changes associated with innovation will precipitate additional changes and responses. Transmission management must be responsive to these changes and must influence them so that performance is improved or at least not degraded.

New Transmission Technologies. Technological changes have marked the entire history of telecommunications. New devices, designs, systems, and modes of operation continue to become available at an increasing rate. They must be controlled and managed as they are introduced in the network.

The effects of these technological innovations are generally re-duced costs, improved performance, added flexibility, and new service capability. For example, Figure 22–1 shows how the transmission grade of service improves as the interoffice portion of the network moves from analog to digital. On the other hand, other technologies such as dynamic routing (discussed in Chapter 7, Part 6) have the potential to degrade transmission perform-ance.

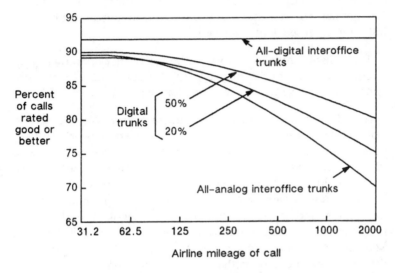

Figure 22–1. Relationship of transmission quality to call mileage with various percentages of digital trunks.

Considerable effort must be expended to ensure that new cir-cuits and systems operate compatibly with existing plant. Fast-acting digital switching systems must operate, at least for a period of transition, with slow–acting step–by–step systems. New digital transmission systems must interface with the existing analog plant. While room must be made for new designs and services, it is uneconomical to scrap all of the existing facility investment. Transmission management must ensure that new designs and technologies are carefully introduced to avoid undue penalties in performance, reliability, and cost. At the same time, methods must be adopted to bring about rapid application in a competitive environment.

Methods. Maintenance, operations, design, and layout methods are changing in response to the need to reduce expenses and to the availability of sophisticated digital computer technology. This has prompted the rapid transition from manual to automated methods and allowed operations to be remotely controlled from centralized computer terminals. The factors of size and speed make computer analysis and control very attractive. In addition, the network has become so large that the evaluation of the effects of changes would be difficult without computer aids. The automated processes that are being introduced must be carefully implemented and managed.

Transmission management must be aware that operators of interconnecting networks may use similar automated testing systems to measure selected transmission parameters.

New Network Services. The integrated services digital network (ISDN) and the public packet–switched network (PPSN) covered in Chapter 3 are examples of innovative responses to service demand and environmental change using new and changing technologies. The introduction of electronic switching systems spawned new services such as speed calling, call waiting, three–way calling, and call forwarding. CLASS[SM] service will allow even more exotic intelligent–network service options and features. Service options that allow customers to control configurations of their networks pose particular transmission challenges.

Other Responses

Transmission management responses to change involve increasing emphasis on several other activities. Transmission input to the facility planning process covered in Chapter 19 must reflect the importance of maintaining transmission quality as well as providing service restoration alternatives. Participation in activities of the standards–setting bodies becomes essential.

With changes taking place so rapidly, it is important that personnel involved in the engineering of the network and in managing transmission receive continuing education and

---

CLASS is a service mark of Bellcore.

technical updates in the form of seminars and training courses. Numerous organizations, educational institutions, and companies are involved in meeting this need. Among these is the Bellcore Technical Education Center [3]. This center conducts an educational program covering all the technology involved in operating and managing the telecommunications network.

22-3 FUTURE TRENDS

In the future, transmission management will be challenged even more by ever-increasing rates of environmental and technological change. Table 22-1 summarizes some of the new transmission networks, services, and technologies discussed in Volumes 2 and 3 of this book that are expected to mature in the 1990s to meet an increasing demand for economical

Table 22-1. Emerging Transmission Networks, Services, and Technologies Noted in *Telecommunications Transmission Engineering*

Network, Service, or Technology	Where Noted
Gb/s fiber optic systems	Vol. 2, Chap. 14
Synchronous transmission at DS3 rate (SYNTRAN)	Vol. 2, Chap. 15
Synchronous Optical Network (SONET)	Vol. 2, Chap. 15
1024-QAM (quadrature amplitude modulation) microwave systems	Vol. 2, Chap. 16
Digital cellular systems	Vol. 2, Chap. 18
Mobile Satellite Service (MSS)	Vol. 2, Chap. 18
Universal digital portable communications (UDPC)	Vol. 2, Chap. 18
Integrated services digital network (ISDN)	Vol. 3, Chap. 3
Public packet-switched network (PPSN)	Vol. 3, Chap. 3
Switched Multi-Megabit Data Service (SMDS)	Vol. 3, Chap. 3
Extensive loop fiber application	Vol. 3, Chap. 5
45-Mb/s DS3 video channels	Vol. 3, Chap. 16
In-service nonintrusive measurement devices (INMDs)	Vol. 3, Chap. 20
Integrated operations systems (OSs)	Vol. 3, Chap. 21

telecommunications. Beyond the 1990s, the expansive development of digital transmission and switching technology suggests the eventual combination of transmission and switching (transswitching). As optical systems move beyond the Gb/s-rate range, fiber capacity may be great enough to allow passive optical networks to eliminate the local central office [4].

References

1. DiBiaso, L. S. "User Reaction to Transmission Quality—A Cost Model," *Conference Record*, IEEE International Conference on Communications (1987), 34.6, pp. 1206–1210.

2. "Computer III Notice of Proposed Rule Making," Federal Communications Commission Mimeo Number 86–252 (June 16, 1986).

3. Stewart, A. "TEC's Mission: Telecommunications," *Telephone Engineer and Management* (Nov. 15, 1986), pp. 148–151.

4. Bain, M. and P. Cochrane. "Future Optical Fiber Transmission Technology and Networks," *IEEE Communications Magazine*, Vol. 26, No. 11 (Nov. 1988), pp. 45–60.

Acronyms

The acronyms listed here reflect usage in this book. They may be used differently in other contexts.

ABBH	Average Bouncing Busy Hour	**AIS**	Automatic Intercept System
ABS	Average Busy Season	**AL**	Acceptance Limit
ABSBH	Average Busy–Season Busy Hour	**ALBO**	Automatic Line Build–Out
ACD	Automatic Call Distributor	**AM**	Amplitude Modulation
ACRS	Accelerated Cost Recovery System	**AMI**	Alternate Mark Inversion
ACXT	Apparatus Case Crosstalk	**AML**	Actual Measured Loss
ADM	Adaptive Delta Modulation	**ANSI**	American National Standards Institute
ADPCM	Adaptive Differential Pulse Code Modulation	**AOS**	Alternate Operator Services
ADR	Asset Depreciation Range	**APC**	Automatic Power Control
ADTS	Automated Digital Termination System	**APD**	Avalanche Photodetector Diode
AGC	Automatic Gain Control	**ARSB**	Automated Repair Service Bureau
AIOD	Automatically Identified Outward Dialing	**ASK**	Amplitude Shift Keying
AIS	Alarm Indication Signal (or)	**AT**	Access Tandem
		AT&T	American Telephone and Telegraph Company

AUTOVON Automatic Voice Network

AWG American Wire Gauge

AXPIC Adaptive Cross–Polar Interference Canceller

B3ZS Bipolar Format with Three–Zero Substitution

B6ZS Bipolar Format with Six–Zero Substitution

B8ZS Bipolar Format with Eight–Zero Substitution

BDDS Basic Dedicated Digital Service

BER Bit Error Ratio

BETRS Basic Exchange Telecommunications Radio Service

BITS Building Integrated Timing Supply

BLER Block Error Ratio

BOC Bell Operating Company (or)

BOC Build–Out Capacitor

BOL Build–Out Lattice

BOR Build–Out Resistor

BPF Bandpass Filter

BRI Basic Rate Interface

BSRF Basic Synchronization Reference Frequency

CAC Compandored Analog Carrier

CAMA Centralized Automatic Message Accounting

CARL Computerized Administrative Route Layouts

CAROT Centralized Automatic Reporting on Trunks

CATV Cable Television

CCC Clear–Channel Capability

CCG Composite Clock Generator

CCIR International Radio Consultative Committee

CCIS Common–Channel Interoffice Signalling

CCITT International Telegraph and Telephone Consultative Committee

CCS Common–Channel Signalling (or)

CCS Hundred Call Seconds (Per Hour)

CCSA Common–Control Switching Arrangement

CDCF Cumulative Discounted Cash Flow

CDMA	Code–Division Multiple Access	**CREG**	Concentrated Range Extension with Gain
CEPT	European Conference of Posts and Telecommunications	**CS**	Channel Switching
		CSA	Carrier Serving Area
CFA	Carrier Failure Alarm	**CSDC**	Circuit Switched Digital Capability
CGA	Carrier Group Alarm	**CSP**	Control Switching Point
CHILL	CCITT High–Level Language	**CSU**	Channel Service Unit
CL	Carrier Liaison	**CTX**	Centrex
CLRC	Circuit Layout Record Card	**CUCRIT**	Capital Utilization Criteria
CMC	Cellular Mobile Carrier	**CX**	Composite (Circuit)
CMD	Circuit–Mode Data		
CMOS	Complementary Metal–Oxide Semiconductor	**DA**	Directory Assistance
		DART	Distribution Area Rehabilitation Tool
CMTT	Joint CCIR–CCITT Study Group on Transmission of Sound Broadcasting and Television Systems Over Long Distances	**DAVID**	Data Above Video
		DBOC	Drop Build–Out Capacitor
		DCE	Data Circuit–Terminating Equipment
		DCF	Discounted Cash Flow
CMV	Joint CCIR–CCITT Study Group for Vocabulary	**DCMS**	Digital Circuit Multiplication System
CO	Central Office	**DCS**	Digital Cross–Connect System
COD	Central–Office District		
COT	Central–Office Terminal	**DCT**	Digital Carrier Terminal (or)
CPE	Customer–Premises Equipment	**DCT**	Digital Carrier Trunk
CRC	Cyclic Redundancy Check	**DDB**	Digital Data Bank

DDD	Direct Distance Dialing	**DSBTC**	Double Sideband with Transmitted Carrier
DDS	Digital Data System		
DEMS	Digital Electronic Message Service	**DSI**	Digital Speech Interpolation
DFB	Distributed Feedback	**DSL**	Digital Subscriber Line
DFSG	Direct–Formed Supergroup	**DSN**	Defense Switched Network
DIC	Direct InterLATA Connecting	**DSS**	Data Station Selector
DILEP	Digital Line Engineering Program	**DSU**	Data Service Unit
		DSX	Digital Signal Cross–Connect
DLC	Digital Loop Carrier	**DTA**	Digitally Terminated Analog
DLP	Decode Level Point	**DTE**	Data Terminal Equipment
DM	Digital Multiplex		
DNHR	Dynamic Nonhierarchical Routing	**DTMF**	Dual–Tone Multifrequency
		DTS	Digital Termination System
DP	Dial Pulse		
DPP	Discounted Payback Period	**DUV**	Data Under Voice
		DX	Duplex (Signalling)
DRS	Digital Reference Signal		
DS	Digital Signal		
DS0	Digital Signal Level 0 (64 kb/s)	**EA**	Equal Access
		EAEO	Equal–Access End Office
DS1	Digital Signal Level 1 (1.544 Mb/s)	**EAS**	Extended Area Service
DS2	Digital Signal Level 2 (6.312 Mb/s)	**ECSA**	Exchange Carriers Standards Association
DS3	Digital Signal Level 3 (44.736 Mb/s)		
		EDD	Envelope Delay Distortion
DS4	Digital Signal Level 4 (274.176 Mb/s)	**EFS**	Error–Free Seconds
DSB	Double Sideband	**EHD**	Expected High Day
DSBSC	Double–Sideband Suppressed Carrier	**EIA**	Electronic Industries Association

EIRP	Effective Isotropic Radiated Power	**FCC**	Federal Communications Commission
ELCL	Equal Level Coupling Loss	**FCOD**	Foreign Central–Office District
ELEPL	Equal Level Echo Path Loss	**FDM**	Frequency–Division Multiplexing
ELERL	Equal Level Echo Return Loss	**FDMA**	Frequency–Division Multiple Access
ELP	Encode Level Point	**FET**	Field–Effect Transistor
ELSRL	Equal Level Singing Return Loss	**FEXT**	Far–End Crosstalk
EML	Expected Measured Loss	**FFM**	First Failure to Match
EO	End Office	**FFT**	Fast Fourier Transform
EPL	Echo Path Loss (or)	**FG**	Feature Group
EPL	Equivalent Peak Level	**FM**	Frequency Modulation
ERL	Echo Return Loss	**FMAC**	Facility Maintenance and Administration Center
ES	Errored Seconds		
ESD	Electrostatic Discharge	**FSK**	Frequency Shift Keying
ESF	Extended Superframe Format	**FSL**	Free Space Loss
ESM	Economic Study Module	**FX**	Foreign Exchange
ET	Exchange Termination	**FXS**	Foreign Exchange Station
ETL	Equipment Test List		
ETN	Electronic Tandem Network	**GDF**	Group Distributing Frame
ETV	Educational Television	**GoB**	Good or Better
EU	End User	**GOSCAL**	Grade–of–Service Calculation
EVE	Extreme Value Engineering	**GS**	Ground Start
		HAIS	Host Automatic Intercept System
FAX	Facsimile Communication	**HC**	High Capacity

HCDS	High–Capacity Digital Service	**ISC**	Intercompany Service Coordination
HDBH	High–Day Busy Hour	**ISDN**	Integrated Services Digital Network
HDTV	High–Definition Television	**ISMX**	Integrated Subrate Multiplexer
HF	High Frequency	**ITC**	International Teletraffic Congress
HU	High Usage	**ITT**	Intertandem Trunk
		ITU	International Telecommunications Union
IAL	Immediate Action Limit		
IC	Interexchange Carrier	**ITV**	Industrial Television
ICCF	Industry Carriers Compatibility Forum	**IXT**	Interaction Crosstalk
		JEG	Joint Expert Group
ICL	Inserted Connection Loss	**JFS**	Jumbogroup Frequency Supply
IDF	Intermediate Distributing Frame	**JIS**	Jurisdictionally Interstate Service
IDLC	Integrated Digital Loop Carrier	**JMX**	Jumbogroup Multiplex
IEEE	Institute of Electrical and Electronics Engineers	**KEMAR**	Knowles Electronic Manikin for Acoustic Research
IEOT	Inter–End–Office Trunk	**KSI**	Key Service Indicator
IF	Intermediate Frequency	**KTS**	Key Telephone System
IMTS	Improved Mobile Telephone Service	**KTU**	Key Telephone Unit
INA	Integrated Network Access	**LAD**	Loop Activity Data
INMD	In–Service Nonintrusive Measurement Device	**LAMA**	Local Automatic Message Accounting
		LAN	Local Area Network
IROR	Internal Rate of Return	**LATA**	Local Access and Transport Area

564

LATIS	Loop Activity Tracking Information System	**MAN**	Metropolitan Area Network
LBO	Line Build–Out	**MARR**	Minimum Attractive Rate of Return
LBR	Local Business Radio	**MAT**	Metropolitan Area Trunk
LBRV	Low–Bit–Rate Voice	**MCVD**	Modified Chemical Vapor Deposition
LCD	Liquid Crystal Display	**MDF**	Main Distributing Frame
LD	Long Distance		
LDR	Local Distribution Radio	**MF**	Multifrequency
		MFD	Mode Field Diameter
LEAD	Loop Engineering Assignment Data	**MFT**	Metallic Facility Terminal
LED	Light–Emitting Diode	**MG**	Mastergroup
LEIM	Loop Electronics Inventory Module	**MGDF**	Mastergroup Distributing Frame
LFACS	Loop Facility Assignment and Control System	**MGT**	Mastergroup Translator
		M–JCP	Multiplexer Jack and Connector Panel
LIU	Line Interface Unit		
LMX	L–Type Multiplex	**MJU**	Multipoint Junction Unit
LOCAP	Low Capacitance		
LPC	Linear Predictive Coding	**ML**	Maintenance Limit
LPF	Low–Pass Filter	**MLDS**	Microwave Local Distribution System
LRD	Long–Route Design		
LRE	Loop Range Extender	**MLRD**	Modified Long–Route Design
LSB	Lower Sideband	**MMGT**	Multimastergroup Translator
LSSGR	LATA Switching Systems Generic Requirements	**MMIC**	Monolithic Microwave Integrated Circuit
LT	LATA Tandem	**MML**	Man–Machine Language
LTEE	Long–Term Economic Evaluator		
		MMX	Mastergroup Multiplex
MAC	Multiplexed Analog Component	**MNRU**	Modulated Noise Reference Unit

MRSELS	Microwave Radio and Satellite Engineering and Licensing System	**NMDG**	Network Management Development Group
MSS	Mobile Satellite Service	**NOF**	Network Operations Forum
MTBF	Mean Time Between Failures	**NOTE**	Network Office Terminal Equipment
MTS	Message Telecommunications Service	**NPA**	Numbering Plan Area
MTSO	Mobile Telephone Switching Office	**NPV**	Net Present Value
		NPWE	Net Present Worth of Expenditures
MTU	Maintenance Terminating Unit	**NRZ**	Non−Return to Zero
		NT	Network Termination
		NT1	Network Termination Type 1
		NTC	Network Transmission Committee
NA	Numerical Aperture		
NANP	North American Numbering Plan	**NTIA**	National Telecommunications & Information Administration
NAS	North American Standard		
NBOC	Network Building−Out Capacitor	**NTS**	Network Technical Support
NCAC	Noncompandored Analog Carrier	**NTSC**	National Television System Committee
NCF	Net Cash Flow	**NTT**	Nippon Telegraph and Telephone
NCTE	Network Channel Terminating Equipment	**NTTMP**	Network Trunk Transmission Measurements Plan
NE	Network Element		
NEBS	Network Equipment Building System	**NXX**	End−Office Code
NEXT	Near−End Crosstalk		
NI	Network Interface		
NIC	Nearly Instantaneously Compandored	**OAM**	Once A Month
		OBF	Ordering and Billing Forum

OCL	Overall Connection Loss	PBX	Private Branch Exchange
OCU	Office Channel Unit	PC	Personal Computer (or)
OCUDP	Office Channel Unit Dataport	PC	Primary Center
OCVD	Outside Chemical Vapor Deposition	PCM	Pulse Code Modulation
OGT	Outgoing Trunk	PDM	Pulse Duration Modulation
OMFS	Office Master Frequency Supply	PEVL	Polyethylene Video Line (16–gauge cable)
ONAL	Off–Network Access Line		
ONI	Operator Number Identification	PFM	Pulse Frequency Modulation
ONS	On–Premises Station	PFS	Primary Frequency Supply
OPS	Off–Premises Station		
OPX	Off–Premises Extension	PIC	Polyethylene–Insulated Conductor
ORB	Office Repeater Bay		
OS	Operations System	PIN	Positive Intrinsic Negative
OSHA	Occupational Safety and Health Administration	PLAR	Private–Line Automatic Ringdown
OSI	Open Switch Interval	PLL	Phase–Locked Loop
OSS	Operations Support System (or)	PM	Phase Modulation
		PMD	Packet–Mode Data
		POT	Point of Termination
OSS	Operator Services System	POTS	Plain Old Telephone Service
OSSGR	Operator Services Systems Generic Requirements	PoW	Poor or Worse
		PPM	Pulse Position Modulation
OTDR	Optical Time–Domain Reflector	PPSN	Public Packet–Switched Network
		PPSNGR	Public Packet–Switched Network Generic Requirements
PAM	Pulse Amplitude Modulation		
PAR	Peak–to–Average Ratio	PR	Protective Relaying

PRR	Project Rate of Return	**REG**	Range Extender with Gain
PSAP	Public Safety Answering Point	**RF**	Radio Frequency
		RG	Ringing Generator
PSDS	Public Switched Digital Service	**RHC**	Regional Holding Company
PSK	Phase Shift Keying	**RL**	Return Loss
PSN	Public Switched Network	**ROH**	Receiver Off Hook
		ROLR	Receive Objective Loudness Rating
PSS	Packet Switching System	**RRD**	Revised Resistance Design
PT	Principal Tandem		
PVN	Private Virtual Network	**RSB**	Repair Service Bureau
PWAC	Present Worth of Annual Charges	**RSM**	Remote Switching Module
PWE	Present Worth of Expenditures	**RSU**	Remote Switch Unit
		RT	Remote Terminal
		RTA	Remote Trunk Arrangement
QAM	Quadrature Amplitude Modulation	**RTU**	Remote Test Unit
		RZ	Return to Zero
QPRS	Quadrature Partial–Response Signalling	**SAGM**	Separated Absorption, Grating, and Multiplication
QSDG	Quality of Service Development Group		
		SAI	Serving Area Interface
RAIS	Remote Automatic Intercept System	**SAS**	Switched Access Service
RC	Regional Center (or)	**SB**	Sideband
RC	Resistor–Capacitor	**SC**	Sectional Center
RD	Resistance Design	**SCA**	Subsidiary Communications Authorization
RDS	Radio Digital System		
RDSN	Region Digital Switched Network	**SCPC**	Single Channel Per Carrier
RDT	Remote Digital Terminal	**SDE**	Synchronization Distribution Expander
REA	Rural Electrification Administration		

SDL	Specification and Descriptive Language	**SRL**	Singing Return Loss
		SSB	Single Sideband
SDM	Standard Error of Mean	**SSBAM**	Single–Sideband Amplitude Modulation
SDN	Switched Digital Network	**SSC**	Special–Services Center
SES	Severely Errored Seconds	**SSMA**	Spread–Spectrum Multiple Access
SF	Single Frequency	**SSN**	Switched Services Network
SG	Study Group		
SGDF	Supergroup Distributing Frame	**ST**	Sector Tandem
		STC	Serving Test Center
SL	System Level	**STDM**	Statistical Time–Division Multiplexer
SMDR	Station Message Detail Recording		
		STD–RL	Standard Return Loss
SMDS	Switched Multi–Megabit Data Service	**STE**	Signalling Terminal Equipment
SMRS	Special Mobile Radio Service	**STL**	Studio–to–Transmitter Link
SMSA	Standard Metropolitan Statistical Area	**SW**	Sync Word
		SX	Simplex
		SYNTRAN	Synchronous Transmission
SOLR	Sidetone Objective Loudness Rating		
SONAD	Speech–Operated Noise–Adjusting Device	**T1**	Carrier System or Standards Committee
SONET	Synchronous Optical Network	**T1DM**	T1 Data Multiplexer
		T1OS	T1 Outstate
SP	Signal Processor	**TAS**	Telephone Answering Service
SPC	Stored Program Control	**TASI**	Time–Assignment Speech Interpolation
SPCS	Stored Program Controlled System		
SR	Sidetone Response	**TC**	Time Consistent (or)
SRDM	Subrate Data Multiplexer	**TC**	Toll Center
SRE	Signalling Range Extender	**TCM**	Time–Compression Multiplexing

TCT	Tandem–Connecting Trunk	**TSPS**	Traffic Service Position System
TD	Terminal Digit (or)	**TT&C**	Tracking, Telemetry, and Control
TD	Test Distributor	**TWT**	Traveling Wave Tube
TDM	Time–Division Multiplexing		
TDMA	Time–Division Multiple Access	**UDPC**	Universal Digital Portable Communications
TE	Transverse Electric		
TEM	Transverse Electromagnetic	**UHF**	Ultra High Frequency
THDBH	Ten High–Day Busy Hour	**UI**	Unit Interval
TIC	Tandem InterLATA Connecting	**UNICCAP**	Universal Cable Circuit Analysis Program
TL	Transducer Loss (or)	**USB**	Upper Sideband
TL	Transmission Level	**USOA**	Uniform System of Accounts
TLP	Transmission Level Point		
TM	Transverse Magnetic	**VBD**	Voiceband Data
TMRS	Telephone Maintenance Radio Service	**VF**	Voice Frequency
		VG	Voice Grade
		VHF	Very High Frequency
TOC	Television Operating Center	**VITS**	Vertical Interval Test Signal
TOLR	Transmit Objective Loudness Rating	**VLSI**	Very–Large–Scale Integration
TRA	Tax Reform Act	**VNL**	Via Net Loss
TSA	Transmission Surveillance Auxiliary	**VNLF**	Via Net Loss Factor
		VOGAD	Voice–Operated Gain–Adjusting Device
TSC	Transmission Surveillance Center	**VSA**	Voice–Switched Attenuator
TSGR	Transport Systems Generic Requirements	**VSAT**	Very–Small–Aperture Terminal
TSI	Time–Slot Interchange	**VSB**	Vestigial Sideband

VSWR	Voltage Standing Wave Ratio	**WEPL**	Weighted Echo Path Loss
VT	Virtual Tributary	**WLEL**	Wire–Line Entrance Link
VU	Volume Unit	**WORD**	Work Order Record and Details
WAL	WATS Access Line		
WAN	Wide Area Network	**WP**	Working Party
WATS	Wide Area Telecommunications Service		
WDM	Wavelength–Division Multiplexing	**ZBTSI**	Zero–Byte Time–Slot Interchange

Index

A

A and B signalling, 287
Absolute delay, 352
ac signalling, 260
Acceptance limit, 432
Access,
  integrated network, 93
  LATA, 226
  nonswitched, 226, 237
  switched, 226
Access line,
  off-network, 234, 240, 303, 310
  SSN, 308
  WATS, 226, 430, 438
Access service, 425
  digital data, 459
  equal, 427
  high-capacity digital, 461
  LATA, 226, 425
  "metallic," 449
  nonequal, 428
  nonswitched, 226, 237
  program audio, 450
  switched, 226, 425
  Technical References, 426
  telegraph grade, 449
  television, 454
  voice-grade, 441
  wideband analog, 456
  wideband data, 458
Access tandem, 176, 433
  connection via, 185
Access transmission, 428
Accounting,
  centralized automatic message,
    124, 200, 219, 220
  local automatic message, 124
Acoustic signal-to-noise ratio, 204
Actual measured loss, 162, 263,
  271, 531
Adaptive differential pulse code
  modulation, 42, 316
Adaptive hybrid, 176
Add-on centrex, 279, 288

Address signal,
  centrex, 287
  DTMF, in AUTOVON/DSN,
    314
Adjustment, loss, 152
Administration, trunk, 143
Administration of Designed
  Services, 243
Administration of local cable
  network,
  connect-through, 83
  permanently connected plant,
    83
  reassignable, 83
Administration of special services,
  243
Alarm, carrier group, 48
Allocation, traffic load, 142
Allocation area, 86
Alternate routing, 14, 139
Alternate routing system, central
  tandem, 25
American National Standards
  Institute, 38
  digital hierarchy interface
    standard, 463
  DSX-1 interconnection
    specification, 464
  network performance standard,
    149, 437
  video standard, 376
Amplifier,
  plug-in, 370
  program, 371
  talk-back, 336
Amplitude, program signal, 373
Amplitude limiting, 206
Analog fiber system, 381
Analog loop-back testing, 339
Analog-to-digital transition, 44
Analog trunk loss objectives, 19
Analysis,
  capacity, 498
  demand, 498
  expenditure, 498
Announcement service, 231

E

F

G

M

N

P

PBX, 43, 150, 225, 257, 277, 291
  cut-through, 261
  main, 232, 292
  satellite, 232, 293, 309
  station lines, 233
  tandem, 233, 293
  tributary, 309
PBX–CO trunk, 233
  standard design of, 253
PBX data-signal considerations, 274
PBX digital loss plan, 298
PBX station line resistance limits, 261
PBX tie trunk, 261, 293, 309
PBX tie-trunk signalling, 261
Performance,
  high-capacity digital service, 317, 469
  voice grade-of-service, 213
Performance, high-speed data, 41
  local and LATA access network, 34
  public packet-switched network, 59
Performance index, transmission, 515, 519
Performance objectives, 551
Performance standard, ANSI, 149, 437
Phase hits, 354
Phase jitter, 353
Phase-locked loop, 410
Pine-tree geometry, 85
Plain old telephone service (POTS), 225
PLAN (module of LEIS), 90
Plan,
  facility analysis, 85
  fixed-loss transmission, 19, 161
  hierarchical switching, 3, 301
  network trunk transmission measurements, 514
  North American Numbering, 10, 302
  outside plant, 84
  region digital switched network transmission, 37, 171,
  transmission,

message network, 15
  private-line, 331
Planned construction, 488
Planning,
  internodal, 489
  loop plant, 81
  transmission facility, 479
Planning function, transmission facility,
  documentation and coordination, 481, 496
  exploratory studies, 480, 486
  service planning, 481, 492
Planning subfunction,
  carrier, 493
  circuit, 492
  equipment, 495
  facility, 494
Plant, out-of-sight, 74, 87
Plant design, multiple, 83
Plug-in amplifier, 370
Point,
  control switching, 6, 21, 201
  decode level, 152, 207
  encode level, 152, 207
  intermediate, 10
  toll, 6
  transmission level, 162, 330, 345
Point of termination, 426
Polygrid network, 240, 311
  facilities, 315
  features of, 314
Poor-or-worse grade of service, 107, 265
POTS, 225
Power,
  data, on private lines, 347
  signal, 273, 347
Preemphasis, video signal, 376
Prequalification of loops, 226
Present worth of annual charges, 107
Present worth of expenditures, 90
Primary center, 6
Primary rate interface, 54
Private branch exchange, see PBX
Private-line bridging arrangements, 331
Private-line channels, 321

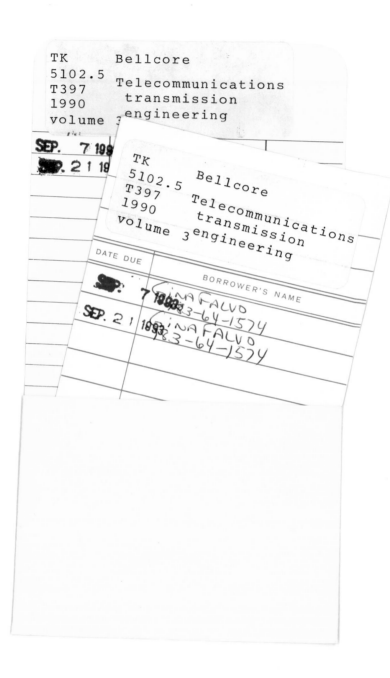